# Mathematical Analysis and its Applications

*Editor*

**Ferit Gürbüz**

Department of Mathematics
Kırklareli University, Turkey

Taylor & Francis Group

Boca Raton London New York

CRC Press is an imprint of the
Taylor & Francis Group, an **informa** business

A SCIENCE PUBLISHERS BOOK

First edition published 2025
by CRC Press
2385 NW Executive Center Drive, Suite 320, Boca Raton FL 33431

and by CRC Press
4 Park Square, Milton Park, Abingdon, Oxon, OX14 4RN



*Library of Congress Cataloging-in-Publication Data (applied for)*

ISBN: 978-1-032-64245-1 (hbk)
ISBN: 978-1-032-64246-8 (pbk)
ISBN: 978-1-032-64247-5 (ebk)

DOI: 10.1201/9781032642475

Typeset in Times New Roman
by Prime Publishing Services

# Preface

This book covers several contemporary topics in the field of mathematical analysis and its applications and relevancies in other areas of research. It focuses on enriching the understanding of several methods, problems, and applications in the area of mathematical analysis. The book also gives applications and uses of operator theory, approximation theory, optimization, variable exponent analysis, inequalities, special functions, functional equations, statistical convergence and some function spaces. Also, it presents various associated problems and ways to solve such problems. That is, each chapter of this book aims to offer readers the understanding of discussed research problems by presenting related developments in reasonable details. We hope these chapters will also inspire further innovation in the field of Analysis. As a result, with this book, we aim to bring academic scientists, researchers and scholars together in order to exchange and share their experiences and research results about all aspects of "Mathematical Analysis and its Applications". Researchers will also find this book useful.

There are 10 chapters in this book, and they are organized as follows.

Chapter "Approximation Results for Stochastic Processes via Statistical Convergence Based on a Power Series (*Emre Taş*)" presents strong Korovkin-type results for stochastic processes with the use of statistical convergence based on a power series. The author provides such examples which illustrate that his results can not be deduced from earlier results. The sequences of operators in these examples not only satisfy his theorems but also do not satisfy the known theorems. It is important to note that this concept of statistical convergence, shortly $P$-statistical convergence, can not be implied by statistical convergence or $A$-statistical convergence where A is a nonnegative regular infinite matrix. Furthermore, he also aims to obtain commutative conformable Korovkin properties for stochastic processes by using fractional calculus and $P$-statistical convergence.

Chapter "Fractional Korovkin-type Results by *P*-statistical Convergence (*Tuğba Yurdakadim*)" obtains fractional Korovkin-type results by statistical convergence which is based on a power series. In approximation theory, Korovkin-type theorems are used frequently since they give the opportunity to determine the uniform convergence of positive linear operators to identity operator by minimum calculations. These theorems are investigated by using different concepts of convergences, changing the function spaces and test functions. These results are also examined by quantum calculus and fractional calculus. In the existing literature there are some results obtained by fractional calculus but rarely. Therefore the author's aim is to obtain fractional Korovkin-type results and illustrate examples which make important contribution to existing literature since statistical convergence and *P*-statistical convergence do not imply each other.

Chapter "Two Approaches for Evaluating the Period Function of Some Hamiltonian Systems (*Daniele Ritelli* and *Giulia Spaletta*)" presents two approaches for the symbolic approximation of the energy-period function: one based on the asymptotic expansion of the period function and the other through the approximation of the fifth order of the restoring force and the consequent closed-form solution of the approximated equation. The approximation quality obtained is verified through different comparison methods illustrated in the work.

Chapter "Continuous Characterizations of Weighted Besov and Triebel-Lizorkin-Type Spaces (*Ahmed Loulit*)" gives continuous characterizations of the recently introduced weighted type homogeneous Besov-Triebel-Lizorkin spaces. The weight function w is assumed to be in the Muckenhoupt's class. The author proves different integrals representations of the quasi-norms of these spaces in terms of continuous Peetre maximal function of local, the tent space (Lusin area function) associated with local means and the Peetre maximal function of local means.

Chapter "Variable Exponent Vanishing Morrey Type Spaces on Unbounded Domains (*Ferit Gürbüz*)", applying some properties of variable exponent analysis, the author mainly focuses on some integral operators in variable exponent generalized Morrey type space on unbounded domains. Precisely, our aim is to characterize the boundedness for rough-admissible and rough -admissible operators in variable exponent vanishing generalized Morrey spaces on unbounded sets, respectively. Moreover, this work is intended to give an introduction to some important inequalities in harmonic analysis and applications. This work is also suitable for graduate students with some knowledge of real analysis and measure theory.

Chapter "Boundedness of the Intrinsic Square Function on Herz Spaces with Variable Exponents (*Liwei Wang*)" proves the boundedness of the intrinsic square function on the generalized Herz spaces, where all the three main indices are variable. Also, the author shows that the commutator of the intrinsic square function with BMO symbols is bounded in these spaces. By applying some properties of variable exponents and the generalized BMO norms.

Chapter "$q$-Deformed and $\lambda$-Parametrized Hyperbolic Tangent Function Relied Complex Valued Trigonometric and Hyperbolic Neural Network High Order Approximations (*George A. Anastassiou*)" researches the univariate quantitative approximation of complex valued continuous functions on a compact interval by complex valued neural network operators. These approximations are derived by establishing Jackson type inequalities involving the modulus of continuity of the engaged function's high order derivatives. The nature of the author's approximations are trigonometric and hyperbolic. His operators are defined by using a density function generated by a $q$-deformed and $\lambda$-parametrized hyperbolic tangent function, which is a sigmoid function. The approximations are pointwise and of the uniform norm. The related complex valued feed-forward neural networks are with one hidden layer.

Chapter "Nonlinear Exponential Sampling: Approximation Results and Applications (*Danilo Costarelli*)" introduces a family of nonlinear exponential sampling operators. The proposed definition represents a nonlinear extension of the classical exponential sampling operators introduced by Bardaro, Faina, and Mantellini in 2017. Here, regularization results are established, as well as pointwise and uniform convergence theorems in the case of the approximation of bounded continuous and uniformly continuous functions, respectively. Moreover, quantitative estimates for the order of approximation are also obtained using suitable moduli of continuity. Finally, some special cases are discussed in detail and numerical examples with graphical illustrations are also given.

Chapter "Refinements and Reverses of Some Inequalities for the Normalized Determinants of Sequences of Positive Operators on Hilbert Spaces (*Silvestru Sever Dragomir*)" presents several refinements and reverses for the normalized determinant of sequences of operators that have the spectra in a positive interval. For this purpose, the author used some Jensen's type discrete inequalities for twice differentiable functions obtained by the author earlier. Some reverses of the celebrated Ky Fan's inequality for determinants of positive operators are given. Upper and lower bounds in terms of the operator variance are also provided.

Chapter "Fuzzy Inference Based Approach of Ant Colony Optimization (ACO) in Fuzzy Transportation Models (*M.K. Sharma, Tarun Kumar, Laxmi Rathour* and *Vishnu Narayan Mishra*)"—An effective and significant approach of optimization is ant colony optimization (ACO). This crowd-based meta-heuristic approach was inspired by the biological feature of social ants. Consequently, the purpose of this study is to create a methodology for fuzzy transportation models using ACO. The transport issue has been approached in a fuzzy setting in this paper. We have opted to describe transportation factors like cost, demand, and availability in the fuzzy logic environment because they may not always be precise and may consequently be ambiguous. Trapezoidal and pentagonal fuzzy numbers are used to represent price, demand, and availability. Then, in order to determine the fuzzy cost, ACO is employed by converting the trapezoidal and pentagonal fuzzy transportation issues into pointwise crisp transportation problems. Numerical calculations are used to explain the effectiveness of the suggested methodology.

Finally, the editor sincerely acknowledges the cooperation and patience of contributors during the entire process of handling their works submitted for this book. Reviewers deserve deep gratitude for selflessly offering their help for successfully bringing out this book. The editor also thankfully acknowledges the support of editorial staff at Taylor & Francis.

**Ferit Gürbüz**
Department of Mathematics
Kırklareli University
Kırklareli 39100, Türkiye

July 2024

# Contents

# Chapter 1

# Approximation Results for Stochastic Processes via Statistical Convergence Based on a Power Series

*Emre Taş*

## 1.1 Introduction and Preliminaries

One of the central concepts in approximation theory is the quest to represent any given function in terms of more amenable or straightforward functions, or even a combination of both. Specifically, when we expand a function into a power series, we effectively express that function using polynomials, which possess both simplicity and elegance. Consequently, a fundamental inquiry arises: What constitutes a favorable category of these amiable and straightforward functions that yield superior approximations, and to what extent can we employ them for ap-

Department of Mathematics, Kırşehir Ahi Evran University, Kırşehir, 40100, Turkey.
Email: emretas86@hotmail.com

proximation purposes? Within this context, the Weierstrass theorem stands as a cornerstone of approximation theory, affirming that the set of algebraic polynomials is dense in the space of continuous functions defined on the closed interval [a, b], denoted as $C[a,b]$. This profound result has been attributed to Karl Weierstrass [71]. The proof of Weierstrass's theorem is known to be intricate and demanding. Nevertheless, it was David Bernstein who first managed to offer a well-recognized, more comprehensible alternative proof [19], subsequently further explored by other mathematicians [26], [49]. This alternative proof opened the door to the investigation of positive, linear operators within the realm of approximation theory [21], [48], [60]. Notably, it was established that one could ascertain the uniform convergence of such operators towards the identity operator with minimal computations, a result known as Korovkin-type theorems. In essence, if the sequence of positive linear operators $U_j$ converges uniformly to $f$ for every $f \in C[a,b]$, it suffices that $U_j(x^i)$ converges uniformly to $x^i$ for $i = 0, 1, 2$, over the interval $[a,b]$.

As a consequence, Korovkin-type theorems have become a ubiquitous tool in approximation theory, and numerous mathematicians have undertaken diverse approaches to investigate the intricacies of this theory. Let us delve into a detailed exploration of these topics:

- Some researchers have opted for alternative notions of convergence, departing from the conventional limit concept, as these alternative concepts offer a broader perspective on convergence beyond the classical limit [11], [29], [31], [70]. It is worth pausing to underscore the significance of summability theory. Summability theory endeavors to assign limits and sums to sequences and series that would otherwise diverge. The study of convergent series dates back to ancient times, with little interest in divergent series and sequences until the era of Leonhard Euler (1707–1783). Prior to Euler, divergent series and sequences were largely disregarded. Euler, however, developed an ingenious method for associating a divergent series, Carl Friedrich Gauss (1777–1855) also introduced infinite processes into mathematical analysis in his early career. His contributions include the development of the binomial theorem and investigations into the convergence of power series. Augustin-Louis Cauchy (1789–1857) introduced rigor into mathematical analysis and formulated ideas regarding the convergence and divergence of power series. Niels Henrik Abel (1802–1829) was another prominent mathematician who delved into the concepts of convergence and divergence. In the latter half of the nineteenth century, interest in divergent series waned significantly but was later revisited. Among those who rekindled investigations into divergent series was Ernesto Cesàro (1859–1906), who introduced $C_1$-convergence. This development allowed mathematicians to discuss the Cauchy product of two infinite series and their convergence properties. Subsequently, extensive studies on summability were conducted [20], [61]. Gadjiev and

Orhan were the first to combine approximation and summability theories [33]. Many results in approximation theory have since been explored by using different convergence concepts from summability theory [25], [63].

■ Recent research has delved into the realm of nonlinear operators. While linear operators naturally possess a linear algebraic structure, it is pertinent to question whether linear operators are the only viable choice. Bede et al. addressed this question by investigating nonlinear operators, particularly pseudo-linear operators, max-product type operators, and max-min type operators within the theory of approximation [12], [16], [17], [18]. Notably, pseudo-linearity relaxes the stringent linearity condition typically associated with operators. In addition to Bede et al.'s approach, another type of max-min approach, introduced by He [40], has gained traction. These operators have demonstrated superior performance in terms of approximation order, computational complexity, image processing, and have found significant applications in deep learning, particularly in feedforward neural networks and ReLU networks [17], [62], [66]. Consequently, various approximation theorems involving these operators have been investigated using different convergence concepts [9], [36], [37], [38].

■ Another intriguing facet of these studies explores the fuzzy setting. A brief historical overview of fuzzy theory reveals its departure from conventional set theory. Concepts like "pretty animals" or "nice cities" have limited applicability in classical set theory. Fuzziness, as a measure of the degree of certainty or ambiguity, was first recognized by Lotfi Zadeh [74], who systematically introduced fuzzy sets and fuzzy numbers. These fuzzy sets are characterized by membership functions that assign each element a membership grade ranging from 0 to 1. Goetschel and Voxman [35] introduced a modified version of fuzzy numbers and defined a metric on this class of sets. The concept of fuzziness and the role it plays in addressing uncertainty has substantial implications in fields such as human cognition, pattern recognition, machine learning, weather forecasting, robotics, stock market analysis, and bio-medicine. Given its effectiveness in mitigating uncertainty and its real-world relevance, mathematicians have sought to adapt fundamental mathematical concepts to the fuzzy setting [3], [22], [23], [53], [54], [57]. Fuzzy topology has been extensively studied and has significant applications in quantum particle physics [27], [28]. Moreover, fuzzy logic has found its way into the theory of approximation, with researchers like Gal, Anastassiou, Duman, Yurdakadim, Tas, Baxhau, and others presenting approximation theorems within the fuzzy framework [5], [15], [34], [73].

■ Some researchers have explored results in weighted spaces, modular spaces, and abstract spaces. For instance, efforts have been made in [1], [10], [13], [14], [41], [51], [68] to identify sets of functions, analogous to $1, x, x^2$, that satisfy the same properties as Korovkin sets. Another line of inquiry involves the investigation of Korovkin-type results in various function spaces or abstract Banach spaces [43], [44], [69].

■ The field of $q$-calculus, which plays a pivotal role in applications across various applied sciences and engineering disciplines, serves as a bridge between mathematics and physics. Thus, it is crucial to bridge gaps in the mathematical theory of $q$-calculus. Concepts such as binomial expansion, factorial, Taylor expansion, and derivatives have been extended to the realm of $q$-calculus. Consequently, numerous results obtained, or previously unattainable, have been examined within this framework [42], [55], [58], [67]. For example, the $q$-analog of Bernstein operators was introduced by Lupaş [50]. Different types of $q$-Bernstein operators have been proposed by Phillips [58], and Ostrovska [56] has investigated the approximation properties of these operators. Karsli and Gupta [45] introduced $q$-Chlodowsky operators, extending the concept of $q$-Bernstein operators to unbounded intervals.

■ Korovkin type results have been stated for stochastic processes [2], [4], [6], [7], [8], [72]. Lastly, the study of Korovkin-type results has been expanded to encompass fractional calculus. Fractional calculus deals with integrals and derivatives of arbitrary order, fundamental concepts in mathematical analysis. The exploration of half-order derivatives of $f(x) = x$, which began with a correspondence between Leibniz and L'Hospital, persisted through the works of Liouville, Grünwald, Letnikov, Riemann, and Caputo. From the latter part of the 20th century, fractional derivatives gained significance in real-world applications, especially for characterizing materials like polymers and rocks [24], [59], [65]. Unfortunately, the mathematical theory of fractional calculus lagged behind the practical demands. Consequently, the examination of analytical results in a fractional setting became imperative.

In this chapter, we aim to present Korovkin-type theorems for stochastic process and commutative conformable fractional Korovkin theorems for stochastic process utilizing $P$-statistical convergence, where $P$ denotes power series. We will also provide illustrative examples that underscore the importance of our results, considering that statistical convergence and $P$-statistical convergence are not mutually inclusive, making our findings a valuable contribution to the literature. Let us review the fundamental concepts, definitions, and established results that will prove valuable throughout this paper.

We begin with the initial definition, that of the natural density of a subset $G \subseteq \mathbb{N}$ and the concept of statistical convergence for a sequence $s = (s_j)$. It is important to emphasize that statistical convergence serves as a powerful extension of classical convergence, addressing the limitations associated with the classical notion of a limit.

**Definition 1.1** The natural density of $G \subseteq \mathbb{N}$ is given by

$$\delta(G) := \lim_{n \to \infty} \frac{1}{n} \#\{j \le n : j \in G\}$$

if the limit exists where $\#E$ denotes the cardinality of $E$ and $\mathbb{N}$ is the set of all nonnegative integers. If $\delta(G_\varepsilon) = 0$ for every $\varepsilon > 0$ where $G_\varepsilon = \{j \in \mathbb{N} : |s_j - l| \ge \varepsilon\}$ then $s = (s_j)$ is said to be statistically convergent to $l$ [29], [31], [64].

As it can be seen that a convergent sequence is also statistically convergent but, in general the converse does not hold. Let us give a couple of examples to examine this situation in details.

**Example 1** *Let $n \in \mathbb{N}$ and $s_j = \begin{cases} n^2 & , \quad j = n^2 \\ 2 & , \quad j \ne n^2. \end{cases}$*

*Of course, $(s_j)$ is not convergent since it is not bounded. On the other hand, $\#\{j \le n : |s_j - 2| \ge \varepsilon\} \le \#\{j \le n : s_j \ne 2\} \le \sqrt{n}$ holds for $\varepsilon > 0$*

*and it implies that*

$$\lim_{n \to \infty} \frac{1}{n} \#\{j \le n : |s_j - 2| \ge \varepsilon\} \le \lim_{n \to \infty} \frac{1}{n} \#\{j \le n : s_j \ne 2\} \le \lim_{n \to \infty} \frac{1}{n} \sqrt{n} = 0.$$

*Then we obtain that $\delta(K_\varepsilon) = 0$, for every $\varepsilon > 0$ where $K_\varepsilon = \{j \in \mathbb{N} : |s_j - 2| \ge \varepsilon\}$. This means that $s = (s_j)$ statistically converges to 2.*

It is evident that a statistically convergent sequence does not necessarily require boundedness, unlike a convergent sequence, which must be bounded. Numerous classical convergence concepts have undergone extensive examination and adaptation within the framework of statistical convergence. Notably, concepts such as statistical boundedness, statistical limit points, statistical cluster points, and the relationships between these statistical notions and their classical counterparts have been the subject of thorough investigation [30].

It has become clear that statistical convergence does not uphold all the properties associated with classical convergence. For instance, while it is a well-established fact that every subsequence of a convergent sequence also converges, Miller has demonstrated that this does not hold true for subsequences of statistically convergent sequences. This divergence arises from the incorporation of binary expansion, measures, and probability theory in the analysis [52].

Now, let $A = (a_{nj})$ $n, j = 1, 2, \cdots$; be an infinite matrix. Then all of the aboved mentioned concepts have all been studied for the extension of statistical convergence which is called as $A$-statistical convergence and recalled below.

**Definition 1.2**   Let $A = (a_{nj})$ $n, j = 1, 2, \cdots$; be an infinite matrix and $s = (s_j)$ be a sequence. Then the $A$-transform of $s$ denoted by $As := ((As)_n)$ is defined by

$$(As)_n = \sum_{j=1}^{\infty} a_{nj} s_j$$

where the series is convergent for every $n$. Also regularity of a matrix is defined by: if the $A$-transform of a convergent sequence is also convergent to the same limit then it is said that $A$ is regular [39].

One of important matrices in summability theory is Cesàro matrix and it is given by
$C_1 := (c_{nj})$

$$c_{nj} = \begin{cases} \frac{1}{n} & , \quad 1 \leq j \leq n \\ 0 & , \quad j > n \end{cases}$$

and explicitly given by

$$C_1 = \begin{pmatrix} 1 & 0 & 0 & \cdots & \cdot & \cdots \\ \frac{1}{2} & \frac{1}{2} & 0 & \cdots & \cdot & \cdots \\ \cdot & \cdot & \cdot & \cdots & \cdot & \cdots \\ \frac{1}{n} & \frac{1}{n} & \cdot & \frac{1}{n} & 0 & \cdots \\ \frac{1}{n+1} & \frac{1}{n+1} & \cdot & \frac{1}{n+1} & \frac{1}{n+1} & \cdots \\ \vdots & \vdots & \vdots & \vdots & \vdots & \ddots \end{pmatrix}$$

The regularity of an infinite matrix $A = (a_{nj})$ is characterized by Silverman-Toeplitz theorem [20]. One can easily notice that Cesàro matrix is regular.

**Definition 1.3**   Let $A = (a_{nj})$ be a nonnegative regular infinite matrix. Then if

$$\delta_A(K) = \lim_n \sum_{j \in K} a_{nj} \tag{1.1.1}$$

exits it is said that the number $\delta_A(K)$ is the $A$-density of $K$ where $K \subset \mathbb{N}$ [32].

**Example 2** *Consider*

$$a_{nj} = \begin{cases} 2 & , \quad j = n^2 \\ 0 & , \quad j \neq n^2 \end{cases}$$

*then for $A = (a_{nj})$, $\delta_A(K) = 1$ where $K = \{1, 4, 9, 16, \cdots, j^2, \cdots\}$. Notice that $\delta_A(K^c) = 0$ for $K^c = \{2, 3, 5, \cdots\}$.*

**Definition 1.4**  Let $A = (a_{nj})$ be a nonnegative regular infinite matrix. If there exists $\ell$ such that

$$\lim_n \sum_{j:|s_j-\ell|\geq\varepsilon} a_{nj} = 0 \tag{1.1.2}$$

holds for every $\varepsilon > 0$ then we say that $s = (s_j)$ $A$-statistically converges to $\ell$. Indeed, $\delta_A(K_\varepsilon) = 0$ for every $\varepsilon > 0$ where $K_\varepsilon = \{j \in \mathbb{N} : |s_j - \ell| \geq \varepsilon\}$.

This is denoted by $st_A - \lim s = \ell$ or $s_j \to \ell(st_A)$ [46], [52].

In Definiton 1.4,

- ■ if $A$ is $I$, the identity matrix then $I$-statistical convergence reduces to classical convergence,

- ■ if $A$ is $C_1$ then $A$-statistical convergence reduces to statistical convergence.

Again many concepts of classical convergence and statistical convergence have been examined for $A$-statistical convergence. Also the relationships between these concepts have been studied by many authors.

**Definition 1.5**  Let $(p_j)$ be sequence of real numbers such that for all $j \geq 1$, $p_j \geq 0$, and $p_0 > 0$, $p(t) := \sum_{j=0}^{\infty} p_j t^j$ with a radius of convergence $R \in (0,\infty]$. Then power series method is defined as follows:

Let also

$$C_p := \left\{ f : (-R,R) \to \mathbb{R}\, \Big|\, \lim_{0<t\to R^-} \frac{1}{p(t)} f(t) \quad \text{exists} \right\}$$

and

$$C_{P_p} := \left\{ s = (s_j)\,|\,p_s(t) := \sum_{j=0}^{\infty} p_j t^j s_j \quad \text{with radius of convergence} \geq R \quad \text{and} \quad p_s \in C_p \right\},$$

$$P - \lim x = \lim_{0<t\to R^-} \frac{1}{p(t)} \sum_{j=0}^{\infty} p_j t^j s_j$$

where the functional $P - \lim : C_{P_p} \to \mathbb{R}$ and then it is said that $s$ is $P-$convergent [20], [47].

Consider the following examples:

**Example 3** *Let* $(s_j) = (1,0,1,0,1,0,...)$, $R = \infty$, *for* $j \geq 0$, $p_j = \dfrac{1}{j!}$, *then* $p(t) = e^t$ *and it is easy to see that*

$$\lim_{t \to \infty} \frac{1}{e^t} \sum_{j=0}^{\infty} \frac{s_j t^j}{j!} = \lim_{t \to \infty} \frac{1}{e^t} \sum_{j=0}^{\infty} \frac{t^{2j}}{(2j)!} = \lim_{t \to \infty} \frac{1}{e^t} \{ \frac{e^t + e^{-t}}{2} \} = \frac{1}{2}.$$

*So the sequence* $s = (s_j)$ *is convergent to* $\frac{1}{2}$ *in the sense of power series method but it is not convergent in the ordinary sense.*

**Example 4** *Let also* $s_j = (-1)^j$, *for* $j \geq 0$, $p_j = 1$ *then* $R = 1$, $p(t) = \frac{1}{1-t}$ *and*

$$\lim_{t \to 1^-} (1 - t) \sum_{j=0}^{\infty} s_j t^j = 0$$

*which means that* $s = (s_j)$ *is convergent to* $0$ *in the sense of power series method, i.e., P-convergent to* $0$ *but s is not convergent.*

Therefore by taking into consideration the above examples, we can say that power series method is more effective.

If a convergent sequence is also $P$-convergent to same limit then it is said that $P$ is regular and it is characterized by the following condition:

$$\lim_{t \to R^-} \frac{p_j t^j}{p(t)} = 0, \text{ for every } j \in \mathbb{N}_0$$

[20]. It is also noteworthy to recall that

- in the case of $R = 1$, $p(t) = \dfrac{1}{1-t}$ and for $j \geq 0$, $p_j = 1$ the power series method coincides with Abel method which is a sequence-to-function transformation,

- in the case of $R = \infty$, $p(t) = e^t$ and for $j \geq 0$, $p_j = \dfrac{1}{j!}$ the power series method coincides with Borel method.

By combining statistical convergence and power series, Ünver and Orhan [70] have recently introduced $P$-statistical convergence and have presented a Korovkin-type theorem for positive linear operators defined on $C[0,1]$.

Now we are ready to recall $P$-statistical convergence which is the main tool of our results.

**Definition 1.6**  Let $P$ be regular and $G \subseteq \mathbb{N}_0$. If

$$\delta_P(G) := \lim_{0<t\to R^-} \frac{1}{p(t)} \sum_{j\in G} p_j t^j$$

exists then it is said to be the *P*-density of *G*.

One can immediately observe that if $\delta_P(G)$ exists then $\delta_P(G) \in [0,1]$ [70].

**Definition 1.7**  Let $s = (s_j)$ be a sequence of real numbers and let $P$ be regular. If for every $\varepsilon > 0$

$$\lim_{0<t\to R^-} \frac{1}{p(t)} \sum_{j\in G_\varepsilon} p_j t^j = 0$$

that is, $\delta_P(G_\varepsilon) = 0$ for every $\varepsilon > 0$, then $s$ is said to be *P*-statistically convergent to $l$ and we denote it by $st_P - \lim s = l$ [70].

In order to show that statistical convergence does not imply *P*-statistical convergence and vice versa, the following examples can be examined. Therefore we can emphasize that our results make an important contribution to the existing literature.

**Example 5**  *Consider the following method defined by the sequence $(p_j)$: $j \in \mathbb{N}_0$,*

$$p_j = \begin{cases} 1 & , \quad j \text{ prime} \\ 0 & , \quad j \text{ otherwise} \end{cases}$$

*and also the sequence $s = (s_j)$ defined by*

$$s_j = \begin{cases} 0 & , \quad j \text{ prime} \\ n & , \quad j \text{ otherwise} \end{cases}$$

*Then it is obvious that $s = (s_j)$ is not statistically convergent. But for every $\varepsilon > 0$, we have that $\delta_P(K_\varepsilon) = 0$ where $K_\varepsilon = \{j \in \mathbb{N} : |s_j - 0| \geq \varepsilon\}$.*
*This means that $s = (s_j)$ is P-statistically convergent to 0.*

**Example 6**  *Consider the following method defined by the sequence $(p_j)$: $j \in \mathbb{N}_0$*

$$p_j = \begin{cases} 1 & , \quad j \text{ prime} \\ 0 & , \quad j \text{ otherwise} \end{cases}$$

*and also the sequence $s = (s_j)$ defined by*

$$s_j = \begin{cases} j & , \quad j \text{ prime} \\ 0 & , \quad j \text{ otherwise.} \end{cases}$$

*Here, it is obvious that the sequence $s = (s_j)$ statistically converges to 0. But it does not converge in the P-statistical sense.*

Numerous generalizations of $\frac{d^n}{dx^n} f(x)$ for cases where $n \notin \mathbb{N}$ exist, and they have been extensively explored by mathematicians. Notable among these generalizations are the Riemann-Liouville, Caputo, Grünwald-Letnikov, Weyl, Riesz, and conformable fractional derivatives. Each of these distinct definitions of fractional derivatives offers the flexibility to choose the most suitable one for a given problem, facilitating the quest for the optimal solution.

It is important to highlight that, despite certain mathematical relationships between these definitions, they possess distinct physical interpretations. For instance, a notable disparity arises in the treatment of constants. The Caputo derivative of a constant is consistently zero, whereas the Riemann-Liouville derivative, bounded by a finite lower limit, is not zero. In cases where it is imperative for the derivative of a constant to be zero to provide a physically meaningful interpretation of a problem, Caputo derivatives have proven to be the preferred choice in existing literature.

## 1.2 Korovkin Theory for Stochastic Process via Power Series Method

$C([a,b])$ is the space of continuous real valued functions, $C^m([a,b])$ is the space of all $m-$th order continuously differentiable functions and $B([a,b])$ is the space of bounded real valued functions on $[a,b]$ endowed with the supremum norm $\|.\|_\infty$. Let $(\Gamma, \mathscr{A}, \mu)$ be a probability space. Along this section we use some assumptions:

**Assumption 1.** $X(t,\gamma)$ is a stochastic process from $[a,b] \times (\Gamma, \mathscr{A}, \mu)$ into $\mathbb{R}$ such that $X(.,\gamma) \in C^m([a,b])$ for each $\gamma \in \Gamma$, and that $X^{(k)}(t,.)$ is a measurable for any $t \in [a,b]$ and for all $k = 0, 1, 2, ..., m$.

**Assumption 2.** Expectation operator defined by

$$(EX)(t) = \int_\Gamma X(t,\gamma)\mu(d\gamma), \; for \; every \; x \in [a,b]$$

**Assumption 3.** $X^{(m)}(t,\gamma)$ is continuous in $t \in [a,b]$ uniformly with respect to $\gamma \in \Gamma$. We write $X^{(m)} \in C^U([a,b])$.

**Assumption 4.** For any $t \in [a,b]$,

$$\left(E\left|X^{(k)}\right|^q\right)(t) < \infty, \; \text{for each } k = 0, 1, 2, ..., m, \; q \in (1,\infty)$$

holds.

**Assumption 5.** For any $t \in [a,b]$,

$$\left( E \left| X^{(k)} \right| \right)(t) < \infty, \text{ for each } k = 0, 1, 2, ..., m$$

holds.

Modulus of continuity of a stochastic process $X$ is defined in integral form by

$$\Omega_1 \left( X; \delta \right)_{L_q, [a,b]} = \left\{ \sup_{\substack{|t-x| \leq \delta \\ t,x \in [a,b]}} \int_{\Gamma} |X(t,\gamma) - X(x,\gamma)|^q \mu(d\gamma) \right\}^{\frac{1}{q}}, \delta > 0 \text{ and } q \geq 1.$$

### *Lemma 1.1*

*If $\{\delta_j\}$ is a sequence of positive real numbers such that $st_P - \lim \delta_j = 0$ then we have, $st_P - \lim \Omega_1 (X; \delta_j)_{L_q} = 0$ for $q \geq 1$ and every stochastic process $X$ satisfying the Assumption 3 from $[a,b] \times (\Gamma, \mathscr{A}, \mu)$ into $\mathbb{R}$.*

**Proof 1.1**    Since $st_P - \lim \delta_j = 0$, for every $\eta > 0$ we have

$$\lim_{t \to R^-} \frac{1}{p(t)} \sum_{j:\delta_j \geq \eta} p_j t^j = 0. \tag{1.2.1}$$

According to Assumption 3, we can write for every $\eta > 0$, there exists $\kappa > 0$ such that for every $x, y \in [a,b]$ satisfying $|x - y| \leq \kappa$,

$$|X(x,\gamma) - X(y,\gamma)| < \eta.$$

Hence we get

$$\left\{ \int_P |X(x,\gamma) - X(y,\gamma)|^q \mu(d\gamma) \right\}^{\frac{1}{q}} < \eta, q \geq 1.$$

By using the last inequality, in case of $0 < \rho < \kappa$ we obtain $\Omega_1 (X;\rho)_{L_q} < \eta$. That is, whenever $\Omega_1 (X;\rho)_{L_q} \geq \eta$, it should be $\rho \geq \kappa$. If we replace $\rho$ with $\delta_j$, for any $\eta > 0$ we have

$$\left\{ j \in \mathbb{N}_0 : \Omega_1 (X;\delta_j)_{L_q} \geq \eta \right\} \subset \left\{ j \in \mathbb{N}_0 : \delta_j \geq \gamma \right\}.$$

From the above relation, we have for any $0 < t < R$,

$$\sum_{j:\Omega_1(X;\delta_j)_{L_q} \geq \eta} p_j t^j \leq \sum_{j:\delta_j \geq \gamma} p_j t^j.$$

Taking limit whenever $t \to R^-$ and considering (1.2.1), for every $\eta > 0$

$$\lim_{t \to R^-} \frac{1}{p(t)} \sum_{j:\Omega_1(X;\delta_j)_{L_q} \geq \eta} p_j t^j = 0$$

holds. This implies $st_P - \lim \Omega_1 (X;\delta_j)_{L_q} = 0$.

### Theorem 1.1

*Consider a sequence of positive linear operators $\left\{U_j\right\}_{j \in \mathbb{N}_0}$ on stochastic processes. Let $\left\{\tilde{U}_j\right\}_{j \in \mathbb{N}_0}$ be a corresponding sequence of positive linear operators which satisfies the following equality for every stochastic process $X$ from $[a,b] \times (\Gamma,\mathscr{A},\mu)$ into $\mathbb{R}$*

$$U_j(X)(t,\gamma) = \tilde{U}_j(X(.,\gamma);t) \text{ for } \gamma \in \Gamma \text{ and } t \in [a,b].$$

*Also suppose that*

$$st_P - \lim \left\|\tilde{U}_j(e_0) - e_0\right\|_\infty = 0 \tag{1.2.2}$$

*and*

$$st_P - \lim \left\|\tilde{U}_j\left(|\varphi_t|^{q(m+1)}\right)\right\|_\infty = 0, \text{ fixed } 1 < q < \infty \text{ and } m \in \mathbb{N}.$$

*Then we have for all $X$ satisfying Assumption 3 and Assumption 4*

$$st_P - \lim \left\|E\left(|U_j(X) - X|^q\right)\right\|_\infty = 0.$$

**Proof 1.2** Fixing $m \in \mathbb{N}$ and $1 < q < \infty$ and by Corollary 2.1 in [4], we know the following inequality for all $X$ satisfying Assumption 3 and Assumption 4 and for each $j \in \mathbb{N}_0$,

$$
\begin{aligned}
\left\|E\left(|U_j(X) - X|^q\right)\right\|_\infty^{\frac{1}{q}} \leq{} & \left\|E(|X^q|)\right\|_\infty^{\frac{1}{q}} \cdot \left\|\tilde{U}_j(e_0) - e_0\right\|_\infty + \sum_{k=1}^m \frac{\left\|E\left(\left|X^{(k)}\right|^q\right)\right\|_\infty^{\frac{1}{q}}}{k!} \left\|\tilde{U}_j\left(\varphi_t^k\right)\right\|_\infty \\
& + \left(\frac{2(q-1)\left\|\tilde{U}_j(e_0)\right\|_\infty}{qm-1}\right)^{1-\frac{1}{q}} \cdot \frac{1}{(m-1)!(q+1)^{\frac{m}{q(m+1)}}} \\
& \times \left(\left\|(\tilde{U}_j(e_0))^{\frac{1}{m+1}}\right\|_\infty (q+1)^{\frac{m}{m+1}} + 1\right)^{\frac{1}{q}} \cdot \left\|\tilde{U}_j\left(|\varphi_t|^{q(m+1)}\right)\right\|_\infty^{\frac{m}{q(m+1)}} \\
& \times \Omega_1\left(X^{(m)}; \frac{1}{(q+1)^{\frac{1}{q(m+1)}}} \left\|\tilde{U}_j\left(|\varphi_t|^{q(m+1)}\right)\right\|_\infty^{\frac{1}{q(m+1)}}\right)_{L_q}.
\end{aligned}
$$

In the meanwhile, by using Hölder's inequality with $\alpha = \frac{q(m+1)}{q(m+1)-k}$ and $\beta = \frac{q(m+1)}{k}$, we can write

$$\left\|\tilde{U}_j\left(\varphi_t^k\right)\right\|_\infty \leq \left\|\tilde{U}_j(e_0)\right\|_\infty^{\frac{q(m+1)-k}{q(m+1)}} \left\|\tilde{U}_j\left(|\varphi_t|^{q(m+1)}\right)\right\|_\infty^{\frac{k}{q(m+1)}}, k = 1,2,...,m.$$

By taking into consideration last inequalities, we have

$$
\left\| E\left(\left|U_j\left(X\right)-X\right|^q\right)\right\|_\infty^{\frac{1}{q}} \;\le\; Z_{m,q}\left\{\sum_{k=1}^m \left\|\tilde{U}_j\left(e_0\right)\right\|_\infty^{\frac{q(m+1)-k}{q(m+1)}} \left\|\tilde{U}_j\left(\left|\varphi_t\right|^{q(m+1)}\right)\right\|_\infty^{\frac{k}{q(m+1)}}\right.
$$

$$
+\left\|\tilde{U}_j\left(e_0\right)-e_0\right\|_\infty + \left\|\tilde{U}_j\left(e_0\right)\right\|_\infty^{1-\frac{m}{q(m+1)}}\left\|\tilde{U}_j\left(\left|\varphi_t\right|^{q(m+1)}\right)\right\|_\infty^{\frac{m}{q(m+1)}}
$$

$$
\times\,\Omega_1\left(X^{(m)};\delta_j\left(m,q\right)\right)_{L_q}
$$

$$
\left.+\left\|\tilde{U}_j\left(e_0\right)\right\|_\infty^{1-\frac{1}{q}}\left\|\tilde{U}_j\left(\left|\varphi_t\right|^{q(m+1)}\right)\right\|_\infty^{\frac{m}{q(m+1)}}\Omega_1\left(X^{(m)};\delta_j\left(m,q\right)\right)_{L_q}\right\}
$$

where

$$
Z_{m,q}=\max\left\{\frac{1}{(m-1)!}\left(\frac{2\left(q-1\right)}{qm-1}\right)^{1-\frac{1}{q}},\left\|E\left(\left|X^q\right|\right)\right\|_\infty^{\frac{1}{q}},\frac{\left\|E\left(\left|X'\right|^q\right)\right\|_\infty^{\frac{1}{q}}}{1!},...,\frac{\left\|E\left(\left|X^{(m)}\right|^q\right)\right\|_\infty^{\frac{1}{q}}}{m!}\right\}
$$

and

$$
\delta_j\left(m,q\right)=\frac{1}{\left(q+1\right)^{\frac{1}{q(m+1)}}}\left\|\tilde{U}_j\left(\left|\varphi_t\right|^{q(m+1)}\right)\right\|_\infty^{\frac{1}{q(m+1)}}.
$$

our hypothesis implies that $Z_{m,q}$ is a finite positive number for each fixed $m\in\mathbb{N}$ and $1<q<\infty$. On the other hand, we have

$$
\left\| E\left(\left|U_j\left(X\right)-X\right|^q\right)\right\|_\infty^{\frac{1}{q}} \;\le\; Z_{m,q}\left\{\sum_{k=1}^m \left\|\tilde{U}_j\left(e_0\right)-e_0\right\|_\infty^{\frac{q(m+1)-k}{q(m+1)}} \left\|\tilde{U}_j\left(\left|\varphi_t\right|^{q(m+1)}\right)\right\|_\infty^{\frac{k}{q(m+1)}}\right.
$$

$$
+\sum_{k=1}^m \left\|\tilde{U}_j\left(\left|\varphi_t\right|^{q(m+1)}\right)\right\|_\infty^{\frac{k}{q(m+1)}} + \left\|\tilde{U}_j\left(e_0\right)-e_0\right\|_\infty
$$

$$
+\left\|\tilde{U}_j\left(e_0\right)-e_0\right\|_\infty^{1-\frac{m}{q(m+1)}}\left\|\tilde{U}_j\left(\left|\varphi_t\right|^{q(m+1)}\right)\right\|_\infty^{\frac{m}{q(m+1)}}
$$

$$
\times\,\Omega_1\left(X^{(m)};\delta_j\left(m,q\right)\right)_{L_q}
$$

$$
+\left\|\tilde{U}_j\left(e_0\right)-e_0\right\|_\infty^{1-\frac{1}{q}}\left\|\tilde{U}_j\left(\left|\varphi_t\right|^{q(m+1)}\right)\right\|_\infty^{\frac{m}{q(m+1)}}\Omega_1\left(X^{(m)};\delta_j\left(m,q\right)\right)_{L_q}
$$

$$
\left.+\,2\left\|\tilde{U}_j\left(\left|\varphi_t\right|^{q(m+1)}\right)\right\|_\infty^{\frac{m}{q(m+1)}}\Omega_1\left(X^{(m)};\delta_j\left(m,q\right)\right)_{L_q}\right\}.
$$

Consider the sets for given $\varepsilon>0$:

$$
D\left(\varepsilon\right)=\left\{j\in\mathbb{N}_0:\ \left\|E\left(\left|U_j\left(X\right)-X\right|^q\right)\right\|_\infty\ge\varepsilon^q\right\}
$$

$$
A_1\left(\varepsilon\right)=\left\{j\in\mathbb{N}_0:\ \left\|\tilde{U}_j\left(e_0\right)-e_0\right\|_\infty\ge\frac{\varepsilon}{2(m+2)Z_{m,q}}\right\}
$$

$$
A_2\left(\varepsilon\right)=\left\{j\in\mathbb{N}_0:\ \left\|\tilde{U}_j\left(e_0\right)-e_0\right\|_\infty^{1-\frac{m}{q(m+1)}}\left\|\tilde{U}_j\left(\left|\varphi_t\right|^{q(m+1)}\right)\right\|_\infty^{\frac{m}{q(m+1)}}\Omega_1\left(X^{(m)};\delta_j\left(m,q\right)\right)_{L_q}\ge\frac{\varepsilon}{2(m+2)Z_{m,q}}\right\}
$$

$$
A_3\left(\varepsilon\right)=\left\{j\in\mathbb{N}_0:\ \left\|\tilde{U}_j\left(e_0\right)-e_0\right\|_\infty^{1-\frac{1}{q}}\left\|\tilde{U}_j\left(\left|\varphi_t\right|^{q(m+1)}\right)\right\|_\infty^{\frac{m}{q(m+1)}}\Omega_1\left(X^{(m)};\delta_j\left(m,q\right)\right)_{L_q}\ge\frac{\varepsilon}{4(m+2)Z_{m,q}}\right\}
$$

$$A_4(\varepsilon) = \left\{ j \in \mathbb{N}_0 : \left\| \tilde{U}_j \left( |\varphi_t|^{q(m+1)} \right) \right\|_\infty^{\frac{m}{q(m+1)}} \geq \frac{\varepsilon}{2(m+2)Z_{m,q}} \right\}$$

$$B_k(\varepsilon) = \left\{ j \in \mathbb{N}_0 : \left\| \tilde{U}_j(e_0) - e_0 \right\|_\infty^{\frac{q(m+1)-k}{q(m+1)}} \left\| \tilde{U}_j \left( |\varphi_t|^{q(m+1)} \right) \right\|_\infty^{\frac{k}{q(m+1)}} \geq \frac{\varepsilon}{2(m+2)Z_{m,q}} \right\}, \; k = 1, 2, ..., m$$

$$C_k(\varepsilon) = \left\{ j \in \mathbb{N}_0 : \left\| \tilde{U}_j \left( |\varphi_t|^{q(m+1)} \right) \right\|_\infty^{\frac{k}{q(m+1)}} \geq \frac{\varepsilon}{2(m+2)Z_{m,q}} \right\}, \; k = 1, 2, ..., m.$$

Hence we obtain

$$D(\varepsilon) \subset \left( \bigcup_{i=1}^{4} A_i(\varepsilon) \right) \cup \left( \bigcup_{k=1}^{m} B_k(\varepsilon) \right) \cup \left( \bigcup_{k=1}^{m} C_k(\varepsilon) \right)$$

which implies

$$\delta_P(D(\varepsilon)) \leq \sum_{i=1}^{4} \delta_P(A_i(\varepsilon)) + \sum_{k=1}^{m} \delta_P(B_k(\varepsilon)) + \sum_{k=1}^{m} \delta_P(C_k(\varepsilon)).$$

From the hypothesis, we obtain for each fixed $m \in \mathbb{N}$ and $q \in (1, \infty)$ and for each $k = 1, 2, ..., m$

$$st_P - \lim \left\| \tilde{U}_j(e_0) - e_0 \right\|_\infty^{\frac{q(m+1)-k}{q(m+1)}} = 0, \tag{1.2.3}$$

$$st_P - \lim \left\| \tilde{U}_j \left( |\varphi_t|^{q(m+1)} \right) \right\|_\infty^{\frac{k}{q(m+1)}} = 0$$

and

$$st_P - \lim \delta_j(m, q) = 0.$$

Hence by using Lemma 1.1, we get

$$st_P - \lim \Omega_1 \left( X^{(m)}; \delta_j(m, q) \right)_{L_q} = 0.$$

Consider the following sets: $H(\varepsilon) =$

$$\left\{ j \in \mathbb{N}_0 : \left\| \tilde{U}_j(e_0) - e_0 \right\|_\infty^{\frac{q(m+1)-k}{q(m+1)}} \left\| \tilde{U}_j \left( |\varphi_t|^{q(m+1)} \right) \right\|_\infty^{\frac{k}{q(m+1)}} \Omega_1 \left( X^{(m)}; \delta_j(m, q) \right)_{L_q} \geq \varepsilon \right\}$$

$$H_1(\varepsilon) = \left\{ j \in \mathbb{N}_0 : \left\| \tilde{U}_j(e_0) - e_0 \right\|_\infty^{\frac{q(m+1)-k}{q(m+1)}} \geq \sqrt[3]{\varepsilon} \right\}$$

$$H_2(\varepsilon) = \left\{ j \in \mathbb{N}_0 : \left\| \tilde{U}_j \left( |\varphi_t|^{q(m+1)} \right) \right\|_\infty^{\frac{k}{q(m+1)}} \geq \sqrt[3]{\varepsilon} \right\}$$

$$H_3(\varepsilon) = \left\{ j \in \mathbb{N}_0 : \Omega_1 \left( X^{(m)}; \delta_j(m, q) \right)_{L_q} \geq \sqrt[3]{\varepsilon} \right\}.$$

For every $\varepsilon > 0$, since

$$H(\varepsilon) \subset H_1(\varepsilon) \cup H_2(\varepsilon) \cup H_3(\varepsilon)$$

we have for every $t \in (0,R)$,

$$\frac{1}{p(t)} \sum_{j \in H(\varepsilon)} p_j t^j \leq \frac{1}{p(t)} \sum_{j \in H_1(\varepsilon)} p_j t^j + \frac{1}{p(t)} \sum_{j \in H_2(\varepsilon)} p_j t^j + \frac{1}{p(t)} \sum_{j \in H_3(\varepsilon)} p_j t^j.$$

Taking limit whenever $t \to R^-$ in the above inequality, we observe that

$$st_P - \lim \left\| \tilde{U}_j(e_0) - e_0 \right\|^{\frac{q(m+1)-k}{q(m+1)}} \left\| \tilde{U}_j \left( |\varphi_t|^{q(m+1)} \right) \right\|_\infty^{\frac{k}{q(m+1)}} \Omega_1 \left( X^{(m)}; \delta_j(m,q) \right)_{L_q} = 0.$$

One can observe that for every $\varepsilon > 0$,

$$\begin{aligned}
\delta_P(A_i(\varepsilon)) &= 0, \, i = 1,2,3,4 \\
\delta_P(B_k(\varepsilon)) &= \delta_P(C_k(\varepsilon)) = 0, \, k = 1,2,...,m
\end{aligned}$$

hold. These imply that for every $\varepsilon > 0$,

$$\delta_P(D(\varepsilon)) = 0$$

which completes the proof.

### Theorem 1.2

*Consider a sequence of positive linear operators $\left\{ U_j \right\}_{j \in \mathbb{N}_0}$ on stochastic process. Let $\left\{ \tilde{U}_j \right\}_{j \in \mathbb{N}_0}$ be a corresponding sequence of positive linear operators which satisfies the following equality for every stochastic process $X$ from $[a,b] \times (\Gamma, \mathscr{A}, \mu)$ into $\mathbb{R}$*

$$U_j(X)(t,\gamma) = \tilde{U}_j(X(.,\gamma);t) \text{ for } \gamma \in \Gamma \text{ and } t \in [a,b].$$

*Also suppose that*

$$st_P - \lim \left\| \tilde{U}_j(e_0) - e_0 \right\|_\infty = 0 \tag{1.2.4}$$

*and*

$$st_P - \lim \left\| \tilde{U}_j \left( |\varphi_t|^{m+1} \right) \right\|_\infty = 0, \text{ fixed } m \in \mathbb{N}.$$

*Then we have for all $X$ satisfying Assumption 3 and Assumption 5*

$$st_P - \lim \left\| E \left| U_j(X) - X \right| \right\|_\infty = 0.$$

The proof can be obtained similarly to the above theorem by the help of Corollary 2.2 of [4].

**Example 7** *The classical Bernstein operator is defined by for every $f \in C([0,1])$,*

$$B_j(f)(t) = \sum_{k=0}^{j} f\left(\frac{k}{j}\right) \binom{j}{k} t^k (1-t)^{j-k}, \, t \in [0,1], \, j \in \mathbb{N}_0.$$

*Consider the following operator*

$$\tilde{B}_j(X)(t,\gamma) = B_j(X(.,\gamma))(t)$$

$$= \sum_{k=0}^{j} X\left(\frac{k}{j},\gamma\right)\binom{j}{k} t^k (1-t)^{j-k}, \ t \in [0,1].$$

*By using the operators $B_j$ and $\tilde{B}_j$, we can write*

$$U_j(X(.,\gamma))(t) = (1+s_j)B_j(X(.,\gamma))(t)$$

*and*

$$\tilde{U}_j(X)(t,\gamma) = (1+s_j)\tilde{B}_j(X)(t,\gamma),$$

*where $(s_j)$ is P-convergent to 1 and not convergent in the classical sense. These operators $U_j$ and $\tilde{U}_j$ satisfy the assumptions of Theorem 1.1 but doesn't satisfy classical theorem. This shows that our theorem is more general.*

## 1.3  Korovkin Theory for Stochastic Process via Commutative Conformable Fractional Derivative

**Definition 1.8**  Let $a,b \in \mathbb{R}$. The left conformable fractional derivative starting from $a$ of a function $f : [a,\infty) \to \mathbb{R}$ of order $0 < \alpha \le 1$ is defined by

$$(T_\alpha^a f)(t) = \lim_{\varepsilon \to 0} \frac{f\left(t + \varepsilon(t-a)^{1-\alpha}\right) - f(t)}{\varepsilon}.$$

If $(T_\alpha^a f)(t)$ exists on $(a,b)$, then

$$(T_\alpha^a f)(a) = \lim_{t \to a^+} (T_\alpha^a f)(t).$$

The right conformable fractional derivative of order $0 < \alpha \le 1$ terminating at $b$ of $f : (-\infty, b] \to \mathbb{R}$ is defined by

$$\left(_\alpha^b T f\right)(t) = \lim_{\varepsilon \to 0} \frac{f\left(t + \varepsilon(b-t)^{1-\alpha}\right) - f(t)}{\varepsilon}.$$

If $\left(_\alpha^b T f\right)(t)$ exists on $(a,b)$, then

$$\left(\left(_\alpha^b T f\right)(b)\right) = \lim_{t \to b^-} \left(_\alpha^b T f\right)(t).$$

**Definition 1.9**    Let $\alpha \in (n, n+1]$ and set $\beta = \alpha - n$. Then, the left conformable fractional derivative starting from $a$ of a function $f : [a, \infty) \to \mathbb{R}$ of order $\alpha$, where $f^{(n)}(t)$ exists, is defined by

$$(T_\alpha^a f)(t) = \left(T_\beta^a f^{(n)}\right)(t).$$

The right conformable fractional derivative of order $\alpha$ terminating at $b$ of $f : (-\infty, b] \to \mathbb{R}$, where $f^{(n)}(t)$ exists, is defined by

$$\left({}_\alpha^b T f\right)(t) = (-1)^{n+1}\left({}_\alpha^b T f^{(n)}\right)(t).$$

Consider $(\Gamma, \mathscr{A}, \mu)$ being probability space and $L_1(\Gamma, \mathscr{A}, \mu)$ being the space of all real valued random variables $Y = Y(\gamma)$ with

$$\int_\Gamma |Y(\gamma)|\, \mu(d\gamma) < \infty.$$

Let $X = X(t, \gamma)$ be a stochastic process from $[a, b] \times (\Gamma, \mathscr{A}, \mu)$ into $\mathbb{R}$. $C_\Gamma([a, b]) := C([a, b], L_1(\Gamma, \mathscr{A}, \mu))$ is the space of $L_1$-continuous stochastic process in $t$.
$B_\Gamma([a, b]) = \left\{X : \sup_{t \in [a,b]} \int_\Gamma |X(t, \gamma)|\, \mu(d\gamma) < \infty\right\}$. Observe that $B_\Gamma([a, b]) \subset$ $C_\Gamma([a, b])$. Let $\alpha \in (n, n+1)$, $n \in \mathbb{Z}_+$ and consider the subspace of stochastic processes satisfying the following conditions:

■   $X(., \gamma) \in C^{n+1}([a, b])$ for any $\gamma \in \Gamma$,

■   There exists a positive real number $M$ such that $\left|X^{(n+1)}(t, \gamma)\right| \leq M$ for any $(t, \gamma) \in [a, b] \times \Gamma$,

■   $X^{(k)}(t, \gamma) \in C_\Gamma([a, b])$, $k = 0, 1, 2, \ldots, n$,

■   $X^{(n+1)}(t, \gamma)$ is a stochastic process for $(t, \gamma) \in [a, b] \times \Gamma$.

We use the following definition of modulus of continuity in [8]

$$\omega_1\left(E(T_\alpha^v X); \sigma\right)_{[v,b]} = \sup_{\substack{|t-x| \leq \sigma \\ t, x \in [v,b]}} \left|(E(T_\alpha^v X))(t) - (E(T_\alpha^v X))(x)\right|, \ \sigma > 0$$

$$\omega_1\left(E({}_\alpha^v T X); \sigma\right)_{[a,v]} = \sup_{\substack{|t-x| \leq \sigma \\ v, x \in [a,v]}} \left|(E({}_\alpha^v T X))(t) - (E({}_\alpha^v T X))(x)\right|, \ \delta > 0$$

$$\omega_1\left(E({}^\alpha T^v); \sigma\right) = \max\left\{\omega_1\left(E(T_\alpha^v X); \sigma_1\right)_{[v,b]}, \omega_1\left(E({}_\alpha^v T X); \sigma_2\right)_{[a,v]}\right\}, \ \sigma, \sigma_1, \sigma_2 > 0$$

### Theorem 1.3

*Let $\{U_j\}_{j \in \mathbb{N}_0}$ be a sequence of positive E-commutative linear operators from $C_\Gamma([a,b])$ into $B_\Gamma([a,b])$. Also assume that*

$$st_P - \lim \left\| U_j(e_0) - e_0 \right\|_\infty = 0 \tag{1.3.1}$$

*and*

$$st_P - \lim \left\| U_j\left( |.-v|^{\alpha+1} \right)(v) \right\|_\infty = 0 \tag{1.3.2}$$

*we have for all $X \in C_\Gamma^{\alpha,n+1}([a,b])$*

$$st_P - \lim \left\| E(U_jX)(v) - (EX)(v) \right\|_\infty = 0$$

*where $\alpha \in (m, m+1)$, $m \in \mathbb{Z}_+$.*

**Proof 1.3**  Following Theorem 11 of [8], we know that for all $u \in [a,b]$

$$\left\| E(U_jX) - (EX) \right\|_\infty \leq \left\| EX \right\|_\infty \left\| U_j(e_0) - e_0 \right\|_\infty$$

$$+ \sum_{k=1}^m \frac{\left\| EX^{(k)} \right\|_\infty}{k!} \left\| U_j(.-v)^k(v) \right\|_\infty$$

$$+ \frac{\displaystyle\sup_{v \in [a,b]} \omega_1\left( E\left( {}^\alpha T^v X \right), \sigma \right)}{\displaystyle\prod_{n=1}^m (\alpha + n - m)}$$

$$\times \left\| U_j\left( |.-v|^{\alpha+1} \right)(v) \right\|_\infty^{\frac{\alpha}{\alpha+1}} \left[ \frac{\left\| U_j(e_0) \right\|_\infty^{\frac{1}{\alpha+1}}}{\alpha - m} - \frac{\left\| U_j\left( |.-v|^{\alpha+1} \right)(v) \right\|_\infty^{\frac{\alpha}{\alpha+1}}}{\delta(\alpha+1)} \right]$$

$$< \infty.$$

On the other hand the following inequality

$$\left\| U_j(s-v)^k(v) \right\|_\infty \leq \left\| U_j\left( |.-v|^k \right)(v) \right\|_\infty$$

$$\leq \left\| U_j\left( |.-v|^{\alpha+1} \right)(v) \right\|_\infty^{\frac{k}{\alpha+1}} \cdot \left\| U_j(e_0) \right\|_\infty^{\frac{\alpha+1-k}{\alpha+1}}$$

holds for all $k = 1, 2, ..., m$.

$$\left\| E\left(U_j X\right) - (EX) \right\|_\infty \leq \left\| EX \right\|_\infty \left\| U_j\left(e_0\right) - e_0 \right\|_\infty$$

$$+ \sum_{k=1}^{m} \frac{\left\| EX^{(k)} \right\|_\infty}{k!} \left\| U_j \left| . - v \right|^{\alpha+1}(v) \right\|_\infty^{\frac{k}{\alpha+1}} \left\| U_j\left(e_0\right) \right\|_\infty^{\frac{\alpha+1-k}{\alpha+1}}$$

$$+ \frac{\displaystyle\sup_{v\in[a,b]} \omega_1\left(E\left({}^{\alpha}T^v X\right), \sigma\right)}{\displaystyle\prod_{n=1}^{m}\left(\alpha+n-m\right)} \cdot \left\| U_j\left(\left| . - v \right|^{\alpha+1}\right)(v) \right\|_\infty^{\frac{\alpha}{\alpha+1}}$$

$$\times \left[ \frac{\left\| U_j\left(e_0\right) \right\|_\infty^{\frac{1}{\alpha+1}}}{\alpha-m} - \frac{\left\| U_j\left(\left| . - v \right|^{\alpha+1}\right)(v) \right\|_\infty^{\frac{1}{\alpha+1}}}{\sigma\left(\alpha+1\right)} \right].$$

By the hypothesis, the sequence $\left\{ \left\| U_j\left(e_0\right) \right\|_\infty \right\}_{j\in\mathbb{N}}$ is $P$-statistical bounded, i.e., $\left\| U_j\left(e_0\right) \right\|_\infty \leq K$ for every $j \in K$ having $P$-density 1. Also $\omega_1\left(E\left({}^{\alpha}T^v X\right), \sigma\right)$ is bounded. Consider the sets:

$$D(\varepsilon) = \left\{ j : \left\| E\left(U_j X\right) - (EX) \right\|_\infty \geq \varepsilon \right\},$$

$$A_1(\varepsilon) = \left\{ j : \left\| U_j\left(e_0\right) - e_0 \right\|_\infty \geq \frac{\varepsilon}{\left\| EX \right\|_\infty (m+3)} \right\},$$

$$A_2^{(k)}(\varepsilon) = \left\{ j : \left\| U_j \left| . - v \right|^{\alpha+1}(v) \right\|_\infty^{\frac{k}{\alpha+1}} \geq \frac{\varepsilon}{\displaystyle\max_{1\leq k\leq m} \frac{\left\| EX^{(k)} \right\|_\infty}{k!} \left\| U_j(e_0) \right\|_\infty^{\frac{\alpha+1-k}{\alpha+1}} (m+3)} \right\}, \quad k = 1, 2, ..., m,$$

$$A_3(\varepsilon) = \left\{ j : \left\| U_j \left(\left| . - v \right|^{\alpha+1}\right)(v) \right\|_\infty^{\frac{\alpha}{\alpha+1}} \geq \frac{\varepsilon}{\dfrac{\displaystyle\sup_{v\in[a,b]} \omega_1(E({}^{\alpha}T^v X),\sigma)}{\displaystyle\prod_{n=1}^{m}(\alpha+n-m)} \dfrac{\left\| U_j(e_0) \right\|_\infty^{\frac{1}{\alpha+1}}}{\alpha-m} (m+3)} \right\},$$

$$A_4(\varepsilon) = \left\{ j : \left\| U_j \left(\left| . - v \right|^{\alpha+1}\right)(v) \right\|_\infty \geq \frac{\varepsilon}{\dfrac{\displaystyle\sup_{v\in[a,b]} \omega_1(E({}^{\alpha}T^v X),\sigma)}{\displaystyle\prod_{n=1}^{m}(\alpha+n-m)\sigma(\alpha+1)} (m+3)} \right\}.$$

Hence we obtain

$$D(\varepsilon)\cap K \subset \left[ A_1(\varepsilon) \cup \left( \bigcup_{k=1}^{m} A_2^{(k)}(\varepsilon) \right) \cup A_3(\varepsilon) \cup A_4(\varepsilon) \right] \cap K.$$

which implies

$$\frac{1}{p(t)} \sum_{j\in D(\varepsilon)\cap K} p_j t^j \leq \frac{1}{p(t)} \sum_{j\in A\cap K} p_j t^j \leq \frac{1}{p(t)} \sum_{j\in A} p_j t^j. \tag{1.3.3}$$

where $A := A_1(\varepsilon) \cup \left( \bigcup_{k=1}^{m} A_2^{(k)}(\varepsilon) \right) \cup A_3(\varepsilon) \cup A_4(\varepsilon)$. By using (1.3.1) we get

$$\lim_{t \to R^-} \frac{1}{p(t)} \sum_{j \in A_1(\varepsilon)} p_j t^j = 0$$

and by the help of (1.3.2) we have

$$\lim_{t \to R^-} \frac{1}{p(t)} \sum_{j \in A_2^{(k)}(\varepsilon)} p_j t^j = 0, \, k = 1, 2, \dots, m$$

$$\lim_{t \to R^-} \frac{1}{p(t)} \sum_{j \in A_3(\varepsilon)} p_j t^j = 0$$

and

$$\lim_{t \to R^-} \frac{1}{p(t)} \sum_{j \in A_4(\varepsilon)} p_j t^j = 0.$$

Taking limit whenever $t \to R^-$ in (1.3.3), the following limit

$$\lim_{t \to R^-} \frac{1}{p(t)} \sum_{j \in D(\varepsilon) \cap K} p_j t^j = 0$$

exists. We have the following inequality

$$\frac{1}{p(t)} \sum_{j \in D(\varepsilon)} p_j t^j = \frac{1}{p(t)} \sum_{j \in D(\varepsilon) \cap K} p_j t^j + \frac{1}{p(t)} \sum_{j \in D(\varepsilon) \cap (\mathbb{N}_0 \setminus K)} p_j t^j$$

$$\leq \frac{1}{p(t)} \sum_{j \in D(\varepsilon) \cap K} p_j t^j + \frac{1}{p(t)} \sum_{j \in \mathbb{N}_0 \setminus K} p_j t^j$$

which implies that $st_P - \lim \left\| E(U_j X)(v) - (EX)(v) \right\|_\infty = 0$.

**Example 8** *Consider the operator $U_j$ in the Example 7, observe that $U_j$ : $C_\Gamma([0,1]) \to C_\Gamma([0,1])$ and*

$$(EU_j(X))(t) = \sum_{k=0}^{j} (EX) \left( \frac{k}{j} \right) \binom{j}{k} t^k (1-t)^{j-k}, \, t \in [0,1] \text{ and } f \in C_\Gamma([0,1]).$$

*This shows that $U_j$ is an E-commutative positive linear operator, i.e., $EU_j = U_j E$. This example satisfies the conditions of Theorem 1.3 but doesn't satisfy classical theorem. This shows that our theorem is more general.*

# References


[1] Altomare, F. and Diomede, S. 2001. Contractive Korovkin subsets in weighted spaces of continuous functions. Rend. Circ. Mat. Palermo. 50: 547–568.

[2] Anastassiou, G.A. 1991. Korovkin inequalities for stochastic process. J. Math. Anal. Appl. 157: 366–384.

[3] Anastassiou, G.A. 2005. On basic fuzzy Korovkin theory. Stud. Univ. Babes-Bolyai Math. 50: 3–10.

[4] Anastassiou, G.A. 2007. Stochastic Korovkin theory given quantitatively. Facta. Univ. Ser. Math. Inform. 22: 43–60.

[5] Anastassiou, G.A. and Duman, O. 2008. Statistical fuzzy approximation by fuzzy positive linear operators. Comput. and Math. Appl. 55: 573–580.

[6] Anastassiou, G.A., Duman, O. and Erkuş-Duman, E. 2009. Statistical approximation for stochastic processes. Stoch. Anal. Appl. 27: 460–474.

[7] Anastassiou, G.A. and Duman, O. 2010. Statistical Korovkin theory for multivariate stochastic processes. Stoch. Anal. Appl. 28: 648–661.

[8] Anastassiou, G.A. 2021. Commutative conformable fractional Korovkin properties for stochastic process. Acta Math. Univ. Comenianae 90: 259–276.

[9] Arif, A. and Yurdakadim, T. 2022. Approximation results on nonlinear operators by $P_p$-statistical convergence. Adv. Studies: Euro-Tbilisi Math. J. 15(3): 1–10.

[10] Atlıhan, Ö.G., Yurdakadim, T. and Taş, E. 2022. A new approach to approximation by positive linear operators in weighted spaces. Ukrainian J. Math. 74(11): 1447–1453.

[11] Balaz, V. and Salat, T. 2006. Uniform density $u$ and corresponding $I_u$-convergence. Math. Commun. 11: 1–7.

[12] Bardaro, C., Musielak, J. and Vinti, G. 2003. Nonlinear Integral Operators and Applications. De Gruyter Ser. Nonlinear Anal. Appl. 9, Walter de Gruyter, Berlin.

[13] Bardaro, C., Boccuto, A., Dimitriou, X. and Mantellini, I. 2009. A Korovkin theorem in multivariate modular function spaces. J. Func. Spaces Appl. 7: 105–120.

[14] Bardaro, C., Boccuto, A., Dimitriou, X. and Mantellini, I. 2013. Abstract korovkin type theorems in modular spaces and applications. Cent. Eur. J. Math. 11: 1774–1784.

[15] Baxhau, B., Agrawal, P.N. and Shukla, R. 2022. Some fuzzy Korovkin type approximation theorems via power series summability method. Soft Comput. 26: 11373–11379.

[16] Bede, B., Nobuhara, H., Fodor, J. and Hirota, K. 2006. Max-product Shepard approximation operators. J. Adv. Comput. Intelligence Intelligent Informatics 10: 494–497.

[17] Bede, B., Nobuhara, H., Dankova, M. and Nola, A.D. 2008. Approximation by pseudo-linear operators. Fuzzy Sets Syst. 159: 804–820.

[18] Bede, B., Coroianu, L. and Gal, S.G. 2009. Approximation and shape preserving properties of the Bernstein operators of max-product kind. Int. J. Math. Sci. Art. ID 590589, 22 pp.

[19] Bernstein, S.N. 1912. Demonstration du theoreme de Weierstrass fondee. Communications of the Kharkov Mathematical Society 13: 1–2.

[20] Boos, J. 2000. Classical and Modern Methods in Summability. Oxford University Press.

[21] Bohman, H. 1952. On approximation of continuous and of analytic functions. Ark. Mat. 2(1): 43–56.

[22] Buckley, J.J. 1991. Solving systems of linear fuzzy equations. Fuzzy Sets and Syst. 43: 33–43.

[23] Coker, D. 1997. An introduction to intuitionistic fuzzy topological spaces. Fuzzy Sets and Syst. 88: 81–89.

[24] Diethelm, K. 2004. The Analyis of Fractional Differential Equations. Lecture Notes in Mathematics. Springer.

[25] Duman, O., Khan, M.K. and Orhan, C. 2003. A-statistical convergence of approximating operators. Math. Inequal. Appl. 6: 689–699.

[26] Dzyadyk, V.K. 1959. On the best trigonometric approximation in the $L$ metric of some functions. Dokl. Akad. Nauk. SSSR 129: 19–22.

[27] El Naschie, M.S. 1998. On uncertainty of Cantorian geometry and two-slit experiment. Chaos Solitons and Fractals 9: 517–529.

[28] El Naschie, M.S. 2004. A review of E-infinity theory and mass spectrum of high energy particle physics. Chaos Solitons and Fractals 19: 209–236.

[29] Fast, H. 1951. Sur la convergence statistique. Colloq. Math. 2: 241–244.

[30] Fridy, J.A. and Orhan, C. 1997. Statistical limit superior and limit inferior. Proc. Amer. Mat. Soc. 125(12): 3625–3631.

[31] Fridy, J.A. 1985. On statistical convergence. Analysis 5: 301–313.

[32] Freedman, A.R. and Sember, J.J. 1981. Densities and summability. Pacific J. Math. 95: 293–305.

[33] Gadjiev, A.D. and Orhan, C. 2002. Some approximation theorems via statistical convergence. Rocky Mountain Journal of Math. 32: 129–137.

[34] Gal, S. 2000. Approximation theory in fuzzy setting. Chapter 13 in Handbook of Analytic-Computational Methods in Applied Mathematic. Chapman and Hall/CRC, Boca Raton, 617–666.

[35] Goetschel, R.J. and Voxman, W. 1986. Elementary fuzzy calculus. Fuzzy Sets and Syst. 18: 31–43.

[36] Gökçer, T.Y. and Duman, O. 2020. Approximation by max-min operators: A general theory and its application. Fuzzy Sets and Systems 394: 146–161.

[37] Gökçer, T.Y. and Duman, O. 2016. Summation process by max-product operators. Comp. Anal. Springer Proceedings in Math. and Stat. 155, DOI 10.1007/978-3-319-28443-9-4.

[38] Gökçer, T.Y. and Aslan, İ. 2022. Approximation by Kantorovich-type max-min operators. Appl. Math. Comput. 423: 127011.

[39] Hardy, G.H. 1949. Divergent Series. Oxford University Press, London.

[40] He, J.H. 2008. Max-min approach to nonlinear oscillators. Int. J. Nonlinear Sci. Numer. Simul. 9(2): 207–210.

[41] Izuchi, K., Takagi, H. and Watanabe, S. 1996. Sequential BKW-operators and function algebras. J. Approx. Theory 85: 185–200.

[42] Kac, V. and Cheung, P. 2002. Quantum Calculus. Springer-Verlag.

[43] Karakus, S., Demirci, K. and Duman, O. 2010. Statistical approximation by positive linear operators on modular spaces. Positivity 14: 321–334.

[44] Karakus, S. and Demirci, K. 2010. Matrix summability and Korovkin type approximation theorem on modular spaces. Acta Math. Univ. Commenianae. 2: 281–292.

[45] Karsli, H. and Gupta, V. 2008. Some approximation properties of $q$-Chlodowsky operators. Applied Mathematics and Computation 195: 220–229.

[46] Kolk, E. 1993. Matrix summability of statistically convergent sequence. Analysis 13: 77–83.

[47] Kratz, W. and Stadtmüller, U. 1989. Tauberian theorems for $J_p$-summability. J. Math. Anal. Appl. 139: 362–371.

[48] Korovkin, P.P. 1953. On convergence of linear positive operators in the space of continuous functions. Doklady Akad. Nauk SSR. 90: 961–964.

[49] Lorentz, G.G. 1986. Bernstein Polynomials. AMS Chelsea Publishing.

[50] Lupaş, A. 1987. A $q$-analogue of the Bernstein operator. Seminar on Numerical and Statistical Calculus, University of Cluj-Napoca 9: 85–92.

[51] Micchelli, C.A. 1975. Convergence of positive linear operators on C(X). J. Approx. Theory 13: 305–315.

[52] Miller, H.I. 1995. A measure theoretical subsequences characterization of statistical convergence. Trans. Amer. Math. Soc. 347(5): 1811–1819.

[53] Nuray, F. and Savaş, E. 1995. Statistical convergence of sequences of fuzzy numbers. Math. Slovaca. 45: 269–273.

[54] Olgun, M., Ünver, M. and Yardımcı, Ş. 2019. Pythagorean fuzzy topological spaces. Complex and Intelligent Systems 5: 177–183.

[55] Oruc, H. and Tuncer, N. 2002. On the convergence and iterates of $q$-Bernstein polynomials. Journal of Approximation Theory 117: 301–313.

[56] Ostrovska, S. 2003. $q$-Bernstein polynomials and their iterates. J. Approx. Th. 123: 232–255.

[57] Peeva, K. 1992. Fuzzy linear systems. Fuzzy Sets and Syst. 49: 339–355.

[58] Phillips, G.M. 1997. Bernstein polynomials based on the $q$-integers. Annals of Numerical Mathematics 4: 511–518.

[59] Podlubny, I. 2001. Geometric and physical interpretation of fractional integration and fractional differentiation. arXiv preprint math/0110241.

[60] Popoviciu, T. 1950. Asupra demonstratiei teoremei lui weierstrass cu ajutorul polinoamelor de interpolare. Lucrarile Ses, Gen. St. Acad. Române din 1–4.

[61] Powell, R. E. and Shah, S. H. 1972. Summability Theory and its Applications. Van Nostrand-Reinhold.

[62] Qian, Y. and Yu, D. 1980. Rates of approximation by neural network interpolation operators. Appl. Math. Comput. 418: 126781.

[63] Sakaoğlu, I. and Orhan, C. 2013. Strong summation process in $L_p$ spaces. Nonlinear Analysis 86: 89–94.

[64] Salat, T. 1980. On statistically convergent sequences of real numbers. Mat. Slovaca. 30: 139–150.

[65] Samko, S.G., Kilbas, A.A. and Marichev, O.I. 1993. Fractional integrals and derivatives. Gordon and Breach Science Publishers, Yverdon Yverdon-les-Bains, Switzerland, Vol. 1.

[66] Shen, Z., Yang, H. and Zhang, S. 2022. Optimal approximation rate of ReLU networks in terms of width and depth. J. Math. Pures Appl. 157: 101–135.

[67] Taş, E., Orhan, C. and Yurdakadim, T. 2013. The Stancu-Chlodowsky operators based on $q$-Calculus. In: Proceedings of the 11th International Conference of Numerical Analysis and Applied Mathematics (ICNAAM), Sept 21–27 2013, Rhodes Island, GREECE. AIP Conference Proceeding 1558: 1152–1155.

[68] Taş, E. and Yurdakadim, T. 2019. Korovkin theory for extraordinary test functions by A-statistical convergence. Palestine Journal of Mathematics 8(1): 86–90.

[69] Taş, E. and Yurdakadim, T. 2017. Approximation by positive linear operators in modular spaces by power series method. Positivity 21: 1293–1306.

[70] Ünver, M. and Orhan, C. 2019. Statistical convergence with respect to power series methods and applications to approximation theory. Numer. Func. Anal. Opt. 40: 535–547.

[71] Weierstrass, K.G. 1885. Über die analytische Dorstellbarkeit sogenannter willkürlicher Funktionen einer reellen. Veranderlichen. Sitzungsber, Akad. Berlin.

[72] Weba, M. 1986. Korovkin systems of stochastic processes. Math. Z. 192: 73–80.

[73] Yurdakadim, T. and Taş, E. 2022. Effects of fuzzy setting in Korovkin theory via $P_p$-statistical convergence. Roman. J. Math. 2(12): 1–8.

[74] Zadeh, L.A. 1965. Fuzzy sets. Inform. and Control 8: 338–353.

# Chapter 2

# Fractional Korovkin-type Results by $P$-statistical Convergence

*Tuğba Yurdakadim*

## 2.1　Introduction and Preliminaries

One of the main themes of approximation theory is to express an arbitrary function in terms of nicer or simpler functions or both. In fact, while expanding a function to a power series, we are expressing the function in terms of polynomials which are both nice and simple. Therefore, one of the trivial questions is what is a good class of nice and simple functions which approximates better and what kind of functions we can approximate. In this context, the theorem of Weierstrass lies in the heart of approximation theory and states that the class of algebraic polynomials is dense in $C[a,b]$ where $C[a,b] = \{f | f : [a,b] \longrightarrow \mathbb{R}, continuous\}$ [78]. The proof of this theorem is not easy to follow and Bernstein is the first who has achieved to give a well-known, more understandable alternative proof [19], [27], [51]. Then this proof has lead to study positive, linear operators in approximation theory [21], [50], [65]. It has been obtained that one can determine the

Department of Mathematics, Bilecik Şeyh Edebali University, Bilecik, 11100, Turkey.
Email: tugbayurdakadim@hotmail.com

uniform convergence of such operators to identity operator by minimum calculations which are also known as Korovkin-type results, i.e., $T_n(f)$ converges uniformly to $f$ for every $f \in C[a,b]$ if $T_n(x^j)$ converges to $x^j$, $j = 0, 1, 2$ uniformly on $[a,b]$ where $(T_n)$ is a sequence of positive linear operators. Therefore Korovkin-type theorems are used frequently and many mathematicians have studied such problems of this theory in various ways. Let us mention them in details:

(1) Some of them have used different types of convergences instead of classical limit since these concepts generalize the ordinary convergence [9], [11], [30], [32], [34], [72], [77]. It will be nice to pause in order to mention the importance of summability theory. The aim of summability theory is to assign a limit and a sum to divergent sequences and series, respectively. Studying the convergence of series is an ancient art. Before L. Euler (1707–1783), divergent series and sequences were out of interest, indeed they were considered to be illeginate. Therefore they were all left alone up to Euler. He then developed a way that a divergent series $\sum_{k=0}^{\infty} a_k = a$ if $\sum_{k=0}^{\infty} a_k z^k$ converges to $f(z)$ for small values of $z$ and $f(1) = a$. In this way, $\frac{1}{1+z} = \sum_{k=0}^{\infty}(-1)^k z^k, |z| < 1$ and hence $\sum_{k=0}^{\infty}(-1)^k = \frac{1}{2}$. However, $\frac{1-z^2}{1-z^3} = 1 - z^2 + z^3 - z^5 + z^6 \ldots$ and $\frac{1-z^2}{1-z^3} = \frac{1+z}{1+z+z^2}$ for $z \neq 1$. Therefore, $\frac{1+z}{1+z+z^2} = 1 - z^2 + z^3 - z^5 + z^6 \ldots$ and for $z = 1$, $\sum_{k=0}^{\infty}(-1)^k = \frac{1+1}{1+1+1} = \frac{2}{3}$. But in this way, it seems that one can assign any value to $\sum_{k=0}^{\infty}(-1)^k$ with Euler's notation. C. F. Gauss (1777–1855) has also introduced the concept of the use of infinite processes into mathematical analysis. At a young age he obtained the binomial theorem and this made him interested in the convergence of power series. He has also made contributions to the studies on hypergeometric series and their convergence properties. A. L. Cauchy (1789–1857) has introduced rigor to mathematical analysis, for example, he has formulated the ideas on the convergence and divergence of power series. N. H. Abel (1802–1829) is a third important mathematician who has also dealt with the concepts of convergence and divergence. In the second half of nineteenth century, the interest on divergent series has declined seriously, but only to be rehandled at a later date. Among those to start reinvestigations of divergent series was E. Cesàro (1859–1906) who has introduced $C_1$-convergence. This has allowed mathematicians to argue the Cauchy product of two infinite series and its convergence properties. Up untill this time, many extensive studies on summability have been made [20], [66]. Gadjiev and Orhan are the first who have combined approximation and summability theories [34] and then many results of approximation theory have been investigated with the use of different concepts of convergences of summability theory [26], [68].

(2) Recently, some of these have considered nonlinear operators. Notice that while we are working with linear operators we have a linear structure in the algebraic sense. But we should ask that whether we have to use linear operators or not. Bede et al. have answered this question by considering nonlinear operators,

in particular pseudo-linear operators, operators of max-product type, max-min type in the theory of approximation [12], [16], [17], [18]. It is worth mentioning that pseudo-linearity weakens the ordinary linearity condition. Besides the approach of Bede et al., there is also another type of max-min approach presented by He [41]. These operators give better results in the order of approximation, in computational complexity or in image processing and also these results have significant applications in deep learning which depends on feedforward neural networks, ReLU networks [17], [67], [71]. Therefore some approximation theorems dealing with these operators have been investigated by using different concepts of convergences [8], [37], [38], [39].

(3) Another perspective of these studies is fuzzy setting. Let us give a brief history of fuzzy theory. The class of pretty animals, or the class of nice cities in the world doesn't make sense from the perspective of set theory. Fuzziness is a popular way to measure the completeness of a certain supposition or concern and Zadeh [80] is the first who noticed this and introduced the fuzzy sets and fuzzy numbers in a systemathical way. He has described fuzzy sets with the use of membership functions by assigning to each element a grade of membership from 0 to 1. There is a slight modification of fuzzy numbers which has been presented by Goetschel and Voxman [36] and they have also introduced a function which is a metric on this family of sets. Furthermore, the uncertainty in the expressions of such classes plays a key role in human thinking, pattern recognition, machine learning, weather forecast, robotics, stock market and bio-medicines. Since it is effective to use membership functions to overcome the uncertainty and the impact of this situation in the problems of real world, all of these observations have motivated mathematicians to adapt the fundamental concepts of mathematics in a fuzzy setting [5], [22], [23], [56], [57], [61]. Fuzzy topology has been deeply studied and it has a significant application to quantum particle physics [28], [29]. In addition to these studies, fuzzy logic has also been considered in the theory of approximation, for example, Gal, Anastassiou and Duman, Yurdakadim and Tas, Baxhau et. al have presented some approximation theorems in fuzzy setting [2], [15], [35], [79].

(4) Some of them have investigated such results in weighted spaces, modular spaces and abstract spaces. For example, in [1], [10], [13], [14] [42], [53], [73], [75] to find other sets of functions which satisfy the same property as $\{1, x, x^2\}$ also called as Korovkin sets is one of the problems and the other problem is the investigation of Korovkin-type results in different function spaces or abstract Banach spaces [45], [46], [58], [76].

(5) $q$-calculus which plays an increasingly important role in applications to many branches of applied sciences and engineering acts as a bridge between mathematics and physics. Therefore it is important to fill the gaps in the mathematical the-

ory of $q$-calculus. Binomial expansion, factorial, Taylor expansion, derivatives have all been introduced in the sense of $q$-calculus and then many results which are obtained or unable to obtain have been examined with this setting [44], [59], [62], [74]. For example, the $q$-analog of Bernstein operators have been introduced by Lupaş [52]. Different type of $q$-Bernstein operators has also been introduced by Phillips [62] and Ostrovska [60] have investigated the approximation properties of these operators. Karsli and Gupta [47] have introduced $q$-Chlodowsky operators which extend $q$-Bernstein operators to an unbounded interval.

The last way to study Korovkin-type results is combining approximation theory and fractional calculus. Fractional calculus deals with integrals and derivatives of arbitrary order which are the fundamental concepts of mathematical analysis. The starting point of this problem is a letter between Leibniz and L'Hospital which is the discussion of half order derivative of $f(x) = x$. This interest has hesitated up to Liouville, Grünwald, Letnikov, Riemann and Caputo. From the last decades of 1900's, fractional derivatives have found their meaning in the real world since they are more appropriate to prescribe the problem. For example, they allow us to define the properties of polymers, rocks and different materials [25], [64], [70]. Unfortunately, the mathematical theory of fractional calculus falls behind the necessities of real world problems. Therefore it is worth examining the results of analysis in a fractional setting.

In this chapter our goals are to present fractional Korovkin-type theorems by $P$-statistical convergence where $P$ denotes power series and to provide examples which assert the importance of our results. Since statistical convergence and $P$-statistical convergence do not imply each other, our results make enough contribution to the literature. Now we can remember the basic concepts, definitions and also the known results which will be useful along the paper.

The first definition is the natural density of a subset $G \subseteq \mathbb{N}$ and the statistical convergence of a sequence $s = (s_n)$. It worths noting that statistical convergence is an effective generalization of classical convergence since it overcomes the lack of a classical limit.

**Definition 2.1**   The natural density of $G \subseteq \mathbb{N}$ is given by

$$\delta(G) := \lim_{k \to \infty} \frac{1}{k} \#\{n \le k : n \in G\}$$

if the limit exists where $\#E$ denotes the cardinality of $E$ and $\mathbb{N}$ is the set of all nonnegative integers. If $\delta(G_\varepsilon) = 0$ for every $\varepsilon > 0$ where $G_\varepsilon = \{n \in \mathbb{N} : |s_n - l| \ge \varepsilon\}$ then $s = (s_n)$ is said to be statistically convergent to $l$ [30], [32], [69].

As it can be seen that a convergent sequence is also statistically convergent but, in general the converse does not hold. Let us give a couple of examples to examine this situation in details.

**Example 1** *Let $m \in \mathbb{N}$ and*

$$s_n = \begin{cases} \frac{2n+1}{n+2} & , \quad n = m^2 \\ 0 & , \quad n \neq m^2. \end{cases}$$

*Then it is obvious that $s = (s_n)$ is not convergent and for every $\varepsilon > 0$,*

*$\{n \leq k : |s_n - 0| \geq \varepsilon\} \subset \{n \leq k : s_n \neq 0\}$ holds. Therefore one can obtain that $\#\{n \leq k : |s_n| \geq \varepsilon\} \leq \#\{n \leq k : s_n \neq 0\} \leq \sqrt{k}$ which implies*

*$\lim_{k \to \infty} \frac{1}{k} \#\{n \leq k : |s_n| \geq \varepsilon\} \leq \lim_{k \to \infty} \frac{1}{k} \#\{n \leq k : s_n \neq 0\} \leq \lim_{k \to \infty} \frac{1}{k} \sqrt{k} = 0$. This means that $\varepsilon > 0$, $\delta(K_\varepsilon) = 0$ where $K_\varepsilon = \{n \in \mathbb{N} : |s_n - 0| \geq \varepsilon\}$ and $s = (s_n)$ is statistically convergent to 0.*

**Example 2** *Let $m \in \mathbb{N}$ and*

$$s_n = \begin{cases} \sqrt{n} & , \quad n = m^2 \\ 6 & , \quad n \neq m^2. \end{cases}$$

*Of course, $(s_n)$ is not convergent since it is not bounded. On the other hand, $\#\{n \leq k : |s_n - 6| \geq \varepsilon\} \leq \#\{n \leq k : s_n \neq 6\} \leq \sqrt{k}$ holds for $\varepsilon > 0$*

*and it implies that*

*$\lim_{k \to \infty} \frac{1}{k} \#\{n \leq k : |s_n - 6| \geq \varepsilon\} \leq \lim_{k \to \infty} \frac{1}{k} \#\{n \leq k : s_n \neq 6\} \leq \lim_{k \to \infty} \frac{1}{k} \sqrt{k} = 0$. Then we obtain that $\delta(K_\varepsilon) = 0$, for every $\varepsilon > 0$ where $K_\varepsilon = \{n \in \mathbb{N} : |s_n - 6| \geq \varepsilon\}$. This means that $s = (s_n)$ statistically converges to 6.*

It can be seen that a statistically convergent sequence does not need to be bounded, while a convergent sequence needs to be bounded. Many concepts of classical convergence have been investigated and introduced for statistical convergence. For example, statistical boundedness, statistical limit points, statistical cluster points and the relations between these concepts and classical definitions have all been studied [31]. It has been understood that not all properties of classical convergence are true for statistical convergence. For instance, we know that every subsequence of a convergent sequence is also convergent but Miller has shown that this is not true for subsequences of a statistically convergent sequences with the use of binary expansion, measure and probability theory [54].

Now, let $A = (a_{jn}) \, j, n = 1, 2, \cdots$; be an infinite matrix. Then all of the above mentioned concepts have all been studied for the extension of statistical convergence which is referred to as $A$-statistical convergence and recalled below.

**Definition 2.2**    Let $A = (a_{jn})$ $j, n = 1, 2, \cdots$; be an infinite matrix and $s = (s_n)$ be a sequence. Then the $A$-transform of $s$ denoted by $As := ((As)_j)$ is defined by

$$(As)_j = \sum_{n=1}^{\infty} a_{jn} s_n$$

where the series is convergent for every $j$. Also regularity of a matrix is defined by: if the $A$-transform of a convergent sequence is also convergent to the same limit then it is said that $A$ is regular [40].

One of important matrices in summability theory is Cesàro matrix and it is given by

$$C_1 := (c_{jn})$$
$$c_{jn} = \begin{cases} \frac{1}{j} & , \quad 1 \leq n \leq j \\ 0 & , \quad n > j \end{cases}$$

and explicitly given by

$$C_1 = \begin{pmatrix} 1 & 0 & 0 & \cdots & \cdot & \cdots \\ \frac{1}{2} & \frac{1}{2} & 0 & \cdots & \cdot & \cdots \\ \cdot & \cdot & \cdot & \cdots & \cdot & \cdots \\ \cdot & \cdot & \cdot & \cdots & \cdot & \cdots \\ \frac{1}{j} & \frac{1}{j} & \cdot & \frac{1}{j} & 0 & \cdots \\ \frac{1}{j+1} & \frac{1}{j+1} & \cdot & \frac{1}{j+1} & \frac{1}{j+1} & \cdots \\ \vdots & \vdots & \vdots & \vdots & \vdots & \vdots \end{pmatrix}.$$

The regularity of an infinite matrix $A = (a_{jn})$ is characterized by the Silverman-Toeplitz theorem. The theorem is recalled below.

### Theorem 2.1 Silverman-Toeplitz
$A = (a_{jn})$ *is regular if and only if*

    *i.*    $\sup_j \sum_{n=1}^{\infty} |a_{jn}| < \infty,$

    *ii.*   $\forall n \in \mathbb{N}, \ a_n := \lim_j a_{jn} = 0,$

    *iii.*  $\lim_j \sum_{n=1}^{\infty} a_{jn} = 1$

*hold [40].*

One can easily note that the Cesàro matrix is regular since

    i.    $\sup_j \sum_{n=1}^{\infty} |a_{jn}| = 1 < \infty,$

    ii.   $\forall n \in \mathbb{N} \ a_n := \lim_j \frac{1}{j} = 0,$

    iii.  $\lim_j \sum_{n=1}^{\infty} a_{jn} = \lim_j 1 = 1.$

**Definition 2.3** Let $A = (a_{jn})$ be a nonnegative regular infinite matrix. Then if

$$\delta_A(K) = \lim_j \sum_{n \in K} a_{jn} \tag{2.1.1}$$

exits it is said that the number $\delta_A(K)$ is the $A$-density of $K$ where $K \subset \mathbb{N}$ [33].

**Example 3** *For $j = 1, 2, \cdots$ define*

$$a_{jn} = \begin{cases} 1 & , & n = j^2 \\ 0 & , & n \neq j^2 \end{cases}$$

*then for $A = (a_{jn})$, $\delta_A(K) = 1$ where $K = \{1, 4, 9, 16, \cdots, n^2, \cdots\}$. Notice that $\delta_A(K^c) = 0$ for $K^c = \{2, 3, 5, \cdots\}$.*

**Definition 2.4** Let $A = (a_{jn})$ be a nonnegative regular infinite matrix. If there exists $\ell$ such that

$$\lim_j \sum_{n : |s_n - \ell| \geq \varepsilon} a_{jn} = 0 \tag{2.1.2}$$

holds for every $\varepsilon > 0$ then we say that $s = (s_n)$ $A$-statistically converges to $\ell$. Indeed, $\delta_A(K_\varepsilon) = 0$ for every $\varepsilon > 0$ where $K_\varepsilon = \{n \in \mathbb{N}_0 : |s_n - \ell| \geq \varepsilon\}$.

This is denoted by $st_A - \lim s = \ell$ or $s_n \to \ell(st_A)$ [48], [54].

In Definiton 2.4,

- if $A$ is $I$, the identity matrix then $I$-statistical convergence reduces to classical convergence,

- if $A$ is $C_1$ then $A$-statistical convergence reduces to statistical convergence.

Again many concepts of classical convergence and statistical convergence have been examined for $A$-statistical convergence. Also the relationships between these concepts have been studied by many authors.

**Definition 2.5** Let $(p_n)$ be sequence of real numbers such that for all $n \geq 2$, $p_n \geq 0$, and $p_1 > 0$, $p(t) := \sum_{n=1}^{\infty} p_n t^{n-1}$ with a radius of convergence $R \in (0, \infty]$. Then the power series method is defined as follows:

Let also

$$C_p := \left\{ f : (-R, R) \to \mathbb{R} \Big| \lim_{0 < t \to R^-} \frac{1}{p(t)} f(t) \text{ exists} \right\}$$

and

$$C_{p_p} := \left\{ s = (s_n) \Big| p_s(t) := \sum_{n=1}^{\infty} p_n t^{n-1} s_n \text{ with radius of convergence} \geq R \text{ and } p_s \in C_p \right\},$$

$$P - \lim x = \lim_{0 < t \to R^-} \frac{1}{p(t)} \sum_{n=1}^{\infty} p_n t^{n-1} s_n$$

where the functional $P - \lim : C_{P_p} \to \mathbb{R}$ and then it is said that $s$ is $P$-convergent [20], [49].

Consider the following examples:

**Example 4** *Let* $(s_n) = (1, 0, 1, 0, 1, 0, ...)$, $R = \infty$, *for* $n \geq 1$, $p_n = \dfrac{1}{(n-1))!}$, *then* $p(t) = e^t$ *and it is easy to see that*

$$\lim_{t \to \infty} \frac{1}{e^t} \sum_{n=1}^{\infty} \frac{s_n t^{n-1}}{(n-1)!} = \lim_{t \to \infty} \frac{1}{e^t} \sum_{n=0}^{\infty} \frac{t^{2n}}{(2n)!} = \lim_{t \to \infty} \frac{1}{e^t} \left\{ \frac{e^t + e^{-t}}{2} \right\} = \frac{1}{2}.$$

*So the sequence* $s = (s_n)$ *is convergent to* $\frac{1}{2}$ *in the sense of power series method but it is not convergent in the ordinary sense.*

**Example 5** *Let also* $s_n = (-1)^n$, *for* $n \geq 1$, $p_n = 1$ *then* $R = 1$, $p(t) = \frac{1}{1-t}$ *and*

$$\lim_{t \to 1^-} (1 - t) \sum_{n=1}^{\infty} s_n t^{n-1} = 0$$

*which means that* $s = (s_n)$ *is convergent to* 0 *in the sense of power series method, i.e., P-convergent to* 0 *but s is not convergent.*

Therefore by taking into consideration the above examples, we can say that the power series method is more effective.

If a convergent sequence is also $P$-convergent to same limit then it is said that $P$ is regular and it is characterized by the following condition:

$$\lim_{t \to R^-} \frac{p_n t^{n-1}}{p(t)} = 0, \, for \, every \, n \in \mathbb{N}$$

[20]. It is also noteworthy to recall that

- ■ in the case of $R = 1$, $p(t) = \dfrac{1}{1-t}$ and for $j \geq 1$, $p_j = 1$ the power series method coincides with the Abel method which is a sequence-to-function transformation,

- ■ in the case of $R = \infty$, $p(t) = e^t$ and for $j \geq 1$, $p_j = \dfrac{1}{(j-1)!}$ the power series method coincides with Borel method.

By combining statistical convergence and power series, Ünver and Orhan [77] have recently introduced $P$-statistical convergence and have presented a Korovkin-type theorem for positive linear operators defined on $C[0,1]$.

Now we are ready to recall $P$-statistical convergence and Caputo derivative which are the main tools of our results.

**Definition 2.6**  Let $P$ be regular and $G \subseteq \mathbb{N}$. If

$$\delta_P(G) := \lim_{0 < t \to R^-} \frac{1}{p(t)} \sum_{n \in G} p_n t^{n-1}$$

exists then it is said to be the $P$-density of $G$.

One can immediately observe that if $\delta_P(G)$ exists then $\delta_P(G) \in [0,1]$ [77].

**Definition 2.7**  Let $s = (s_n)$ be a sequence of real numbers and let $P$ be regular. If for every $\varepsilon > 0$

$$\lim_{0 < t \to R^-} \frac{1}{p(t)} \sum_{n \in G_\varepsilon} p_n t^{n-1} = 0$$

that is, $\delta_P(G_\varepsilon) = 0$ for every $\varepsilon > 0$, then $s$ is said to be $P$-statistically convergent to $l$ and we denote it by $st_P - \lim s = l$ [77].

In order to show that statistical convergence does not imply $P$-statistical convergence and vice versa, the following examples can be examined. Therefore we can emphasize that our results make an important contribution to the existing literature.

**Example 6**  *Consider the following method defined by the sequence $(p_n)$: $n \in \mathbb{N}$,*

$$p_n = \begin{cases} 1 & , \quad n\ prime \\ 0 & , \quad n\ otherwise \end{cases}$$

*and also the sequence $s = (s_n)$ defined by*

$$s_n = \begin{cases} 0 & , \quad n\ prime \\ n & , \quad n\ otherwise \end{cases}$$

*Then it is obvious that $s = (s_n)$ is not statistically convergent.*

*But for every $\varepsilon > 0$, we have that $\delta_P(K_\varepsilon) = 0$ where*

$$K_\varepsilon = \{n \in \mathbb{N} : |s_n - 0| \geq \varepsilon\}.$$

*This means that $s = (s_n)$ is $P$-statistically convergent to 0.*

**Example 7** *Consider the following method defined by the sequence* $(p_n)$*:* $n \in \mathbb{N}$

$$p_n = \begin{cases} 1 & , \quad n\ prime \\ 0 & , \quad n\ otherwise \end{cases}$$

*and also the sequence* $s = (s_n)$ *defined by*

$$s_n = \begin{cases} n & , \quad n\ prime \\ 0 & , \quad n\ otherwise \end{cases}$$

*Here, it is obvious that the sequence* $s = (s_n)$ *statistically converges to* 0. *But it does not converge in the P-statistical sense.*

There are many possible generalizations of $\frac{d^n}{dx^n} f(x)$ in the case of $n \notin \mathbb{N}$. Some of them are Riemann-Liouville, Caputo, Grünwald-Letnikov, Weyl, Riesz and they are well studied by many mathematicians. These different definitions of fractional derivatives give us the opportunity to study with the most suitable one with the problem and to obtain the best solution. Although there are important relations between these definitions, the physical meanings of them differ from each other. For example, an attractive difference between Riemann-Liouville and Caputo is the derivative of a constant. The Caputo derivative of a constant is zero but for a finite lower bound Riemann-Liouville derivative is not zero. In order to give a physical comment of a problem, it is necessary to be satisfied that the derivative of a constant is zero. Hence, Caputo derivatives are well used in the existing literature and here we consider them.

Throughout the paper we consider the closed interval $I = [a,b]$, let $\mu$ be a positive real number, $m$ is the ceiling of the number $\mu$, i.e., $m = \lceil \mu \rceil$, $\Gamma$ is the Gamma function and $AC(I) = \{f : I \longrightarrow \mathbb{R}, f\ is\ absolutely\ continuous\}$,

$$AC^m(I) = \{f : I \longrightarrow \mathbb{R}, f^{(m-1)} \in AC(I)\}.$$

Then, the left and right Caputo fractional derivatives of $f \in AC^m(I)$ are defined by

$$D_{*a}^{\mu} f(y) := \frac{1}{\Gamma(m-\mu)} \int_a^y (y-t)^{m-\mu-1} f^{(m)}(t)dt$$

for $y \in I$,

$$D_{b-}^{\mu} f(y) := \frac{(-1)^m}{\Gamma(m-\mu)} \int_y^b (\xi - y)^{m-\mu-1} f^{(m)}(\xi)d\xi$$

for $y \in I$, respectively. Here, we also let $D_{*a}^0 f = f, D_{b-}^0 f = f$ on $I$ and suppose for $y < a, D_{*(a)}^{\mu} f(y) = 0$ and for $y > b, D_{b-}^{\mu} f(y) = 0$. We can present a few examples on the Caputo derivatives of some functions.

**Example 8** *Let* $f(t) = t^{\alpha} \in C[0,1]$*,* $t \in [0,1]$*,* $\mu = \frac{1}{3}$ *then* $m = 1$*. Notice that*

$$D_{*0}^{\frac{1}{3}} f(y) := \frac{1}{\Gamma(\frac{2}{3})} \int_0^y (y-t)^{-\frac{1}{3}} \alpha t^{\alpha-1} dt$$

*and by substituting, $t = yx$ we obtain that*

$$D_{*0}^{\frac{1}{3}}f(y) := \frac{\alpha y^{\alpha - \frac{1}{3}}}{\Gamma(\frac{2}{3})} \int_0^1 (1-x)^{-\frac{1}{3}} \alpha x^{\alpha - 1} dx.$$

*Using the properties of Beta function, we have*

$$D_{*0}^{\frac{1}{3}}f(y) = \frac{\Gamma(\alpha + 1)y^{\alpha - \frac{1}{3}}}{\Gamma(\alpha + \frac{2}{3})}.$$

*If $\alpha \in (0, \frac{1}{3})$ then $D_{*0}^{\frac{1}{3}}f(0) = \infty$, and if $\alpha = \frac{1}{3}$ then $D_{*0}^{\frac{1}{3}}f(0) = \Gamma(\frac{4}{3}) > 0$. As it can be seen we do not know what $D_{*a}^{\mu}f(a)$ is, it can be infinite, or finite not zero or zero.*

**Example 9** *Let $f(y) = y$ and $\mu = \frac{1}{2}$. Then*

$$D_{*0}^{\frac{1}{2}}y = \frac{1}{\Gamma(\frac{1}{2})} \int_0^y (y-t)^{-\frac{1}{2}} dt$$

$$= \frac{y^{\frac{1}{2}}}{\Gamma(\frac{1}{2})} \int_0^1 (1-s)^{-\frac{1}{2}} ds$$

$$= \frac{2}{\sqrt{\pi}} y^{\frac{1}{2}}.$$

*If we again take $\frac{1}{2}$ derivative of this function, we obtain that*

$$D_{*0}^{\frac{1}{2}} \frac{2}{\sqrt{\pi}} y^{\frac{1}{2}} = 1$$

*which means that two times $\frac{1}{2}$ derivative of $f(y) = y$ is the ordinary first derivative of it.*

Also the Caputo derivative of $e^{\lambda x}$ is $\lambda^n x^{n-\alpha} E_{1, n-\alpha+1}(\lambda x)$ where $E_{\alpha, \beta}(z)$ is the Mittag-Leffler function defined by $E_{\alpha, \beta}(z) = \sum_{k=0}^{\infty} \frac{z^k}{\Gamma(\alpha k + \beta)}$. The following are well known from [3], [4], [6]:

1. If $\mu > 0, \mu \notin \mathbb{N}, m = \lceil \mu \rceil, f \in C^{m-1}(I)$ and $f^{(m)} \in L_\infty(I)$ then $D_{*a}^{\mu}f(a) = 0, D_{b-}^{\mu}f(b) = 0.$

2. Let $y \in I$ be fixed. For $\mu > 0, m = \lceil \mu \rceil, f \in C^{m-1}(I)$ and $f^{(m)} \in L_\infty(I)$ take into consideration the following Caputo fractional derivatives:

$$U_f(x,y) := D_{*x}^{\mu}f(y) = \frac{1}{\Gamma(m-\mu)} \int_x^y (y-t)^{m-\mu-1} f^{(m)}(t) dt, \; for \; y \in [x,b]$$

and

$$V_f(x,y) := D^{\mu}_{x-} f(y) := \frac{(-1)^m}{\Gamma(m-\mu)} \int_y^x (\xi - y)^{m-\mu-1} f^{(m)}(\xi) d\xi, \ for \ y \in [a,x].$$

Then for each fixed $x \in I$, $U_f(x,\cdot)$ and $V_f(x,\cdot)$ are continuous on $[x,b]$ and $[a,x]$, respectively. Furthermore, $U_f(\cdot,\cdot), V_f(\cdot,\cdot)$ are continuous on $I \times I$ in the case $f \in C^m(I)$.

3.   If $g \in C(I \times I)$, then $s(x) := w(g(x,\cdot),\delta)_{[a,x]}$ and $t(x) := w(g(x,\cdot),\delta)_{[x,b]}$ are continuous for any $\delta > 0$ at the point $x \in I$ where $w(f,\delta), \delta > 0$ is the modulus of continuity in the classical sense.

4.   For any $\delta > 0$,

$$\sup_{x \in I} w(U_f(x,.),\delta)_{[x,b]} < \infty$$

and

$$\sup_{x \in I} w(V_f(x,.),\delta)_{[a,x]} < \infty$$

if $f \in C^{m-1}(I)$ with $f^{(m)} \in L_\infty(I)$.

5.   By setting $\rho_{n,\mu} := \|T_n(\phi^{\mu+1})\|^{\frac{1}{\mu+1}}$, we can write

$$\|T_n(f) - f\| \leq K_{\mu,m}\{\|T_n(e_0) - e_0\| + \sum_{k=1}^{m-1} \|T_n(|\psi|^k)\|$$
$$+ \rho^{\mu}_{n,\mu}(\sup_{x \in I} w(U_f(x,.),\rho_{n,\mu})_{[x,b]}) + \rho^{\mu}_{n,\mu}(\sup_{x \in I} w(V_f(x,.),\rho_{n,\mu})_{[a,x]})$$
$$+ \rho^{\mu}_{n,\mu}\|T_n(e_0) - e_0\|^{\frac{1}{\mu+1}}(\sup_{x \in I} w(U_f(x,.),\rho_{n,\mu})_{[x,b]} + \sup_{x \in I} w(V_f(x,.),\rho_{n,\mu})_{[a,x]}\}$$

where

$$K_{\mu,m} = \max\left\{\frac{1}{\Gamma(\mu+1)}, \frac{\mu+1}{\Gamma(\mu+2)}, \|f\|, \|f'\|, \frac{\|f''\|}{2!}, ..., \frac{\|f^{(m-1)}\|}{(m-1)!}\right\},$$

$\psi(y) := \psi_x(y) = y - x$, $e_0(y) = 1$ on $I$ and $\{T_n\}$ is a sequence of positive linear operators. Notice that the sum in the above inequality collapses if $\mu \in (0,1)$.

## 2.2   Fractional Calculus and *P*-statistical Convergence

In this section, we present our main results. With these results we deal with the fractional trigonometric approximation in different function spaces by *P*-statistical convergence. As mentioned earlier, *P*-statistical convergence is effective to use since there is no implication between it and other concepts of convergences. Throughout the section we let $\mu > 0, \mu \notin \mathbb{N}, m = \lceil \mu \rceil$.

**Theorem 2.2**

*[4] Let $T_n : C(I) \to C(I)$ be positive linear operators. If the sequence $\rho_{n,\mu}$ converges to 0 as $n \longrightarrow \infty$ and $(T_n(e_0))$ is uniformly convergent to $e_0$ on $I$, then $(T_n(f))$ converges uniformly to $f$ on $I$ for every $f \in AC^m(I)$ with $f^{(m)} \in L_\infty(I)$. Also, this uniform convergence is still true on $I$ when $f \in C^m(I)$.*

In order to obtain $P$-statistical version of above theorem, we first need the following lemma.

**Lemma 2.1**

*Let $P$ be regular and $T_n : C(I) \to C(I)$ be positive linear operators. If $st_P - \lim \|T_n(e_0) - e_0\| = 0$ and $st_P - \lim \rho_{n,\mu} = 0$ then $st_P - \lim \|T_n(|\psi|^k)\| = 0$ for every $k = 1, 2, ..., m-1$.*

**Proof 2.1**

Let $k \in \{1, 2, ..., m-1\}$ be fixed. With the use of Hölder inequality for positive linear operators which has been obtained in [63],

$$\|T_n(|\psi|^k)\| \le (2\pi)^k \left[ (\rho_{n,\mu})^k \|T_n(e_0) - e_0\|^{\frac{\mu+1-k}{\mu+1}} + (\rho_{n,\mu})^k \right]$$

has been obtained in [6]. Now let us define the following sets:

$$G = \left\{ n \in \mathbb{N} : \|T_n(|\psi|^k)\| \ge \varepsilon \right\}$$

$$G_1 = \left\{ n \in \mathbb{N} : (\rho_{n,\mu})^k \|T_n(e_0) - e_0\|^{\frac{\mu+1-k}{\mu+1}} \ge \frac{\varepsilon}{2(2\pi)^k} \right\}$$

$$G_2 = \left\{ n \in \mathbb{N} : \rho_{n,\mu} \ge \frac{1}{2\pi} (\frac{\varepsilon}{2})^{\frac{1}{k}} \right\}.$$

Then it is immediate that $G \subseteq G_1 \cup G_2$. Also we define the following sets as

$$G_1' = \left\{ n \in \mathbb{N} : \rho_{n,\mu} \ge \frac{1}{\sqrt{2\pi}} (\frac{\varepsilon}{2})^{\frac{1}{2k}} \right\}$$

$$G_2'' = \left\{ n \in \mathbb{N} : \|T_n(e_0) - e_0\| \ge (\frac{\varepsilon}{2(2\pi)^k})^{\frac{\mu+1}{2(\mu+1-k)}} \right\}.$$

Then it follows that $G \subseteq G_1' \cup G_2'' \cup G_2$.

Hence,

$$\frac{1}{p(t)} \sum_{n \in G} p_n t^{n-1} \le \frac{1}{p(t)} \sum_{n \in G_1'} p_n t^{n-1} + \frac{1}{p(t)} \sum_{n \in G_2''} p_n t^{n-1} + \frac{1}{p(t)} \sum_{n \in G_2} p_n t^{n-1}$$

holds and

$$st_P - \lim \|T_n(|\psi|^k)\| = 0, \, for \, each \, k = 1, 2, ..., m-1$$

by the hypotheses. Thus we complete the proof.

Now we can present the first fractional approximation result via *P*-statistical convergence.

### Theorem 2.3

*Let $P$ be regular and $T_n : C(I) \to C(I)$ be positive linear operators. If $st_P - \lim \|T_n(e_0) - e_0\| = 0$ and $st_P - \lim \rho_{n,\mu} = 0$ then $st_P - \lim \|T_n(f) - f\| = 0$ for every $f \in AC^m(I)$ such that $f^{(m)} \in L_\infty(I)$.*

**Proof 2.2**   Let $f \in AC^m(I)$ with $f^{(m)} \in L_\infty(I)$. It is known that

$$\|T_n(f) - f\| \leq H_{m,\mu}\{\|T_n(e_0) - e_0\| + \sum_{k=1}^{m-1} \|T_n(|\psi|^k)\|$$
$$+ 2\rho_{n,\mu}^r + 2\rho_{n,\mu}^r\|T_n(e_0) - e_0\|\},$$

where

$H_{m,\mu} = \max\left\{K_{m,\mu}, \sup_{x\in I} w(U_f(x,.),\rho_{n,\mu})_{[x,b]}, \sup_{x\in I} w(V_f(x,.),\rho_{n,\mu})_{[a,x]}\right\}$. Again
define the following:

$$F = \{n \in \mathbb{N} : \|T_n(f) - f\| \geq \varepsilon\}$$

$$F_k = \left\{n \in \mathbb{N} : \|T_n(|\psi|^k)\| \geq \frac{\varepsilon}{(m+2)H_{m,\mu}}\right\}, k = 1, 2, ..., m-1$$

$$F_m = \left\{n \in \mathbb{N} : \|T_n(e_0) - e_0\| \geq \frac{\varepsilon}{(m+2)H_{m,\mu}}\right\}$$

$$F_{m+1} = \left\{\rho_{n,\mu} \geq (\frac{\varepsilon}{(m+2)H_{m,\mu}})^{\frac{1}{\mu}}\right\}$$

$$F_{m+2} = \left\{\rho_{n,\mu}^\mu\|T_n(e_0) - e_0\|^{\frac{1}{\mu+1}} \geq \frac{\varepsilon}{2(m+2)H_{m,\mu}}\right\}.$$

Then it follows that $B \subseteq \cup_{i=1}^{m+2} B_i$. If we also define the following sets,

$$F_{m+3} = \left\{\|T_n(e_0) - e_0\| \geq (\frac{\varepsilon}{2(m+2)H_{m,\mu}})^{\frac{\mu+1}{2}}\right\}$$

and

$$F_{m+4} = \left\{\rho_{n,\mu} \geq (\frac{\varepsilon}{2(m+2)H_{m,\mu}})^{\frac{1}{2\mu}}\right\}$$

then we have $F_{m+2} \subseteq F_{m+3} \cup F_{m+4}$ and $F \subseteq \cup_{i=1}^{m+4} B_i$. From the hypotheses, we obtain $\delta_P(F) = 0$ and this completes the proof.

If we consider $C^m(I)$ instead of $AC^m(I)$, then we can slightly modify the above theorem. For this, let us prove the next lemma.

**Lemma 2.2**
*Let $P$ be regular and $T_n : C(I) \to C(I)$ be positive linear operators. If $st_P - \lim \rho_{n,\mu} = 0$ then we have*

$$st_P - \limsup_{x \in I} w(U_f(x, \cdot), \rho_{n,\mu})_{[x,b]} = 0,$$

*and*

$$st_P - \limsup_{x \in I} w(V_f(x, \cdot), \rho_{n,\mu})_{[a,x]} = 0.$$

**Proof 2.3**    It is already known from [3], [4], that there exists $x_0, x_1 \in I$ such that

$$\sup_{x \in I} w(U_f(x, \cdot), \rho_{n,\mu})_{[x,b]} = w(U_f(x_0, \cdot), \rho_{n,\mu})_{[x_0,b]} =: p(\rho_{n,\mu})$$

and

$$\sup_{x \in I} w(V_f(x, \cdot), \rho_{n,\mu})_{[a,x]} = w(V_f(x_1, \cdot), \rho_{n,\mu})_{[a,x_1]} =: q(\rho_{n,\mu}).$$

By the hypotheses, we get $\delta_P(\{n \in \mathbb{N} : \rho_{n,\mu} \geq \delta\}) = 0$ for any $\delta > 0$. Then, by following the similar arguments from [6] we have that

$$\{n \in \mathbb{N} : p(\rho_{n,\mu}) \geq \varepsilon\} \subseteq \{n \in \mathbb{N} : \rho_{n,\mu} \geq \delta_1\}$$

and

$$\{n \in \mathbb{N} : q(\rho_{n,\mu}) \geq \varepsilon\} \subseteq \{n \in \mathbb{N} : \rho_{n,\mu} \geq \delta_2\}.$$

which implies

$$\frac{1}{p(t)} \sum_{n:p(\rho_{n,\mu}) \geq \varepsilon} p_n t^{n-1} \leq \frac{1}{p(t)} \sum_{n:\rho_{n,\mu} \geq \delta_1} p_n t^{n-1},$$

$$\frac{1}{p(t)} \sum_{n:q(\rho_{n,\mu}) \geq \varepsilon} p_n t^{n-1} \leq \frac{1}{p(t)} \sum_{n:\rho_{n,\mu} \geq \delta_2} p_n t^{n-1}.$$

Then by taking in the limit on both sides and using the hypotheses, we complete the proof.

Now we can present the following result in $C^m(I)$. Since the technic of the proof is similar in earlier results, we omit the proof here.

**Theorem 2.4**
*Let $P$ be regular and $T_n : C(I) \to C(I)$ be positive linear operators. If $st_P - \lim \|T_n(e_0) - e_0\| = 0$ and $st_P - \lim \rho_{n,\mu} = 0$ then $st_P - \lim \|T_n(f) - f\| = 0$ for every $f \in C^m(I)$.*

## 2.3  Applications

This section is devoted to the construction of special sequences of operators which support our results. Here, it is worthy to note that it is not possible to have approximation by earlier results which use different concepts of convergences. But we overcome the critical weakness of ordinary convergence with the use of our method.

**Example 10** *Define the sequences $(p_n)$, $u = (u_n)$ and $s = (s_n)$ as follows:*

$$p_n = \begin{cases} 1 & , & n = 2k \\ 0 & , & n = 2k+1 \end{cases}, \quad u_n = \begin{cases} 1 & , & n = 2k \\ \frac{1}{2} & , & n = 2k+1 \end{cases}, \quad s_n = \begin{cases} 0 & , & n = 2k \\ \sqrt{n} & , & n = 2k+1 \end{cases}.$$

*One can obtain that the method $P$ is regular and also observe that*

$$K_\varepsilon = \{ n \in \mathbb{N} : |s_n - 0| \geq \varepsilon \} \subseteq \{ n = 2k+1 : k \in \mathbb{N} \}$$

*holds for every $\varepsilon > 0$. Then we have*

$$\delta_P(K_\varepsilon) = \lim_{0 < t \to R^-} \frac{1}{p(t)} \sum_{n \in K_\varepsilon} p_n t^{n-1} = 0$$

*i.e., $st_P - \lim s = 0$, $st_P - \lim u = 1$.*
    *Also define*

$$T_n(f;x) = (1 + s_n) \sum_{k=0}^{n} \binom{n}{k} f(\frac{k}{n}) (u_n x)^k (1 - u_n x)^{n-k}, x \in I, n \in \mathbb{N}.$$

    *Let $\mu = \frac{1}{2}, I = [0,1]$, $f \in AC^m(I)$ with $f^{(m)} \in L_\infty(I)$. Then $m = \lceil \mu \rceil = 1$, $st_P - \lim \| T_n(e_0) - e_0 \| = 0$,*

$$\| T_n(|\psi|^{\frac{3}{2}}) \| \leq (1 + s_n) [\frac{u_n x (1 - u_n)}{n}]^{\frac{3}{4}}$$

*and*

$$\rho_{n,\frac{1}{2}}^{\frac{3}{2}} = \| T_n(\phi^{\frac{3}{2}}) \| \leq (1 + s_n) [\frac{1}{4n}]^{\frac{3}{4}}.$$

    *Then $T_n$ satisfies the conditions of our theorem. Again it is not possible to approximate $f$ by using $T_n(f)$ since the sequence $(s_n)$ is not convergent or statistically convergent. Furthermore it is still possible to approximate $f$ by using $T_n(f)$ for every $f \in AC^m(I)$ with $f^{(m)} \in L_\infty(I)$ via $P-$statistical convergence since $(s_n)$ is $P-$statistically convergent to 0.*

**Example 11** *Construct $T_n$ by*

$$T_n(f;x) = s_n B_n(f;x)$$

*where $B_n(f;x)$ is given by*

$$B_n(f;x) = \sum_{k=0}^{n} \binom{n}{k} f(\frac{k}{n})(x)^k (1-x)^{n-k}, \; x \in I, n \in \mathbb{N}$$

*and $s = (s_n)$,*

$$s_n = \left\{ \begin{array}{ll} 1 & , \quad n = m^2 \\ n! & , \quad \text{otherwise} \end{array} \right. , \quad p_n = \left\{ \begin{array}{ll} 1 & , \quad n = m^2 \\ 0 & , \quad \text{otherwise} \end{array} \right. .$$

*One can obtain that the method $P$ is regular and $st_P - \lim s = 1$. Now let $\mu = \frac{1}{4}, f \in AC(I)$ with $f' \in L_\infty(I)$. Then $m = \lceil \mu \rceil = 1$, $st_P - \lim \|T_n(e_0) - e_0\| = 0$, by applying Hölder inequality for $p = \frac{8}{5}, q = \frac{8}{3}$ we obtain*

$$\rho_{n,\frac{1}{4}}^{\frac{5}{4}} = \|T_n(\phi^{\frac{5}{4}})\|(1+s_n)\left[\frac{1}{4n}\right]^{\frac{5}{8}}$$

*which implies $st_P - \lim \rho_{n,\frac{1}{4}} = 0$. As in the earlier examples we notice that $T_n$ satisfies our conditions.*

**Example 12** *Construct $T_n$ by*

$$T_n(f;x) = s_n B_n(f;x)$$

*where $B_n(f;x)$ is given above and $s = (s_n)$*

$$s_n = \left\{ \begin{array}{ll} 1 & , \quad n = 2k \\ \sqrt{n} & , \quad n = 2k+1 \end{array} \right. , \quad p_n = \left\{ \begin{array}{ll} 1 & , \quad n = 2k \\ 0 & , \quad n = 2k+1 \end{array} \right. .$$

*One can obtain that the method $P$ is regular and $st_P - \lim s = 1$. As in the earlier examples we notice that $T_n$ satisfies our conditions for $\mu = \frac{1}{2}, f \in AC^m(I)$ with $f^{(m)} \in L_\infty(I)$.*

**Example 13** *Define the sequences $(p_n)$, $u = (u_n)$ and $s = (s_n)$ as follows:*

$$p_n = \left\{ \begin{array}{ll} 1 & , \quad n = 2k \\ 0 & , \quad n = 2k+1 \end{array} \right. , \quad u_n = \left\{ \begin{array}{ll} 1 & , \quad n = 2k \\ \frac{1}{2} & , \quad n = 2k+1 \end{array} \right. , \quad s_n = \left\{ \begin{array}{ll} 0 & , \quad n = 2k \\ \sqrt{n} & , \quad n = 2k+1 \end{array} \right. .$$

*One can obtain that the method P is regular and also observe that*

$$K_\varepsilon = \{n \in \mathbb{N} : |s_n - 0| \geq \varepsilon\} \subseteq \{n = 2k+1 : k \in \mathbb{N}\}$$

*holds for every $\varepsilon > 0$. Then we have*

$$\delta_P(K_\varepsilon) = \lim_{0 < t \to R^-} \frac{1}{p(t)} \sum_{n \in K_\varepsilon} p_n t^{n-1} = 0$$

*i.e., $st_P - \lim s = 0$, $st_P - \lim u = 1$.*
   *Then define*

$$T_n(f;x) = (1+s_n) \sum_{k=0}^{n} \binom{n}{k} f(\frac{k}{n})(u_n x)^k (1 - u_n x)^{n-k}, \ x \in I, n \in \mathbb{N}.$$

*Now let $\mu = \frac{1}{4}, f \in AC(I)$ with $f' \in L_\infty(I)$. Then $m = \lceil \mu \rceil = 1$, $st_P - \lim \|T_n(e_0) - e_0\| = 0$, by applying Hölder inequality for $p = \frac{8}{5}, q = \frac{8}{3}$ we obtain*

$$\rho_{n,\frac{1}{4}}^{\frac{5}{4}} = \|T_n(\phi^{\frac{5}{4}})\|(1+s_n)\left[\frac{u_n x(1 - u_n)}{4n}\right]^{\frac{5}{8}}$$

*which implies $st_P - \lim \rho_{n,\frac{1}{4}} = 0$.*
   *Then $T_n$ satisfies the conditions of our theorem. Again it is not possible to approximate $f$ by using $T_n(f)$ since the sequence $(s_n)$ is not convergent or statistically convergent. Furthermore it is still possible to approximate $f$ by using $T_n(f)$ for every $f \in AC^m(I)$ with $f^{(m)} \in L_\infty(I)$ via $P-$statistical convergence since $(s_n)$ is $P-$statistically convergent to 0.*

**Example 14** *Define the sequences $(p_n)$, $u = (u_n)$ and $s = (s_n)$ as follows:*

$$p_n = \begin{cases} 1 & , & n = m^2 \\ 0 & , & \text{otherwise} \end{cases}, \quad u_n = \begin{cases} 1 & , & n = m^2 \\ \frac{1}{2} & , & \text{otherwise} \end{cases}, \quad s_n = \begin{cases} 0 & , & n = m^2 \\ n! & , & \text{otherwise} \end{cases}.$$

*P is regular and again one can obtain that $st_P - \lim s = 0$, $st_P - \lim u = 1$.*
   *Define*

$$T_n(f;x) = (1+s_n) \sum_{k=0}^{n} \binom{n}{k} f(\frac{k}{n})(u_n x)^k (1 - u_n x)^{n-k}, \ x \in I, n \in \mathbb{N}.$$

*Now let* $\mu = \frac{1}{4}, f \in AC(I)$ *with* $f' \in L_\infty(I)$. *Then* $m = \lceil \mu \rceil = 1$, $st_P - \lim \|T_n(e_0) - e_0\| = 0$, *by applying Hölder inequality for* $p = \frac{8}{5}, q = \frac{8}{3}$ *we obtain*

$$\rho_{n,\frac{1}{4}}^{\frac{5}{4}} = \|T_n(\phi^{\frac{5}{4}})\|(1 + s_n)[\frac{u_n x(1 - u_n)}{4n}]^{\frac{5}{8}}$$

*which implies* $st_P - \lim \rho_{n,\frac{1}{4}} = 0$.

*Then* $T_n$ *satisfies the conditions of our theorem. Again it is not possible to approximate* $f$ *by using* $T_n(f)$ *since the sequence* $(s_n)$ *is not convergent or statistically convergent. Furthermore it is still possible to approximate* $f$ *by using* $T_n(f)$ *for every* $f \in AC^m(I)$ *with* $f^{(m)} \in L_\infty(I)$ *via* $P-$*statistical convergence since* $(s_n)$ *is* $P-$*statistically convergent to* 0.

**Example 15** *Define the sequences* $(p_n)$, $u = (u_n)$ *and* $s = (s_n)$ *as follows:*

$$p_n = \begin{cases} 1 & , \quad n = m^2 \\ 0 & , \quad otherwise \end{cases}, \quad u_n = \begin{cases} 1 & , \quad n = m^2 \\ \frac{1}{2} & , \quad otherwise \end{cases}, \quad s_n = \begin{cases} 1 & , \quad n = m^2 \\ n! & , \quad otherwise \end{cases}.$$

*P is regular and again one can obtain that* $st_P - \lim s = 0$, $st_P - \lim u = 1$.
*Define*

$$T_n(f;x) = s_n \sum_{k=0}^{n} \binom{n}{k} f(\frac{k}{n})(u_n x)^k (1 - u_n x)^{n-k}, \ x \in I, n \in \mathbb{N}.$$

*Now let* $\mu = \frac{1}{4}, f \in AC(I)$ *with* $f' \in L_\infty(I)$. *Then* $m = \lceil \mu \rceil = 1$, $st_P - \lim \|T_n(e_0) - e_0\| = 0$, *by applying Hölder inequality for* $p = \frac{8}{5}, q = \frac{8}{3}$ *we obtain*

$$\rho_{n,\frac{1}{4}}^{\frac{5}{4}} = \|T_n(\phi^{\frac{5}{4}})\|s_n[\frac{u_n x(1 - u_n)}{4n}]^{\frac{5}{8}}$$

*which implies* $st_P - \lim \rho_{n,\frac{1}{4}} = 0$.

*Then* $T_n$ *satisfies the conditions of our theorem. Again it is not possible to approximate* $f$ *by using* $T_n(f)$ *since the sequence* $(s_n)$ *is not convergent or statistically convergent. Furthermore it is still possible to approximate* $f$ *by using* $T_n(f)$ *for every* $f \in AC^m(I)$ *with* $f^{(m)} \in L_\infty(I)$ *via* $P-$*statistical convergence since* $(s_n)$ *is* $P-$*statistically convergent to* 0.

## 2.4  Conclusion

One of the main themes of approximation theory is to express an arbitrary function in terms of nicer and simpler functions. In fact, while expanding a function to a power series, the main motivation is to express the function in terms of polynomials which are both nice and simple. Therefore, one of the obvious questions is what is a good class of nice and simple functions which approximates better and what kind of functions can we approximate. In this context, the theorem of Weierstrass is at the heart of approximation theory and states that the class of algebraic polynomials is dense in $C[a,b]$. As it is well-known that the proof of this theorem is not easy to follow and Bernstein is the first who has achieved this to give us a more understandable alternative proof [19], [51]. Then this proof has lead to studying positive, linear operators in approximation theory [21], [50], [65]. It has been obtained that one can determine the uniform convergence of such operators to identity operator by minimum calculations which are also known as Korovkin-type results, i.e., $T_n(f)$ converges uniformly to $f$ for every $f \in C[a,b]$ if $T_n(x^j)$ converges to $x^j$, $j = 0, 1, 2$ uniformly on $[a,b]$ where $\{T_n\}$ is a sequence of positive linear operators. Therefore Korovkin-type theorems are used frequently and many mathematicians have studied the problems of this theory in various ways such as: by changing the concept of convergence, by using nonlinear operators instead of linear operators, by taking into consideration the uncertainy with the use of fuzzy theory, by investigating these results in different function spaces, by using $q$-calculus which acts as bridge between mathematics and physics, and by using fractional calculus which deals with integrals and derivatives of arbitrary order.

In this chapter, we have presented fractional Korovkin-type theorems by $P$-statistical convergence where $P$ denotes power series and have provided examples which assert the importance of our results. Since statistical convergence and $P$-statistical convergence do not imply each other, our results make enough of a contribution to the literature wherein there are only a few papers combining approximation theory and fractional calculus [3], [4], [7], [24], [43], [55].

# References


[1] Altomare, F. and Diomede, S. 2001. Contractive Korovkin subsets in weighted spaces of continuous functions. Rend. Circ. Mat. Palermo. 50: 547–568.

[2] Anastassiou, G.A. and Duman, O. 2008. Statistical fuzzy approximation by fuzzy positive linear operators. Comput. and Math. Appl. 55: 573–580.

[3] Anastassiou, G.A. 2009. Fractional Korovkin theory. Chaos, Solitons & Fractals 42(4): 2080–2094.

[4] Anastassiou, G.A. 2010. Fractional trigonometric Korovkin theory. Communications in Applied Analysis, 4(1): 39–58.

[5] Anastassiou, G.A. 2005. On basic fuzzy Korovkin theory. Stud. Univ. Babes-Bolyai Math. 50: 3–10.

[6] Anastassiou, G.A. and Duman, O. 2010. Fractional trigonometric Korovkin theory in statistical sense. Serdica Math. J. 36: 121–136.

[7] Anastassiou, G.A. and Duman, O. 2011. Fractional Korovkin-type approximation based on statistical sense. Towards Intelligent Modeling, ISRL 14(36), Springer-Verlag, 157–167.

[8] Arif, A. and Yurdakadim, T. Approximation results on nonlinear operators by $P_p$-statistical convergence. Adv. Studies: Euro-Tbilisi Math. J. 15(3): 1–10.

[9] Atlıhan, Ö.G. and Orhan, C. 2008. Summation process of positive linear operators. Computers and Mathematics with Appl. 56: 1188–1195.

[10] Atlıhan, Ö.G., Yurdakadim, T. and Taş, E. 2022. A new approach to approximation by positive linear operators in weighted spaces. Ukrainian J. Math. 74(11): 1447–1453.

[11] Balaz, V. and Salat, T. 2006. Uniform density $u$ and corresponding $I_u$-convergence. Math. Commun. 11: 1–7.

[12] Bardaro, C., Musielak, J. and Vinti, G. 2003. Nonlinear Integral Operators and Applications. De Gruyter Ser. Nonlinear Anal. Appl. 9, Walter de Gruyter, Berlin.

[13] Bardaro, C., Boccuto, A., Dimitriou, X. and Mantellini, I. 2009. A Korovkin theorem in multivariate modular function spaces. J. Func. Spaces Appl. 7: 105–120.

[14] Bardaro, C., Boccuto, A., Dimitriou, X. and Mantellini, I. 2013. Abstract korovkin type theorems in modular spaces and applications. Cent. Eur. J. Math. 11: 1774–1784.

[15] Baxhau, B., Agrawal, P.N. and Shukla, R. 2022. Some fuzzy Korovkin type approximation theorems via power series summability method. Soft Comput. 26: 11373–11379.

[16] Bede, B., Nobuhara, H., Fodor, J. and Hirota, K. 2006. Max-product Shepard approximation operators. J. Adv. Comput. Intelligence Intelligent Informatics 10: 494–497.

[17] Bede, B., Nobuhara, H., Dankova, M. and Nola, A.D. 2008. Approximation by pseudo-linear operators. Fuzzy Sets Syst. 159: 804–820.

[18] Bede, B., Coroianu, L. and Gal, S.G. 2009. Approximation and shape preserving properties of the Bernstein operators of max-product kind. Int. J. Math. Sci. Art. ID 590589, 22 pp.

[19] Bernstein, S.N. 1912. Demonstration du theoreme de Weierstrass fondee. Communications of the Kharkov Mathematical Society 13: 1–2.

[20] Boos, J. 2000. Classical and Modern Methods in Summability. Oxford University Press.

[21] Bohman, H. 1952. On approximation of continuous and of analytic functions. Ark. Mat. 2(1): 43–56.

[22] Buckley, J.J. 1991. Solving systems of linear fuzzy equations. Fuzzy Sets and Syst. 43: 33–43.

[23] Coker, D. 1997. An introduction to intuitionistic fuzzy topological spaces. Fuzzy Sets and Syst. 88: 81–89.

[24] Demjanovic, J. 1975. Approximation by local functions in a space with fractional derivatives. Differencial'nye Uravnenija i Primenen, Trudy Sem. Processy 103: 35–49.

[25] Diethelm, K. 2004. The Analysis of Fractional Differential Equations. Lecture Notes in Mathematics. Springer.

[26] Duman, O., Khan, M.K. and Orhan, C. 2003. A-statistical convergence of approximating operators. Math. Inequal. Appl. 6: 689–699.

[27] Dzyadyk, V.K. 1959. On the best trigonometric approximation in the $L$ metric of some functions. Dokl. Akad. Nauk. SSSR 129: 19–22.

[28] El Naschie, M.S. 1998. On uncertainty of Cantorian geometry and two-slit experiment. Chaos Solitons and Fractals 9: 517–529.

[29] El Naschie, M.S. 2004. A review of E-infinity theory and mass spectrum of high energy particle physics. Chaos Solitons and Fractals 19: 209–236.

[30] Fast, H. 1951. Sur la convergence statistique. Colloq. Math. 2: 241–244.

[31] Fridy, J.A. and Orhan, C. 1997. Statistical limit superior and limit inferior. Proc. Amer. Mat. Soc. 125(12): 3625–3631.

[32] Fridy, J.A. 1985. On statistical convergence. Analysis 5: 301–313.

[33] Freedman, A.R. and Sember, J.J. 1981. Densities and summability. Pacific J. Math. 95: 293–305.

[34] Gadjiev, A.D. and Orhan, C. 2002. Some approximation theorems via statistical convergence. Rocky Mountain Journal of Math. 32: 129–137.

[35] Gal, S. 2000. Approximation theory in fuzzy setting. Chapter 13 in Handbook of Analytic-Computational Methods in Applied Mathematic. Chapman and Hall/CRC, Boca Raton, 617–666.

[36] Goetschel, R.J. and Voxman, W. 1986. Elementary fuzzy calculus. Fuzzy Sets and Syst. 18: 31–43.

[37] Gökçer, T.Y. and Duman, O. 2020. Approximation by max-min operators: A general theory and its application. Fuzzy Sets and Systems 394: 146–161.

[38] Gökçer, T.Y. and Duman, O. 2016. Summation process by max-product operators. Comp. Anal. Springer Proceedings in Math. and Stat. 155, DOI 10.1007/978-3-319-28443-9-4.

[39] Gökçer, T.Y. and Aslan, İ. 2022. Approximation by Kantorovich-type max-min operators. Appl. Math. Comput. 423: 127011.

[40] Hardy, G.H. 1949. Divergent Series. Oxford University Press, London.

[41] He, J.H. 2008. Max-min approach to nonlinear oscillators. Int. J. Nonlinear Sci. Numer. Simul. 9(2): 207–210.

[42] Izuchi, K., Takagi, H. and Watanabe, S. 1996. Sequential BKW-operators and function algebras. J. Approx. Theory 85: 185–200.

[43] Jaskolski, M. 1989. Contributions to fractional calculus and approximation theory on the square. Func. Approx. Comment. 18: 77–89.

[44] Kac, V. and Cheung, P. 2002. Quantum Calculus. Springer-Verlag.

[45] Karakus, S., Demirci, K. and Duman, O. 2010. Statistical approximation by positive linear operators on modular spaces. Positivity 14: 321–334.

[46] Karakus, S. and Demirci, K. 2010. Matrix summability and Korovkin type approximation theorem on modular spaces. Acta Math. Univ. Commeni-anae. 2: 281–292.

[47] Karsli, H. and Gupta, V. 2008. Some approximation properties of $q$-Chlodowsky operators. Applied Mathematics and Computation 195: 220–229.

[48] Kolk, E. 1993. Matrix summability of statistically convergent sequence. Analysis 13: 77–83.

[49] Kratz, W. and Stadtmüller, U. 1989. Tauberian theorems for $J_p$-summability. J. Math. Anal. Appl. 139: 362–371.

[50] Korovkin, P.P. 1953. On convergence of linear positive operators in the space of continuous functions. Doklady Akad. Nauk SSR 90: 961–964.

[51] Lorentz, G.G. 1986. Bernstein Polynomials. AMS Chelsea Publishing.

[52] Lupaş, A. 1987. A $q$-analogue of the Bernstein operator. Seminar on Numerical and Statistical Calculus, University of Cluj-Napoca 9: 85–92.

[53] Micchelli, C.A. 1975. Convergence of positive linear operators on C(X). J. Approx. Theory 13: 305–315.

[54] Miller, H.I. 1995. A measure theoretical subsequences characterization of statistical convergence. Trans. Amer. Math. Soc. 347(5): 1811–1819.

[55] Nasibov, F.G. 1962. On the degree of best approximation of functions having a fractional derivative in the Riemann-Liouville sense. SSR Ser. Fiz-Mat. Tehn. Nauk. 3: 51–57.

[56] Nuray, F. and Savaş, E. 1995. Statistical convergence of sequences of fuzzy numbers. Math. Slovaca. 45: 269–273.

[57] Olgun, M., Ünver, M. and Yardımc ı, Ş. 2019. Pythagorean fuzzy topological spaces. Complex and Intelligent Systems 5: 177–183.

[58] Orhan, S. and Demirci, K. 2014. Statistical $\mathscr{A}$-summation process and Korovkin type approximation theorem on modular spaces. Positivity 18: 669–686.

[59] Oruc, H. and Tuncer, N. 2002. On the convergence and iterates of $q$-Bernstein polynomials. Journal of Approximation Theory 117: 301–313.

[60] Ostrovska, S. 2003. $q$-Bernstein polynomials and their iterates. J. Approx. Th. 123: 232–255.

[61] Peeva, K. 1992. Fuzzy linear systems. Fuzzy Sets and Syst. 49: 339–355.

[62] Phillips, G.M. 1997. Bernstein polynomials based on the $q$-integers. Annals of Numerical Mathematics 4: 511–518.

[63] Pitul, P. Evaluation of the approximation order by positive linear operators. Ph.D. Thesis, Cluj-Napoca, Babeş-Bolyai University.

[64] Podlubny, I. 2001. Geometric and physical interpretation of fractional integration and fractional differentiation. arXiv preprint math/0110241.

[65] Popoviciu, T. 1950. Asupra demonstratiei teoremei lui weierstrass cu ajutorul polinoamelor de interpolare. Lucrarile Ses, Gen. St. Acad. Române din. 1–4.

[66] Powell, R.E. and Shah, S.H. 1972. Summability Theory and its Applications. Van Nostrand-Reinhold.

[67] Qian, Y. and Yu, D. 1980. Rates of approximation by neural network interpolation operators. Appl. Math. Comput. 418: 126781.

[68] Sakaoğlu, I. and Orhan, C. 2013. Strong summation process in $L_p$ spaces. Nonlinear Analysis 86: 89–94.

[69] Salat, T. 1980. On statistically convergent sequences of real numbers. Mat. Slovaca. 30: 139–150.

[70] Samko, S.G., Kilbas, A.A. and Marichev, O.I. 1993. Fractional integrals and derivatives Gordon and Breach Science Publishers, Yverdon Yverdon-les-Bains, Switzerland, Vol. 1.

[71] Shen, Z., Yang, H. and Zhang, S. 2022. Optimal approximation rate of ReLU networks in terms of width and depth. J. Math. Pures Appl. 157: 101–135.

[72] Şahin Bayram, N. Criteria for statistical convergence with respect to power series methods. Positivity 25: 1097–1105.

[73] Takahasi, S.E. 1993. Bohman-Korovkin-Wulbert operators on normed spaces. J. Approx. Theory 72: 174–184.

[74] Taş, E., Orhan, C. and Yurdakadim, T. 2013. The Stancu-Chlodowsky operators based on $q$-Calculus. In: Proceedings of the 11th International Conference of Numerical Analysis and Applied Mathematics (ICNAAM), Sept 21–27 2013, Rhodes Island, GREECE. AIP Conference Proceeding 1558: 1152–1155.

[75] Taş, E. and Yurdakadim, T. 2019. Korovkin theory for extraordinary test functions by A-statistical convergence. Palestine Journal of Mathematics 8(1): 86–90.

[76] Taş, E. and Yurdakadim, T. 2017. Approximation by positive linear operators in modular spaces by power series method. Positivity 21: 1293–1306.

[77] Ünver, M. and Orhan, C. 2019. Statistical convergence with respect to power series methods and applications to approximation theory. Numer. Func. Anal. Opt. 40: 535–547.

[78] Weierstrass, K.G. 1885. Über die analytische Dorstellbarkeit sogenannter willkürlicher Funktionen einer reellen. Veranderlichen. Sitzungsber, Akad. Berlin.

[79] Yurdakadim, T. and Taş, E. 2022. Effects of fuzzy setting in Korovkin theory via $P_p$-statistical convergence. Roman. J. Math. 2(12): 1–8.

[80] Zadeh, L.A. 1965. Fuzzy sets. Inform. and Control 8: 338–353.

# Chapter 3

# Two Approaches for Evaluating the Period Function of Some Hamiltonian Systems

*Daniele Ritelli* and *Giulia Spaletta**

## 3.1 Introduction

Analysis of non–linear autonomous conservative oscillators of the form $\ddot{x} + g(x) = 0$ (where $x = x(t)$ and $\dot{x}$ denotes as usual, times differentiation) is one of the most active branches of research in ordinary differential equations (ODEs): this topic continues to be studied for decades, both for the great interest in itself and for the numerous physical and engineering real–world applications, and also thanks to the progress of computer algebra, which allows the rapid and accurate use of a wide variety of tools for the fast computation of elementary functions, as well as special functions, such as Jacobian elliptic and hypergeometric functions of one or more variables. For this reason, researchers can tackle models of considerable computational complexity.

Department of Statistical Sciences, University of Bologna
Email: daniele.ritelli@unibo.it
* Corresponding author: giulia.spaletta@unibo.it

Our contribution is devoted to the study of the energy–period function of Hamiltonian systems:

$$\begin{cases} \dot{p} = -\dfrac{\partial}{\partial q}H(p,q)\,, \\[2mm] \dot{q} = \dfrac{\partial}{\partial p}H(p,q)\,, \end{cases} \tag{3.1}$$

governed by decoupled energy functions of the form:

$$H(p,q) = \frac{1}{2}p^2 + G(q)\,. \tag{3.2}$$

As usual, $p$ is the momentum and $q$ is the Lagrangian coordinate. System (3.1) is represented in scalar form, as a single differential equation of order two:

$$\ddot{q} + g(q) = 0\,, \qquad \text{with} \quad g(q) = G'(q)\,. \tag{3.3}$$

The physical meaning of the representation (3.3) is, as known, that of an oscillatory motion generated by the restoring force $-g(q)$ as a function of the displacement $q$.

In this chapter, we consider exclusively smooth Hamiltonians, i.e., $G \in \mathcal{C}^\infty(\mathbb{R})$. Assuming that $H(p,q)$ has a non–degenerate minimum in the origin, it is well known [27, 8] that system (3.1) has a centre in the origin surrounded by the periodic solutions, each characterised by a curve of level $H(p,q) = h$, where $h$ is the initial energy of the motion determined by the initial conditions.

The energy–period function is:

$$\mathbb{T}(h) = \sqrt{2} \int_{q_m}^{q_M} \frac{\mathrm{d}q}{\sqrt{h - G(q)}}\,, \tag{3.4}$$

where the two extremes of integration are the minimum and maximum Lagrangian coordinates reached in the periodic orbit, as illustrated in Fig. 3.1.

The existence of the non–degenerate minimum of the Hamiltonian may be obtained assuming, for example, that $G(0) = G'(0) = 0$ and $G''(0) > 0$. Many contributions are devoted to the study of the monotonicity of the period function $\mathbb{T}(h)$. For instance, in [8, Theorem A], after introducing the function:

$$N(q) = (G'(q))^4 \left(\frac{G(q)}{(G'(q))^2}\right)''\,,$$

it is shown that $N(q) \geq 0$ implies that $\mathbb{T}(h)$ is increasing, and $N(q) \leq 0$ implies that $\mathbb{T}(h)$ is decreasing. Similar contributions in this field are given in [32, Theorem 1]. Also worthy of mention is the article [10] dedicated to quadratic systems, which are related to cubic Hamiltonians.

More general Hamiltonian systems $H(p,q) = F(p) + G(q)$ have been studied, with $F(p)$ and $G(q)$ both vanishing at the origin, where they have a

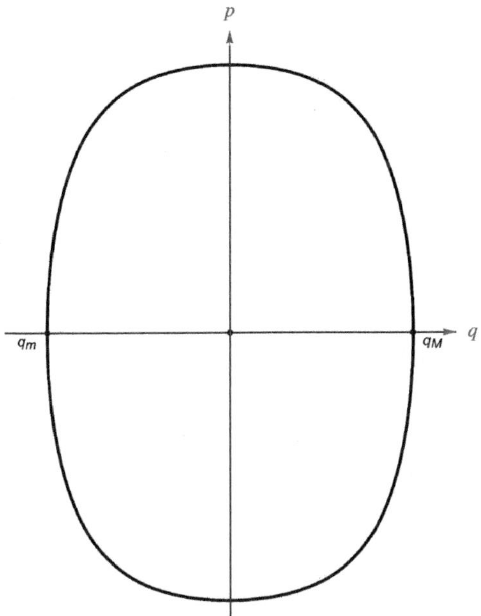

**Figure 3.1:** Phase plot relative to $H(p,q) = \dfrac{p^2}{2} + \dfrac{q^2}{2} + \dfrac{q^4}{4} + \dfrac{q^6}{6}$ .

non–degenerate minimum, inspired by the fact that the famous Lotka–Volterra predator–prey system [36] can be represented in this way, with $F(p) = c\,(e^p - p - 1)$, $a > 0$ , and $G(q) = a\,(e^q - q - 1)$, $c > 0$ . Articles [17, 37] consider monotonicity in the case of the Lotka–Volterra system, and [30, 31] provide the convexity property and asymptotic expansion for the energy–period function of the Lotka–Volterra function. In contrast, asymptotic expansions of the period function are given for general Hamiltonians in [15].

Our main contribution concerns the following *anharmonic* oscillatory systems. Anharmonicity can be seen as the deviation of an ideal oscillator driven by the sole first–power restoring force generated by its displacement. The first group is related to the nonlinear differential equations:

$$\ddot{x} + x + x^7 = 0, \tag{3.5a}$$

$$\ddot{x} + x + x^3 + x^7 = 0, \tag{3.5b}$$

$$\ddot{x} + x + x^3 + x^5 + x^7 = 0. \tag{3.5c}$$

We also consider the generalised Duffing oscillator:

$$\ddot{x} + x^3 + x^7 = 0, \tag{3.6}$$

studied in [1, 7] and more recently in [23, 25].

We present two approaches for the symbolic approximation of the energy—period function: one based on the results in [15] and the other through a fifth—order approximation of the restoring force and the consequent closed—form solution of the approximated equation [18, 9, 24, 12, 13, 3, 2, 4, 5, 6]. In this way, we are able to provide explicit closed—form solutions of initial value problems (IVPs) associated with fifth—degree (odd powers) polynomial oscillators: we discuss an example of this type of problems and then solve the general case. Subsequently, after normalisation of the initial displacement $a$, we approximate the restoring force using fifth—order Chebyshev polynomials, obtaining what some authors call *quantification* [18, 13, 2, 4], and we then proceed to solve the approximate equation. This approach, validated by the study proposed in [14], produces extremely accurate (approximate) analytic solutions, expressed in terms of Jacobian elliptic functions. We exploit closed—form evaluation, using complete elliptic integrals of the first kind of the period of the approximate quintic solutions, to estimate the period of the oscillators object of our study.

## 3.2  Asymptotic Expansion of the Period Function

In this section, for the systems of our interest, we determine the asymptotic representation of the energy–period functions, using the results in [15], that we recall for completeness. We assume that both components $F(p)$ and $G(q)$ of the Hamiltonian $H(p,q) = F(p) + G(q)$ are analytic, with power series representation

$$F(p) = \sum_{n=1}^{\infty} a_n p^n , \qquad G(q) = \sum_{n=1}^{\infty} b_n q^n , \qquad (3.7)$$

where we assume $a_0 = F(0) = 0 = b_0 = G(0)$, $a_1 = F'(0) \neq 0$, $b_1 = G'(0) \neq 0$. In this situation it is possible, in an appropriate neighbourhood of the origin, to invert the series of $F$ and $G$ using the classical method of Lagrange–Bürmann; see for example to [19, paragraph 107] and [38, paragraph 7.3]. Reversion is essential to establish the asymptotic representation theorem of the energy–period function, for the proof of which we refer to [15, Theorem 1.1].

***Theorem 3.1***

*If the sequences $f_n$, $g_n$ are obtained reverting the two power series in (3.7), let:*

$$\alpha_k = (2k - 1) f_{2k-1} \Gamma\left(\frac{2k - 1}{2}\right) , \qquad \beta_k = (2n + 1 - 2k) g_{2n+1-2k} \Gamma\left(\frac{2n + 1 - 2k}{2}\right) ,$$

*where $\Gamma$ denotes the factorial function, and:*

$$\zeta_n = \sum_{k=1}^{n} \alpha_k \beta_k .$$

*Then, if* $\mathbb{T}(h)$ *is the energy–period function of the Hamiltonian system (3.1), it follows:*

$$\lim_{h \to 0^+} \frac{\mathbb{T}(h)}{\displaystyle\sum_{n=1}^{\infty} \zeta_n \frac{h^{n-1}}{(n-1)!}} = 1 \tag{3.8}$$

*Equation (3.8) can be rephrased regarding asymptotics [26, pp.4] as:*

$$\mathbb{T}(h) \backsim \sum_{n=1}^{\infty} \zeta_n \frac{h^{n-1}}{(n-1)!} \qquad as \ h \to 0^+. \tag{3.9}$$

Theorem 3.1 allows one to obtain asymptotic estimates of the energy–period function in relatively short times, even when the period integral cannot be calculated explicitly, since series reversion is implemented in computer algebra systems. The theorem generalises Rothe's contributions [30, 31] involving the Lotka–Volterra predator–prey system. The only limitation lies in the asymptotics expressed by formula (3.8), since formula (3.9) leads to good approximations only for energy levels close to zero.

Given the application of Approximation Theorem 3.1 to the Hamiltonian systems discussed in this chapter, we believe it is useful and instructive to test the level of approximation in the well–known case of the undamped Duffing oscillator [21, chapter 5]:

$$\begin{cases} \ddot{x} + x + cx^3 = 0 \,, \\ x(0) = a > 0, \quad \dot{x}(0) = 0 \,, \end{cases} \tag{3.10}$$

where $c > 0$ due to the physical significance of the model, for which reference is made to the original German source [11] or the recent monograph [20]. The Hamiltonian form of (3.10) is:

$$H(p,q) = \frac{1}{2}p^2 + \frac{1}{2}q^2 + \frac{1}{4}c\,q^4. \tag{3.10h}$$

This is an appropriate time to observe that, when dealing with equations using their geometric formulation, that is:

$$\begin{cases} \ddot{x} + g(x) = 0 \,, \\ x(0) = a > 0 \,, \quad \dot{x}(0) = 0 \,, \end{cases} \tag{3.11}$$

where the restoring force obeys the hypothesis that $g(x) > 0$ for $0 < x \le a$, if we define the potential function $V$ as:

$$V(q) = -2 \int_a^q g(u)\,\mathrm{d}u \,,$$

then the period of the solution of (3.10), expressed in terms of the displacement $a$, is:

$$\mathbb{P}(a) = 2 \int_{-a}^{a} \frac{du}{\sqrt{V(u)}} . \tag{3.12}$$

To compare the two–period expressions (3.4) and (3.12), note that the energy level $h$ in (3.4) is that identified by the initial conditions given in (3.11), since we are in the presence of conservative motion; hence, the energy level is that associated to the initial displacement $a$, that is $h(a) = G(a)$. In other terms:

$$\mathbb{P}(a) = \mathbb{T}(G(a))$$

We further observe that, given the general IVP (3.11), it is possible to normalise the displacement $a$, via the change of dependent variable $u = x/a$, and then deal with the unitary IVP (3.13):

$$\begin{cases} \ddot{u} + \dfrac{1}{a} g(a u) , \\ u(0) = 1 , \quad \dot{u}(0) = 0 . \end{cases} \tag{3.13}$$

Going back to the Duffing equation (3.10), the period (3.12) is well–known and computed in [21, section 5.2]:

$$\mathbb{P}(a) = \frac{4}{\sqrt{1 + c a^2}} K(k_{a,c}) , \tag{3.14}$$

being $K(k)$ the complete elliptic integral of the first kind with modulus $k$ ; here, it is:

$$k_{a,c}^2 = \frac{c a^2}{2 \left(1 + c a^2\right)} .$$

Now, if we approximate the period function using Theorem 3.1, having observed that, in this case, the Hamiltonian function is given by (3.10h), then the reversion of the $q$ component is obtained from $\widetilde{G}(q) = G(q) + G(-q)$ . Using $h$ as the reversion variable, we get:

$$\sqrt{2h} \left( 2 - 2 c h + \frac{7}{2^2} c^2 h^2 - \frac{33}{2^3} c^3 h^3 + \frac{715}{2^6} c^4 h^4 + \cdots \right) . \tag{3.15}$$

Hence, Theorem (3.1) gives the asymptotic expansion for the period related to a generic energy level $h$ :

$$\mathbb{T}(h) \approx \pi \left( 2 - \frac{3}{2} c h + \frac{105}{2^5} c^2 h^2 - \frac{1155}{2^7} c^3 h^3 + \frac{225225}{2^{13}} c^4 h^4 + \cdots \right) . \tag{3.16}$$

At this point, replacing $h = G(a) = \frac{1}{2} a^2 + \frac{1}{4} c a^4$ in (3.16), we obtaine exactly the McLaurin series of (3.14), provided by the Cauchy product between the first factor, which is a binomial series, and the hypergeometric development of the elliptic integral of the first kind. The expression of the approximate evaluation of $\mathbb{P}(a)$ is:

$$\pi \left( 2 - \frac{3}{2^2} a^2 c + \frac{57}{2^7} a^4 c^2 - \frac{315 a^6 c^3}{2^{10}} + \frac{30345 a^8 c^4}{2^{17}} - \frac{193347 a^{10} c^5}{2^{20}} + \cdots \right).$$

At a closer look, Theorem 3.1 seems to be needless in the example considered, since the Duffing oscillator is completely known in the absence of friction and without non–autonomous forcing terms. However, given its use in the study of oscillatory phenomena for which explicit representations of the energy–period function are available, its application confirms that, for low energy levels, the approximation obtained substantially coincides with that of McLaurin.

Conversely, the situation that arises when studying (3.5) oscillators is very different. The Hamiltonians are given respectively by:

$$H_1(p,q) = \frac{p^2}{2} + \frac{q^2}{2} + \frac{q^8}{8}, \tag{3.17a}$$

$$H_2(p,q) = \frac{p^2}{2} + \frac{q^2}{2} + \frac{q^4}{4} + \frac{q^8}{8}, \tag{3.17b}$$

$$H_3(p,q) = \frac{p^2}{2} + \frac{q^2}{2} + \frac{q^4}{4} + \frac{q^6}{6} + \frac{q^8}{8}. \tag{3.17c}$$

The period functions, generated at rest by the initial displacement $a > 0$, in the three cases listed here, are obtained from (3.12) making use of the fact that the integration can be limited to the interval $[0,a]$, due to the symmetry of the restoring forces. We thus obtain:

$$\mathbb{P}_1(a) = \int_0^a \frac{8}{\sqrt{4(a^2 - q^2) + a^8 - q^8}}\, dq, \tag{3.18a}$$

$$\mathbb{P}_2(a) = \int_0^a \frac{8}{\sqrt{4(a^2 - q^2) + 2(a^4 - q^4) + a^8 - q^8}}\, dq, \tag{3.18b}$$

$$\mathbb{P}_3(a) = \int_0^a \frac{8\sqrt{3}}{\sqrt{12(a^2 - q^2) + 6(a^4 - q^4) + 4(a^6 - q^6) + 3(a^8 - q^8)}}\, dq. \tag{3.18c}$$

The integrals (3.18) cannot be expressed in closed form, even with the usual special functions. Certainly, having the quadrature formulas allows us to represent their dependence as a function of $a$, as shown in Fig. 3.2, where the solid line is related to (3.18a), the dotted to (3.18b) and the dashed one to (3.18c).

The first thing that the analysis of the comparative graph suggests is that, for $k = 1, 2, 3$, the following applies:

$$\lim_{a \to 0^+} \mathbb{P}_k(a) = 2\pi.$$

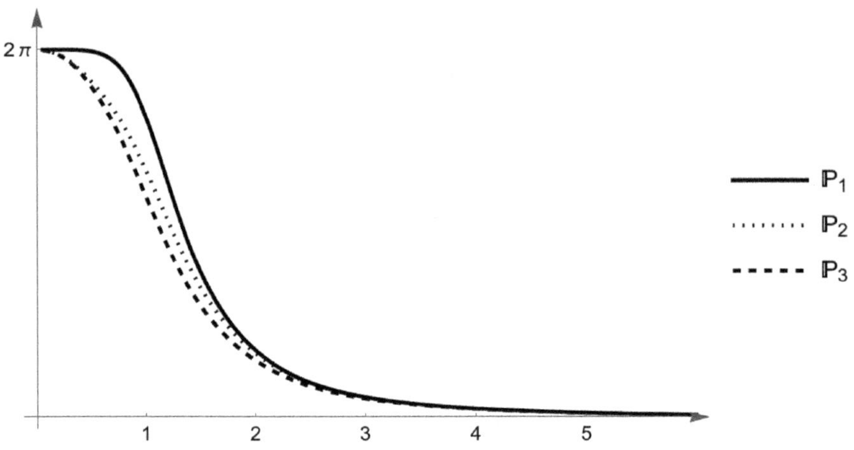


**Figure 3.2:** Comparision between $\mathbb{P}_1$, $\mathbb{P}_2$, $\mathbb{P}_3$ in Formulae (3.18).

This fact is immediately confirmed analytically by applying Theorem 3.1. Such a behaviour has its terse heuristic explanation: when $a$ vanishes, the solution $x = x(t)$ of the IVP, at rest with displacement $a$, also vanishes. This is because, due to the structure of such IVP, the condition $-a \leq x(t) \leq a$ holds, from which it follows that, for $a \to 0^+$, the powers $x^n$ with $n > 1$ become negligible with respect to $x$. This means that the solutions of the anharmonic nonlinear problems are asymptotic to the solution of the harmonic oscillator:

$$\begin{cases} \ddot{x} = -x, \\ x(0) = a > 0, \quad \dot{x}(0) = 0, \end{cases}$$

which is indeed $x(t) = a \cos t$.

As for $a \to +\infty$, Fig. 3.2 suggests that the period of the anharmonic oscillators goes to zero. This fact also has an obvious physical explanation. Again, since the IVP solutions satisfy the condition $-a \leq x \leq a$, it follows that, as $a$ grows towards infinity, the intensity of the restoring force increases by an order equal to $x^7$, reducing the overall time of the oscillation. This can be verified directly using elementary techniques on the integrals (3.18) that define the $\mathbb{P}_k$ periods.

The application of Theorem 3.1 repeats the steps we illustrated in the case of the Duffing oscillator: these are calculations that are elementary in nature, but made more intricate by the reversion of the components of the Hamiltonians in (3.17), which are an integral part of the process, fully manageable with Computer Algebra. We report the first terms that further illustrate the behaviour shown in

Fig. 3.2.

$$\mathbb{P}_1(a) \approx \pi \left( 2 - \frac{35}{2^6} a^6 + \frac{18165}{2^{16}} a^{12} - \frac{6028225}{2^{25}} a^{18} + \cdots \right), \tag{3.19a}$$

$$\mathbb{P}_2(a) \approx \pi \left( 2 - \frac{3}{2^2} a^2 + \frac{57}{2^7} a^4 - \frac{875}{2^{10}} a^6 + \frac{120009}{2^{17}} a^8 + \cdots \right), \tag{3.19b}$$

$$\mathbb{P}_3(a) \approx \pi \left( 2 - \frac{3}{2^2} a^2 - \frac{23}{2^7} a^4 - \frac{91}{2^{10}} a^6 + \frac{171707}{3 \times 2^{16}} a^8 + \cdots \right). \tag{3.19c}$$

The observation that oscillator (3.5a) is the closest to the isochronous behaviour of the harmonic oscillator comes from the fact that the first power of the displacement, appearing in the expansion with order greater than one, is six. In contrast, in the case of the other two oscillators (3.5b) and (3.5c), whose restoring force implies powers of order between one and five, low–order powers of $a$ appear, which significantly distance the behaviour of the oscillation from the harmonic one, leading to anharmonicity.

It is necessary to obtain indications on the quality of the period approximations, at least for small displacements. Not having the feedback of the closed–form expression, available instead for the Duffing oscillator, we plot both the absolute difference between the numerically evaluated integrals $\mathbb{P}_k$ and the polynomials in the left–hand sides of Formulae (3.19), and their quotient, as $a$ varies in $[0,2]$.

## 3.3  Quintic and Septic Oscillators

### 3.3.1  *Perturbed Harmonic Oscillator*

To illustrate how to handle odd–power quintic Duffing oscillators, let us study the IVP:

$$\begin{cases} \ddot{x} = -x - x^5, \\ x(0) = a > 0, \quad \dot{x}(0) = 0. \end{cases} \tag{3.20}$$

Here, we make the simplest choice of considering all coefficients equal to unity; in this way we establish how to treat quintic oscillators in a convenient contest. It is then easy to adapt the procedure to more general situations without unit coefficients. Problem (3.20) was treated in [28] via a time transformation coordinate (therein called *trigonometrification* of time) approximating the periodic solutions to (3.20), using trigonometric functions of the new time coordinate. Here, we present its exact analytic solution following [24].

To compute the period integral $\mathbb{P}(a)$ when $f(x) = -x - x^5$, we recall the elliptic integral tabulated in entry 3.145–2 of [16]:

$$\int_\beta^u \frac{\mathrm{d}x}{\sqrt{(\alpha - x)(x - \beta)[(x - m)^2 + n^2]}} = \frac{1}{\sqrt{pq}} F(\varphi, k), \quad \beta < u < \alpha, \tag{3.21}$$

where $F(\varphi,k)$ is the incomplete elliptic integral of first kind, $p^2 = (m-\alpha)^2 + n^2$, $q^2 = (m-\beta)^2 + m^2$ and:

$$\varphi(u) = 2\operatorname{arccot}\sqrt{\frac{q(\alpha-u)}{p(u-\beta)}}, \qquad k^2 = \frac{1}{4}\frac{(\alpha-\beta)^2 - (p-q)^2}{pq}. \qquad (3.21')$$

The fundamental step in the analysis of a periodic phenomenon is, obviously, to calculate exactly the period and, possibly, the solution in explicit closed–form: for the oscillator governed by (3.20), all this is in fact possible and stated in Theorem 3.2.

### *Theorem 3.2*

*For any $a > 0$, each solution of (3.20) is periodic, with period given by:*

$$\mathbb{P}(a) = \frac{4\sqrt[4]{3}}{\sqrt[4]{(a^4+1)(a^4+3)}}\,\mathbf{K}(k_1(a))\,, \qquad (3.22)$$

*where:*

$$k_1^2(a) = \frac{1}{2} - \frac{\sqrt{3}\,(a^4+2)}{4\,\sqrt{(a^4+3)(a^4+1)}}\,. \qquad (3.23)$$

**Proof**    Given the symmetry of the problem, we can limit ourselves to considering $0 \le x \le a$, which means working in a sort of "first quarter". Referring to the time integral, we have:

$$V(x) = 2\int_a^x f(y)\mathrm{d}y = \frac{1}{3}(a^2 - x^2)(3 + a^4 + a^2 x^2 + x^4)\,.$$

Thus, for $x \le a$, after a straightforward change of variable, we express the time integral associated to (3.20) as:

$$t = \frac{\sqrt{3}}{2}\int_{x^2}^{a^2}\frac{\mathrm{d}s}{\sqrt{s(a^2-s)(3+a^4+a^2s+s^2)}}\,. \qquad (3.24)$$

At this point, in (3.21), we can identify $\alpha$ and $\beta$ with $a^2$ and $0$ respectively; since:

$$3 + a^4 + a^2 s + s^2 = \left(s + \frac{a^2}{2}\right)^2 + 3 + \frac{3}{4}a^4\,,$$

we can also recognise:

$$p^2 = 3\left(a^4+1\right)\,, \qquad q^2 = a^4 + 3\,.$$

Therefore (3.23) follows.    $\square$

The explicit expression (3.22) for the period allows us to verify again that Theorem 3.1 provides the McLaurin series by direct comparison with the hypergeometric expansion of the complete elliptic integral of the first kind:

$$\mathbb{P}(a) = \pi \left( 2 - \frac{5a^4}{2^3} + \frac{515a^8}{3 \times 2^8} - \frac{16285a^{12}}{3^2 \times 2^{13}} + \frac{342265a^{16}}{2^{21}} + \cdots \right).$$

### 3.3.2  General Quintic Oscillator

The oscillator (3.20), studied in Theorem 3.2, is a specific quintic oscillator, representing a particular case in the family of restoring forces of the general form $f(x) = c_1 x + c_3 x^3 + c_5 x^5$, where we assume $c_1, c_3, c_5 > 0$. We thus consider the IVP:

$$\begin{cases} \ddot{x} = -c_1 x - c_3 x^3 - c_5 x^5 \,, \\ x(0) = 1, \quad \dot{x}(0) = 0 \,, \end{cases} \tag{Q}$$

treating only cases leading to periodic solutions $-1 \leq x(t) \leq 1$. We highlight several contributions for this kind of problem [9, 24, 12, 3, 2]. For our purposes, i.e. to approximate oscillators characterised by odd restoring forces with odd quintic oscillators, knowledge of the behaviour of these oscillators is of great importance. The choice to consider unitary initial displacements is not restrictive since, as observed in (3.13), normalisation is always possible through the new dependent variable $u = a^{-1}x$; in the normalisation process, the displacement $a$ modifies the restoring coefficients $c_1, c_3, c_5$; in addition, the unit displacement is related to the fact that the quintic oscillators we use derive from approximations of the restoring force by Chebyshev polynomials.

The period of problem (Q) is obtained by evaluating the (elliptic) Integral:

$$\mathbb{T} = 4\sqrt{\frac{3}{2}} \int_0^1 \frac{ds}{\sqrt{s(1-s)\, h_2(s)}} \,, \tag{3.25}$$

where $h_2(s)$ is the second–degree polynomial:

$$h_2(s) = 2c_5 + 6c_1 + 3c_3 + (2c_5 + 3c_3)\, s + 2c_5 s^2 \,. \tag{3.25'}$$

From (3.25') it follows that, if $h_2$ has no real roots or admits real roots outside the interval $[-1, 1]$, then the solutions $x(t)$ of problem (Q) are periodic and such that $-1 \leq x(t) \leq 1$. Now, the discriminant of $h_2$ is:

$$\Delta = 3\left(3c_3^2 - 4c_5 c_3 - 4c_5^2 - 16c_1 c_5\right) \,, \tag{3.25''}$$

whose sign obviously drives the integration. When $\Delta < 0$, we proceed as in Theorem 3.2, using entry 3.145–2 of [16] again; in this case, the integration is practically identical to that shown in Theorem 3.2, as is the structure of the solution; for this reason, we skip the proof of Theorem 3.3.

**Theorem 3.3**

*Let $\Delta < 0$. If $m,n$ are identified by $h_2(s) = (s-m)^2 + n^2$, then the period of the solution to (Q) is:*

$$\mathbb{P} = \frac{4\sqrt{3}}{\sqrt{c_5}\,\sqrt[4]{\left((m-1)^2 + n^2\right)\left(m^2 + n^2\right)}}\,\mathbf{K}(k)\,, \tag{3.26}$$

*where:*

$$k^2 = \frac{1}{2}\left(1 + \frac{m(1-m) - n^2}{\sqrt{(m^2 + n^2)\left((1-m)^2 + n^2\right)}}\right).$$

We observe that it can be $\Delta \geq 0$ for certain physical systems, depending on the initial displacement $a$. The discriminant of $h_2$ is negative for large values of $a$ but, when $a \to 0^+$, the sign is that of the zero–degree coefficient $3c_3^2 - 16c_1c_5$. There are therefore situations in which integration occurs with a positive or zero discriminant, although, as mentioned, in correspondence with small displacements. Moreover, when $\Delta > 0$, the real roots of $h_2(s)$ are both negative, due to Descartes' rule of signs: in this situation, the relevant integration is provided using entry 3.144-7 of [16]. In Theorem 3.4, we expose the relevant method of integration employed in Section 3.4.

**Theorem 3.4**

*Let $\Delta > 0$ and let $s_1 < s_2$ be the negative roots of $h_2(s)$. Then the period of $x$, solution to problem (Q), is:*

$$\mathbb{P} = \frac{4\sqrt{3c_5}}{\sqrt{s_1(s_2 - 1)}}\,\mathbf{K}(k)\,. \tag{3.27}$$

*As before, $\mathbf{K}(k)$ denotes the complete elliptic integral of first kind, whose modulus $k$ is, here:*

$$k^2 = \frac{s_2 - s_1}{s_1(s_2 - 1)}\,. \tag{3.28}$$

**Proof** To evaluate integral (3.25), we take advantage of entry 3.147-7 of [16], which we recall in full. If $\alpha > v \geq \beta > \gamma > \delta$ then:

$$\int_v^a \frac{ds}{\sqrt{(\alpha - s)(s - \beta)(s - \gamma)(s - \delta)}} = \frac{2}{\sqrt{(a-c)(b-d)}}\,\mathbf{K}(k)\,,$$

with:

$$k^2 = \frac{(\alpha - \beta)(\gamma - \delta)}{(\alpha - \gamma)(\beta - \delta)}\,.$$

Returning to (3.25), we identify $\alpha = 1$, $\beta = 0$, $\gamma = s_2$, $\delta = s_1$, with $k$ given by (3.28). Therefore the motion period is given by (3.27). $\qquad\square$

Finally, if $\Delta = 0$, integral (3.25) degenerates into the elementary integral:

$$\mathbb{T} = 8\sqrt{3\,c_5} \int_0^1 \frac{\mathrm{d}s}{(4c_5 s + 3c_3 + 2c_5)\,\sqrt{s(1-s)}}\,,$$

which, after the (Euler) change of variable $\sqrt{s - s^2} = su$, transforms into:

$$\mathbb{T} = 16\sqrt{3\,c_5} \int_0^\infty \frac{\mathrm{d}u}{3c_3 + 6c_5 + (3c_3 + 2c_5)\,u^2}\,.$$

This case, however, is of little interest due to a very special arrangement of parameters, which change as the initial displacement varies. Thanks to the continuous dependence on data, the relevant solution can be approximated with arbitrary precision, via the solutions obtained in Theorems 3.3 and 3.4.

## 3.4   Chebyshev quintication

Our contribution is based on our previous effort in treating fifth–degree odd–power polynomial oscillators, useful for activating the following procedure, known as *quintication*. First, since Chebyshev polynomials are defined in the interval $[-1, 1]$, we need to normalise the displacement $a$. Then, we approximate the restoring force to order five, using Chebyshev polynomials; see for instance [22, 29]. Finally, the approximated equation is solved explicitly. Several authors use this method; we cite for example [18, 13, 2, 4, 5, 6]. This technique, validated by the numerical analysis presented in [14], produces very accurate approximate analytic period integrals, which are revealed by Theorems 3.3 and 3.4.

### *3.4.1   Generalized Duffing Oscillator*

The behaviour of the septic oscillator (3.6) was studied in [1, 7, 25]. It consists of the following IVP:

$$\begin{cases} \ddot{x} = -x^3 - x^7\,, \\ x(0) = a > 0\,, \quad \dot{x}(0) = 0\,. \end{cases} \tag{3.29}$$

This problem interests us because (since it is possible to determine its period, here in terms of $_2F_1$) it represents a valid test case for the method of approximation using its Chebyshev polynomial of order five. Notably, the asymptotic method illustrated in Section 3.2 cannot be applied to this type of oscillation.

All solutions of (3.29) are periodic; we refer to [25] for details on the computation of the integral:

$$\int_0^a \frac{\mathrm{d}s}{\sqrt{(a^4 - s^4)(a^4 + 2 + s^4)}}\,,$$

providing the oscillation period, which indeed is:

$$\mathbb{P}(a) = \sqrt{\frac{2}{\pi}}\,\Gamma^2\left(\frac{1}{4}\right)\frac{1}{a\sqrt{a^4+2}}\,{}_2\mathrm{F}_1\left(\begin{array}{c}1/4;1/2\\3/4\end{array}\middle|-\frac{a^4}{a^4+2}\right). \tag{3.30}$$

The period function $\mathbb{P}(a)$ is decreasing in this case. Moreover, expanding in Laurent series, we see that:

$$\mathbb{P}(a) = \frac{1}{\sqrt{\pi}}\,\Gamma^2\left(\frac{1}{4}\right)\left(\frac{1}{a}-\frac{1}{3}a^3+\frac{5}{28}a^7-\frac{5}{44}a^{11}+\frac{85}{1056}a^{15}+\cdots\right). \tag{3.31}$$

Explicit knowledge of the period allows us to validate the quality of the approximation obtained using Chebyshev polynomials.

In the case of (3.29), the normalised equation (3.13) becomes:

$$\begin{cases}\ddot{u} = -a^2\,u^3 - a^6\,u^7 := g_a(u)\,,\\ u(0) = 1\,,\quad \dot{u}(0) = 0\,.\end{cases} \tag{3.32}$$

It is worth noting that the structure of (3.32) clearly explains why the initial choice on the sign of the displacement $a$ is not restrictive at all.

Now we expand, in the Chebyshev series truncated to order five, the restoring force $g_a(u)$, arriving at:

$$g_a(u) \simeq c_1(a)\,\check{C}_1(u) + c_3(a)\,\check{C}_3(u) + c_5(u)\,\check{C}_5(u)\,, \tag{3.33}$$

where $\check{C}_i(u)$ denote the Chebyshev polynomial of order $i$, namely:

$$\check{C}_1(u) = u\,,\qquad \check{C}_3(u) = 4u^3 - 3u\,,\qquad \check{C}_5(u) = 16u^5 - 20u^3 + 5u\,, \tag{3.34}$$

and the coefficients $c_i(a)$ are defined by weighted inner product as:

$$c_i(a) = \frac{2}{\pi}\int_{-1}^{1}\frac{g_a(u)\,\check{C}_i(u)}{\sqrt{1-u^2}}\,\mathrm{d}u\,. \tag{3.35}$$

Naturally, having expanded an odd function, there are no even powers. Using the approximate restoring force, we consider our fundamental (normalised) IVP:

$$\begin{cases}\ddot{u} = -\dfrac{7}{64}a^6 u + \dfrac{1}{8}\left(7a^4 - 8\right)a^2 u^3 - \dfrac{7}{4}a^6 u^5\,,\\ u(0) = 1\,,\quad \dot{u}(0) = 0\,.\end{cases} \tag{3.36}$$

Applying to (3.36) the integration procedure indicated by the time equation (3.12), we obtain:

$$\mathbb{P}(a) = \frac{4}{a^3}\sqrt{\frac{3}{7}}\int_0^1\frac{\mathrm{d}z}{\sqrt{(1-z)z\left(z^2 + \left(\dfrac{6}{7a^4}+\dfrac{1}{4}\right)z + \dfrac{6}{7a^4}+\dfrac{7}{16}\right)}}\,. \tag{3.37}$$

For almost any values of the displacement $a$, integral (3.37) is an elliptic integral of the first kind, thus invertible in the Jacobi sense. This explains the great usefulness of the quintication method when applied to restoring forces expressed by odd functions. It should be noted, however, that, unlike the model proposed in [3], the behaviour of the integral is influenced by $a$, since the second–degree polynomial discriminant $\Delta_a$ in (3.37) changes sign according to:

$$\Delta_a < 0 \ \text{ if } a > a^*, \qquad \Delta_a = 0 \ \text{ if } a = a^*, \qquad \Delta_a > 0 \ \text{ if } a < a^*,$$

where:

$$a^* = \sqrt[4]{\frac{16\sqrt{19}}{63} - \frac{8}{9}} \simeq 0.683408 \,.$$

As noted in Theorem 3.4, when $\Delta_a > 0$, the second–degree polynomial, appearing as a factor in the denominator of the integral (3.37), has negative real roots, due to Descartes' rule of signs.

We compare the exact period of the quinticated problem (3.36), obtained using Theorems 3.3 and 3.4, with the exact period computed in (3.30).

### 3.4.1.1 Case $a < a^*$

Here we discuss the impact of Theorem 3.4. We begin by comparing the exact period, indicated in (3.30), with the period of the approximate solution, given by formula (3.27) in terms of the complete elliptic integral of the first kind. Using Theorem 3.4 and taking full advantage of *Mathematica* symbolic calculation [34, 33, 35], we see that the period of the solution of the Chebyshev–approximated problem is:

$$\mathbb{T}_1(a) = \frac{32\sqrt{3}\,\mathbb{K}_1}{a\sqrt{63a^4 + 144 + 2\sqrt{3}\sqrt{192 - 784a^4 - 441a^8}}}, \tag{3.38}$$

where:

$$\mathbb{K}_1 = K\left(\sqrt{\frac{4\sqrt{576 - 2352a^4 - 1323a^8}}{63a^4 + 144 + 2\sqrt{576 - 2352a^4 - 1323a^8}}}\right). \tag{3.39}$$

Expanding expressions (3.38) and (3.39) in Laurent series, we arrive at the approximate representation of the period of the solution (3.29) of the quinticated equation (3.36):

$$\begin{aligned}
\mathbb{T}_1(a) \simeq \frac{1}{\sqrt{\pi}}\Bigg( &\frac{\Gamma^2\left(\frac{1}{4}\right)}{a} + \frac{7\left(5\Gamma^4\left(\frac{1}{4}\right) - 408\pi^2\right)}{2^7\,3\,\Gamma^2\left(\frac{1}{4}\right)}\,a^3 \\
&+ \frac{49\left(1277\,\Gamma^4\left(\frac{1}{4}\right) - 17652\,\pi^2\right)}{2^{13} \times 3^2\,\Gamma^2\left(\frac{1}{4}\right)}\,a^7 \\
&+ \frac{343\left(497740\,\Gamma^4\left(\frac{1}{4}\right) - 9150752\,\pi^2\right)}{2^{23} \times 3^2\,\Gamma^2\left(\frac{1}{4}\right)}\,a^{11} + \cdots \Bigg).
\end{aligned} \tag{3.40}$$

**Table 3.1:** Numerical comparison coefficients in (3.31) and (3.40).

| coefficient of | $a^{-1}$ | $a^3$ | $a^7$ | $a^{11}$ | $a^{15}$ |
|---|---|---|---|---|---|
| exact | 7.4163 | -2.47210 | 1.32434 | -0.842761 | 0.596956 |
| approximate | 7.4163 | -2.47461 | 1.32463 | -0.840173 | 0.591571 |

From the comparison between the power series expansions (3.31) and (3.40), respectively exact and approximated, it can be seen that the solution obtained is particularly effective in this situation: in fact, in both series, the first terms associated with the leading power $a^{-1}$ (recall that, here, $a$ is close to zero) coincide. Moreover, the scan of the successive powers is the same, and the relative coefficients are very close to each other, as shown in Table 3.1, obtained with *Mathematica*.

The maximum difference between the two period expressions, depending on the initial displacement, computed by discretising the interval $(0, a^*)$ with step $10^{-4}$, is $7.53478 \times 10^{-4}$.

### 3.4.1.2 Case $a > a^*$

We discuss here the quality of the approximate solutions given in Theorem 3.3. The explicit formula, as a function of $a$, for the period deduced from (3.26) is:

$$T_2(a) = \frac{32\sqrt[4]{3}}{a\sqrt[4]{(49a^4 + 96)(63a^4 + 64)}} \mathbb{K}_2 , \qquad (3.41)$$

with:

$$\mathbb{K}_2 = K\left( \frac{\sqrt{\sqrt{3(49a^4 + 96)(63a^4 + 64)} - 144 - 63a^4}}{\sqrt[4]{12(49a^4 + 96)(63a^4 + 64)}} \right) . \qquad (3.42)$$

Here, it is neither useful nor meaningful to expand in Laurent series about $a = 0$, since $a$ is away from the origin. To estimate how the two periods (exact and approximate) resemble each other, we expand in Taylor series, for instance, around $a = 1$. The symbolic expressions for both Taylor expansions are quite intricate; here, we limit ourselves to writing the value of both period functions at $a = 1$, observing that their numerical values are very close, namely, $\mathbb{P}(1) \simeq 5.76812$ and $\mathbb{T}_2(1) \simeq 5.76647$ :

$$\mathbb{P}(1) = \sqrt{\frac{2}{3\pi}} \Gamma^2\left(\frac{1}{4}\right) {}_2F_1\left( \begin{array}{c} 1/4; 1/2 \\ 3/4 \end{array} \middle| -\frac{1}{3} \right),$$

$$\mathbb{T}_2(1) = 32\sqrt[4]{\frac{3}{18415}} K\left( \sqrt{\frac{1}{2} - \frac{69}{2}\sqrt{\frac{3}{18415}}} \right) .$$

**Table 3.2:** Numerical comparison of period power series coefficients.

| coefficient of | $(a-1)^0$ | $(a-1)^1$ | $(a-1)^2$ | $(a-1)^3$ | $(a-1)^4$ |
|---|---|---|---|---|---|
| Exact | 5.76647 | -10.2738 | 8.92026 | -4.58378 | 4.05152 |
| Approximate | 5.76812 | -10.2719 | 8.9153 | -4.58693 | 4.06468 |

We then compare numerically the Taylor coefficients, obtained via *Mathematica*, observing again a good quality of approximation, as reported in Table 3.2.

## 3.5  Complete Odd Septic Oscillator

The seventh–degree oscillator (3.29) is governed by only two terms. Here, we consider its complete version, where all odd powers of order up to seven act. Having the objective of comparing the two period approximation techniques, we limit ourselves, once again, to considering the case of powers with unit coefficients:

$$\begin{cases} \ddot{x} = -x - x^3 - x^5 - x^7 \,, \\ x(0) = a > 0 \,, \quad \dot{x}(0) = 0 \,. \end{cases} \tag{3.43}$$

In this case, the normalised equation (3.13) is:

$$\begin{cases} \ddot{u} = -u - a^2 u^3 - a^4 u^5 - a^6 u^7 := g_a(u) \,, \\ u(0) = 1 \,, \quad \dot{u}(0) = 0 \,. \end{cases} \tag{3.44}$$

Then, we expand the restoring force $g_a(u)$ in the Chebyshev series truncated to order five, using (3.33), (3.34), (3.35), so that we arrive at the normalised IVP:

$$\begin{cases} \ddot{u} = -\dfrac{1}{64}(64 + 7 a^6) u - \dfrac{1}{8}(8 - 7 a^4) a^2 u^3 - \dfrac{1}{4}(4 + 7 a^2) a^4 u^5 \,, \\ u(0) = 1 \,, \quad \dot{u}(0) = 0 \,. \end{cases} \tag{3.45}$$

Using the notation adopted in problem **Q**, here the coefficient of the quinticated IVP are:

$$c_1 = \frac{1}{64}(64 + 7 a^6) \,, \quad c_3 = \frac{1}{8}(8 - 7 a^4) a^2 \,, \quad c_5 = \frac{1}{4}(4 + 7 a^2) a^4 \,.$$

It is $c_1, c_5 > 0$ $\forall a$, while $c_3 > 0$ for $0 < a < \dfrac{2^{3/4}}{7^{1/4}} \approx 1.03395$. The discriminant (3.25'') becomes, in this case:

$$\Delta = -\frac{1}{64} a^4 (832 + 2048 a^2 + 1040 a^4 + 784 a^6 + 441 a^8)$$

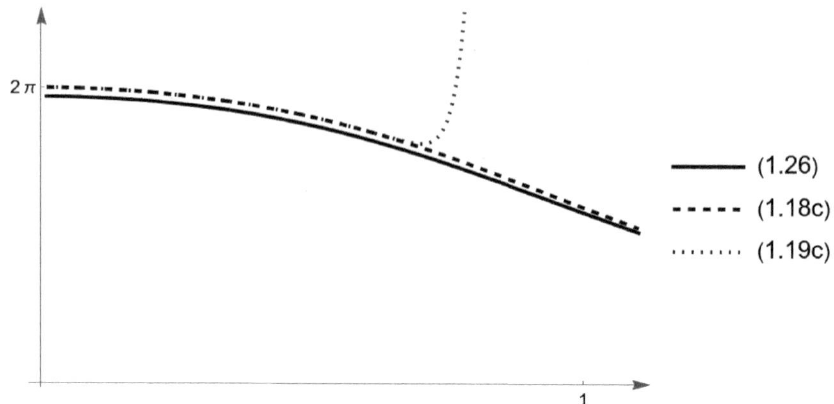

**Figure 3.3:** Period of (3.43) given by (3.26), by (3.18c) and by (3.19c).

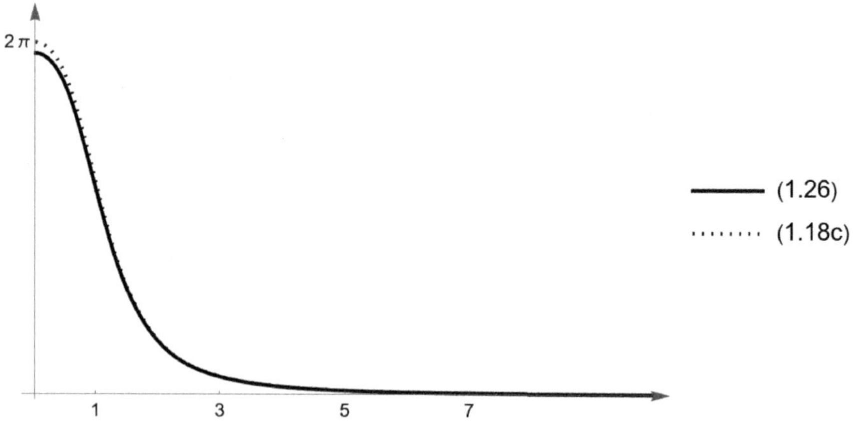

**Figure 3.4:** Period of (3.43) given by (3.26) and by (3.18c).

and is negative for $a \neq 0$. Therefore, Theorem 3.3 applies. We rewrite:

$$\frac{h_2(s)}{2c_5} = (s-m)^2 + n^2, \quad m = -\frac{3c_3}{4c_5} + \frac{1}{2}, \quad n = \sqrt{\frac{3c_1}{c_5} - \left(\frac{3c_3}{4c_5}\right)^2 + \frac{3c_3}{4c_5} + \frac{3}{4}},$$

and apply (3.26). The resulting period expression is not reported here, since it is quite complicated, though manageable via computer algebra. A graphical comparison is instead provided, in Figs. 3.3 and 3.4, among the period of (3.43) obtained via (3.26), the one given by the numerical integral (3.18c) and, finally, $\mathbb{P}_3(a)$ in (3.19c).

## 3.6  Conclusions

The obtained approximation quality is verified through different comparison methods illustrated in this article. Oscillator (3.6) can be approximated immediately (though not with the asymptotic expansion methods of Section 3.2 since not all the conditions required by Theorem 3.1 are satisfied) by employing the approximation method with quintic oscillators introduced in Section 3.3. Note that the choice of restoring forces with coefficients equal to 1 is not restrictive. Indeed, we are interested in highlighting two calculation methods, which are then easily adapted to the case of non–unitary coefficients. What remains essential is that these restoring forces generate periodic oscillations. Furthermore, the choice to study oscillations in which the restoring force only has powers of odd order is motivated here by the fact that, in this way, it is possible to compare the approximations obtainable using Theorem 3.1 with the approximate solutions provided by a special quintic oscillator, constructed by orthogonal projection onto the space of Chebyshev polynomials, as outlined in Section 3.4.

# References


[1] Atkinson, C. P. 1962. On the superposition method for determining frequencies of nonlinear systems. In ASME Procs. of the 4th National Congress of Applied Mechanics, held at the University of California, Berkeley, California, June 18–21, 1962, pp. 57–62, American Society of Mechanical Engineers, New York.

[2] Beléndez, A., Arribas, E., Beléndez, T., Pascual, C., Gimeno, E., and Álvarez, M. L. 2017. Closed–form exact solutions for the unforced quintic nonlinear oscillator. Advances in Mathematical Physics.

[3] Beléndez, A., Beléndez, T., Martínez, F., Pascual, C., Alvarez, M. L., and Arribas, E. 2016. Exact solution for the unforced Duffing oscillator with cubic and quintic nonlinearities. Nonlinear Dynamics 86(3): 1687–1700.

[4] Big-Alabo, A. 2019. Approximate period for large–amplitude oscillations of a simple pendulum based on quintication of the restoring force. European Journal of Physics 41(1): 015001.

[5] Boschi, M., Ritelli, D., and Spaletta, G. 2022. Nonlinear oscillators via Čebyšëv quintic approximations. Arxiv preprint.

[6] Boschi, M., Ritelli, D., and Spaletta, G. 2024. Exact time–integral inversion via Čebyšëv quintic approximations for nonlinear oscillators. Journal of Mathematical Analysis and Applications 533(1): 1–16.

[7] Burton, T. D. and Hamdan, M. N. 1983. Analysis of non-linear autonomous conservative oscillators by a time transformation method. Journal of Sound and Vibration 87(4): 543–554.

[8] Chicone, C. 1987. The monotonicity of the period function for planar Hamiltonian vector fields. Journal of Differential Equations 69(3): 310–321.

[9] Citterio, M. and Talamo, R. 2009. The elliptic core of nonlinear oscillators. Meccanica 44(6): 653.

[10] Coppel, W. A. and Gavrilov, L. 1993. The period function of a Hamiltonian quadratic system. Differential Integral Equations 6(6): 1357–1365.

[11] Duffing, G. 1919. Erzwungene Schwingungen bei veränderlicher Eigenfrequenz und ihre technische Bedeutung. Vieweg, Braunschweig.

[12] Elías-Zúñiga, A. 2013. Exact solution of the cubic–quintic Duffing oscillator. Applied Mathematical Modelling 37(4): 2574–2579.

[13] Elías-Zúñiga, A. 2014. "Quintication" method to obtain approximate analytical solutions of non–linear oscillators. Applied Mathematics and Computation 243: 849–855.

[14] Ershov, A. G. and Kashevarova, T. P. 2005. Interval mathematical library based on Čebyšёv and Taylor series expansion. Reliable Computing 11(5): 359–367.

[15] Foschi, S., Mingari-Scarpello, G., and Ritelli, D. 2004. Higher order approximation of the period–energy function for single degree of freedom Hamiltonian systems. Meccanica 39(4): 357–368.

[16] Gradshteyn, I. S. and Ryzhik, J. M. 2000. Table of Integrals, Series and Products 6th ed. Academic Press, New York.

[17] Hsu, S.-B. 1983. A remark on the period of the periodic solution in the Lotka-Volterra system. Journal of Mathematical Analysis and Applications 95(2): 428–436.

[18] Jonckheere, R. 1971. Determination of the period of nonlinear oscillations by means of Čebyšёv polynomials. Zeitschrift für angewandte Mathematik und Mechanik 51(5): 389–393.

[19] Knopp, K. 1990. Theory and Application of Infinite Series. Dover, New York.

[20] Kovacic, I. and Brennan, M. J. 2011. The Duffing Equation: Nonlinear Oscillators and their Behaviour. Wiley, Hoboken.

[21] Lawden, D. F. 1989. Elliptic Functions and Applications. Springer, New York.

[22] Mason, J. C. and Handscomb, D. C. 2002. Čebyšёv Polynomials. CRC Press, Boca Raton.

[23] Mickens, R. E. 2010. Truly nonlinear oscillations: harmonic balance, parameter expansions, iteration, and averaging methods. World Scientific, Singapore.

[24] Mingari-Scarpello, G. and Ritelli, D. 2010. Exact solution to a first–fifth power nonlinear unforced oscillator. Applied Mathematical Sciences 4(69–72): 3589–3594.

[25] Mingari-Scarpello, G. and Ritelli, D. 2012. Closed form integration of a hyperelliptic, odd powers, undamped oscillator. Meccanica 47(4): 857–862.

[26] Olver, F. 1997. Asymptotics and Special Functions. CRC Press, Boca Raton.

[27] Opial, Z. 1961. Sur les périodes des solutions de l'équation différentielle $x'' + g(x) = 0$. Annales Polonici Mathematici 10(1): 49–72.

[28] Recktenwald, G. and Rand, R. 2007. Trigonometric simplification of a class of conservative nonlinear oscillators. Nonlinear Dynamics 49: 193–201.

[29] Rivlin, T. J. 1969. An introduction to the Approximation of Functions. Blaisdell, Waltham.

[30] Rothe, F. 1985. The periods of the Volterra–Lotka system. Journal für die Reine und Angewandte Mathematik 355: 129–138.

[31] Rothe, F. 1993. Remarks on periods of planar Hamiltonian systems. SIAM Journal on Mathematical Analysis 24(1):129–154.

[32] Schaaf, R. 1985. A class of Hamiltonian systems with increasing periods. Journal für die Reine und Angewandte Mathematik 1985: 96–109.

[33] Sofroniou, M. 1994. Symbolic derivation of Runge–Kutta methods. Journal of Symbolic Computation 18(3): 265–296.

[34] Sofroniou, M. 1994. Symbolic and numerical methods for Hamiltonian system. PhD Thesis, Research Repository, Loughborough University, U.K.

[35] Sofroniou, M. and Spaletta, G. 2002. Symplectic methods for separable Hamiltonian systems. Lecture Notes in Computer Science 2332: 506–515.

[36] Volterra, V. 1931. Thèorie mathèmatique de la lutte pour la vie. Gauthier-Villars, Paris.

[37] Waldvogel, J. 1986. The period in the Lotka–Volterra system is monotonic. Journal of Mathematical Analysis and Applications 114(1): 178–184.

[38] Whittaker, E. T. and Watson, G. N. 1996. A Course of Modern Analysis. Cambridge University Press, Cambridge.

# Chapter 4

# Continuous Characterizations of Weighted Besov and Triebel-Lizorkin-Type Spaces

*Ahmed Loulit*

## 4.1 Introduction

The Besov and Triebel-Lizorkin spaces are a class of function spaces containing many well-known classical function spaces such as Lebesgue spaces $L_p$, Hardy spaces $H_p$, $BMO$ spaces, Sobolev spaces...

A comprehensive treatment of these function spaces and their history can be found in Triebel's monographs [35, 36] and in the fundamental paper by M. Frazier and B. Jawerth [7].

Département de Mathematique, Research Center E. Bernheim Solvay Brussels School, Université Libre de Bruxelles, av F.D. Roosevelt 42, CP 135/01, B-1050 Brussels, Belgium.
Email: ahmed.loulit@ulb.be

Such spaces play an important role in various fields of analysis such as harmonic analysis and partial differential equations. For instance, Morrey [23] study the local regularity of solutions of some partial differential equations in an appropriate space, called Morrey space. This local regularity of solutions is more precise than on the familiar Lebesgue spaces.

During the last decades various classical operators of harmonic analysis, such as maximal, singular, and potential operators were widely investigated both in classical and generalized Morrey spaces, we refer the reader to [11, 12, 13, 14, 15, 16, 17, 18, 19, 21, 27] and the references therein.

In recent years, there has been increasing interest in a new family of function spaces, called New class of Besov and Triebel-Lizorkin spaces. These spaces unify and generalize many classical spaces such as $Q$ spaces, Besov spaces, Triebel-Lizorkin spaces, Morrey spaces and Triebel-Lizorkin-Morrey spaces, see [31, Section 1.4] and references in [38]. From then on, these spaces received a lot of attention, various properties and characterizations of Besov-type and Triebel–Lizorkin type spaces were studied in [41, 42, 43, 45, 46]. In particular, Liang et al. [20, 19] and Wu et al. [40] established continuous characterizations of Besov–Triebel–Lizorkin-type spaces.

In this work we give continuous characterizations of the recently introduced weighted type homogeneous Besov-Triebel-Lizorkin spaces, see [22]. The weight function $w$ is assumed to be in the Muckenhoupt's class. We prove different integrals representations of the quasi-norms of these spaces in terms of continuous Peetre maximal function of local means, the tent space (Lusin area function) associated with local means and the Peetre maximal function of local means.

## 4.2   Some Background Tools

### 4.2.1   *Muckenhoupt Classes and Maximal Function*

In this section $w$ denotes a weight function in $R^n$, i.e., $w$ is an almost every (a.e) positive locally integrable function in $\mathbb{R}^n$. A function $f \in L^p(w)$, $0 < p < \infty$ if and only if

$$||f||_{p,w} = \left( \int_{\mathbb{R}^n} |f(x)|^p w(x) dx \right)^{\frac{1}{p}} < \infty.$$

A weight function $w$ is said to be in the Muckenhoupt classes $A_p$, $1 \leq p < \infty$ if there exists a constant $C_p > 0$ such that for every cube $Q$,

$$\frac{1}{|Q|} \int_Q w\, dy \left( \frac{1}{|Q|} \int_Q w^{1-p'} dy \right)^{p-1} \leq C_p$$

when $1 < p < \infty$, $\frac{1}{p} + \frac{1}{p'} = 1$, well for $p = 1$,

$$\frac{1}{|Q|} \int_Q w\,dy \le C_1 w(x),$$

for, e.g., $x \in Q$, or equivalently $Mw(x) \le C_1 w(x)$ for, e.g., $x \in \mathbb{R}^n$, where $M$ is the Hardy-Littlewood maximal operator defined, for a local integrable function $f$, by

$$Mf(x) = \sup_{x \in Q} \frac{1}{|Q|} \int_Q |f(y)|\,dy.$$

The supremum is taken over all cubes containing $x$.

The classes $A_p$ was introduced by Muckenhoupt, B. [24] in order to characterize the boundedness of the Hardy-Littlewodd maximal operator $M$ on the weighted Lebesgue spaces, see also [4, 8, 32, 33]. The pioneering work of Muckenhoupt, B. [24] showed that

$$M: \quad L_p(w) \to L_p(w)$$

if and only if $w \in A_p$ when $1 < p < \infty$, and

$$M: \quad L_1(w) \to L_{1,\infty}(w),$$

if and only if $w \in A_1$.

$L_{q,\infty}(w)$ denotes the space of all measurable functions $f$ such that

$$\sup_{\lambda > 0} \left( w\left\{ x \in \mathbb{R}^n : f(x) > \lambda \right\} \right)^{\frac{1}{q}} < \infty.$$

Moreover, if $1 < p < \infty$, $1 < q \le \infty$ and $w \in A_p$, then there exists a positive constant C such that for all sequences $\{f_k\}_{k \in \mathbb{Z}}$ of locally integrable functions on $\mathbb{R}^n$,

$$\int_{\mathbb{R}^n} \left( \sum_{k \in \mathbb{Z}} [Mf_k(x)]^q \right)^{p/q} w(x)\,dx \le C \int_{\mathbb{R}^n} \left( \sum_{k \in \mathbb{Z}} |f_k(x)|^q \right)^{p/q} w(x)\,dx. \qquad (4.1)$$

The inequality 4.1 is the well known Fefferman-Stein vector-valued inequality, see for instance [6, 8, 9, 32].

The following important properties of Muckenhoupt weights will be widely used in this work.

**Lemma 4.2.1** *Let $w \in A_p$. Then there exist $\delta > 0$, $C > 0$ s.t, every time we have a measurable subset A of a cube Q, the following "$\delta$-reverse doubling" inequality holds*

$$\frac{w(A)}{w(Q)} \le C \left( \frac{|A|}{|Q|} \right)^{\delta} \qquad (4.2)$$

*and also the following "p-doubling" inequality holds*

$$\frac{w(Q)}{w(A)} \leq C \left( \frac{|Q|}{|A|} \right)^{p}.$$

(4.3)

**Remark 4.2.1** *If $w \in A_p$, then $w^{\frac{-1}{p-1}} \in A_{\frac{p}{p-1}}$ and satisfies the same "$\delta$-reverse doubling" inequality.*

The reverse condition is known as $A_\infty$ - condition and the class of the weights $w$ satisfying $A_\infty$ - condition is denoted by $A_\infty$. It is well known that $A_\infty = \cup_{p \geq 1} A_p$, which motivates the notation $A_\infty$, see for instance [8, Corollary 2.13, pp. 403–404].

**Lemma 4.2.2** *Let $f$ be any measurable function in $\mathbb{R}^n$ and $\lambda > n$. Then*

$$\int_{R^n} f(z) \left( 1 + \frac{|x-z|}{t} \right)^{-\lambda} dz \leq C t^n M f(x)$$

*for any $t > 0$.*

**Lemma 4.2.3** *Let $w \in A_p, 1 < p < \infty$. Then we have the following $B_p$-condition*

$$\int_{R^n} w(z) \left( 1 + \frac{|x-z|}{t} \right)^{-np} dz \leq C \int_{|x-z|<t} w(z) dz$$

*for any $t > 0$ and $x \in \mathbb{R}^n$.*

## 4.3 Tempred Distributions and Reproducing Calderon Formula

Let $\mathscr{S}(\mathbb{R}^n)$ to be the space of all Schwartz functions on $\mathbb{R}^n$ with the classical topology generated by the family of semi-norms

$$||v||_{k,N} = \sup_{x \in \mathbb{R}^n} \sup_{|\beta| \leq N} (1+|x|)^k |\partial^\beta v(x)| \quad k, N \in \mathbb{N}_0, \quad v \in \mathscr{S}(\mathbb{R}^n).$$

The topological dual space, $\mathscr{S}'(\mathbb{R}^n)$ of $\mathscr{S}(\mathbb{R}^n)$ is the set of all continuous linear functional $\mathscr{S}(R^n) \longrightarrow \mathbb{C}$ endowed with the weak $\star$-topology. We denote by $\mathscr{S}_\infty(\mathbb{R}^n)$, the topological subspace of functions in $\mathscr{S}(\mathbb{R}^n)$ having all vanishing moments:

$$\mathscr{S}_\infty(\mathbb{R}^n) = \left\{ v \in \mathscr{S}(\mathbb{R}^n) : \int_{\mathbb{R}^n} x^\beta v(x) dx = 0, \quad \forall \beta \in \mathbb{N}^n \right\}.$$

$\mathscr{S}_\infty'(\mathbb{R}^n)$ denotes the topological dual space of $\mathscr{S}_\infty(\mathbb{R}^n)$, namely, the set of all continuous linear functional on $\mathscr{S}_\infty'(\mathbb{R}^n)$. The space $\mathscr{S}_\infty'(\mathbb{R}^n)$ is also endowed with the weak $\star$-topology. It is well known that $\mathscr{S}'_\infty(\mathbb{R}^n) = \mathscr{S}'(\mathbb{R}^n)/\mathscr{P}(\mathbb{R}^n)$ as topological spaces, where $\mathscr{P}(\mathbb{R}^n)$ denotes the set of all polynomials on $\mathbb{R}^n$; see for example [44, Proposition 8.1].

Similarly, for any $R \in \mathbb{N}$, the space $\mathscr{S}_R(\mathbb{R}^n)$ is defined to be the set of all Schwartz functions having vanishing moments of order $R$ and $\mathscr{S}'_R(\mathbb{R}^n)$ is its topological dual space. We write $\mathscr{S}_{-1}(\mathbb{R}^n) = \mathscr{S}(\mathbb{R}^n)$. The Fourier transform, $\mathscr{F}v = \hat{v}$, of Schwartz function $v$ is defined by

$$\hat{v}(\xi) = (2\pi)^{-n} \int_{\mathbb{R}^n} e^{-i\xi \cdot x} v(x)dx.$$

The convolution of two function $v, \mu \in \mathscr{S}(\mathbb{R}^n)$ is defined by

$$v \star \mu(x) = \int_{\mathbb{R}^n} v(x-y)\mu(y)dy$$

and still belongs to $\mathscr{S}(\mathbb{R}^n)$.

The convolution operator can be extended to $\mathscr{S}(\mathbb{R}^n) \times \mathscr{S}'(\mathbb{R}^n)$ via $v \star f(x) = \langle f, \mu(x-.)\rangle$. It makes sense pointwise and is a $C^\infty$ function in $\mathbb{R}^n$ of at most polynomial growth. To simplify notation, we often write $vf = v \star f$.

For $j \in \mathbb{Z}$ and $k \in \mathbb{Z}^n$, denoted by $Q_{jk}$ the dyadic cube $2^{-j}([0,1]^n + k)$, $l(Q_{jk}) = 2^{-j}$ is its side length, $x_{Q_{jk}} = 2^{-j}k$ is its lower "left-corner" and $c_{Q_{jk}}$ is its center. We set $\mathscr{Q} = \{Q_{jk} : j \in \mathbb{Z}, k \in \mathbb{Z}^n\}$, and $j_Q = -log_2 l(Q)$ for all $Q \in \mathscr{Q}$.

When the dyadic cube $Q$ appears as an index, such as $\sum_{Q \in \mathscr{Q}}$, it is understood that $Q$ runs over all dyadic cubes in $\mathbb{R}^n$.

For a function $v$ and dyadic cube $Q = Q_{jk}$, set

$$v_Q(x) = |Q|^{-1/2} v(2^j x - k) = |Q|^{1/2} v_j(x - x_Q),$$

for all $x \in \mathbb{R}^n$, where $v_j(x) = 2^{nj}v(2^j x)$.

**Definition 4.3.1** *A Schwartz function* $v : \mathbb{R}^n \longrightarrow \mathbb{C}$ *is a Littlewood-Paley function if* $\hat{v}$ *is a real-valued function and satisfies:*

$$supp \quad \hat{v} \subset \{\xi \in \mathbb{R}^n \; : \; 1/2 \leq |\xi| \leq 2\} \tag{4.4}$$

$$|\hat{v}(\xi)| \geq C > 0 \quad if \quad 3/5 \leq |\xi| \leq 5/3. \tag{4.5}$$

*The function* $\hat{\mu}(\xi) = \hat{v}/\eta$ *with* $\eta(\xi) = \sum_{j\in\mathbb{Z}} \hat{v}(2^{-j}\xi)\hat{v}(2^{-j}\xi)$, *is a Littlewood-Paley dual function related to* $v$ *and it is itself a Littlewood-Paley function, satisfying moreover*

$$\sum_{j\in\mathbb{Z}} \hat{\mu}(2^{-j}\xi)\hat{v}(2^{-j}\xi) = 1 \quad for \quad all \quad \xi \neq 0. \tag{4.6}$$

### Lemma 4.3.1 (Reproducing Calderon Formula)

1. *Let $v \in \mathscr{S}(\mathbb{R}^n)$ be such that supp $\hat{v}$ is compact, bounded away from the origin and satisfying $\sum_{j \in \mathbb{Z}} v(2^j \xi) = 1$ for all $\xi \neq 0$. Then for any $f \in \mathscr{S}'_\infty(\mathbb{R}^n)$,*

$$f = \sum_{j \in \mathbb{Z}} \tilde{v}_j \star f. \tag{4.7}$$

2. *Let $\mu, v \in \mathscr{S}(\mathbb{R}^n)$ such that supp $\hat{\mu}$, supp $\hat{v}$ are compact and bounded away from the origin and 4.6 holds. Then for any $f \in \mathscr{S}'_\infty(\mathbb{R}^n)$.*

$$f = \sum_{j \in \mathbb{Z}} 2^{-jn} \sum_{k \in \mathbb{Z}} \tilde{v}_j \star f(2^{-j}k)\mu_j(\,. - 2^{-j}k) = \sum_{Q \in \mathscr{Q}} \langle f, v_Q \rangle \mu_Q \tag{4.8}$$

*where and in what follows $\tilde{v}_j(x) = \overline{v_j(-x)}$.*

## 4.4  Classical Triebel-Lizorkin Spaces

In this section, we recall some characterizations of classical Besov-Triebel-Lizorkin spaces.

**Definition 4.4.1** *Let $w \in A_\infty$, $0 < p, q \leq \infty$, $\gamma \in \mathbb{R}$ and $v \in \mathscr{S}(R^n)$ satisfies 4.4 and 4.5. The homogeneous Triebel-Lizorkin space $\dot{F}_p^{\gamma,q}$ is the set of all distribution $f \in \mathscr{S}'_\infty$ such that*

$$||f||_{\dot{F}_{p,w}^{\gamma,q}} = \left|\left| \left( \sum_{j \in \mathbb{Z}} 2^{j\gamma q}|v_j f|^q \right)^{\frac{1}{q}} \right|\right|_{p,w} < \infty, \qquad 0 < p, q < \infty$$

*and*

$$||f||_{\dot{F}_{\infty,w}^{\gamma,q}} = \sup_Q \left\{ \frac{1}{w(Q)} \int_Q \sum_{j=j_Q}^{\infty} 2^{j\gamma q}|v_j f|^q w(x)dx \right\}^{\frac{1}{q}} < \infty, 0 < q \leq \infty$$

*with the interpretation that when $q = \infty$,*

$$||f||_{\dot{F}_{\infty,w}^{\gamma,\infty}} = \sup_Q \sup_{j \geq j_Q} \frac{1}{w(Q)} \int_Q 2^{j\gamma}|v_j f|w(x)dx < \infty.$$

*The homogeneous Besov space $\dot{B}_{p,w}^{\gamma,q}$ is the set of all distribution $f \in \mathscr{S}'_\infty$ such that*

$$||f||_{\dot{B}_{p,w}^{\gamma,q}} = \left( \sum_{j \in \mathbb{Z}} 2^{j\gamma q}||v_j f||_{p,w}^q \right)^{\frac{1}{q}} < \infty, \quad 0 < p, q < \infty.$$

*Moreover, it is well known that the spaces $\dot{B}_p^{\gamma,q}$ and $\dot{F}_p^{\gamma,q}$ are independent of the choices of $v$ (see, for example, [1, 2, 3, 7, 10]).*

We recall the following continuous characterizations of Triebel-Lizorkin spaces involving integrals, see for instance [2, 37].

**Theorem 4.4.1** *Let $\gamma \in \mathbb{R}$,   $0 < p,q < \infty$, and $0 < r < min(p,q)$. Suppose $w \in A_{p/r}$ and $v \in S$ satisfying the Tauberian condition, i.e., for each $\xi \neq 0$ there exists a $t > 0$ s.t $\hat{v}(t\xi) \neq 0$ and having all vanishing moments, then*

$$||f||_{\dot{F}_{p,w}^{\gamma,q}} \simeq \left|\left| \left( \int_0^\infty t^{-\gamma q} |v_t f|^q \frac{dt}{t} \right)^{\frac{1}{q}} \right|\right|_{p,w} \simeq \left| \left( \int_0^\infty t^{-\gamma q} (v_{t,\lambda}^\star f)^q \frac{dt}{t} \right)^{\frac{1}{q}} \right|_{p,w}$$

$$\simeq || g_{\lambda,\gamma,q}^\star f ||_{p,w}$$

*for all $f \in \dot{F}_{p,w}^{\gamma,q}$ and some large $\lambda$.*

*Where $v_{t,\lambda}^\star f(x)$ is the the Peetre maximal function and $g_{\lambda,\gamma,q}^\star$ is the Littlewood-Paley g function defined respectively by*

$$v_{t,\lambda}^\star f(x) = \sup_{y \in \mathbb{R}^n} |v_t f(y)| \left( 1 + \frac{|x-y|}{t} \right)^{-\lambda}$$

*and*

$$g_{\lambda,\gamma,q}^\star f(x) = \left( \int_0^\infty \int_{\mathbb{R}^n} s^{-\gamma q} |v_s f(y)|^q \left( 1 + \frac{|x-y|}{s} \right)^{-\lambda q} \frac{dyds}{s^{n+1}} \right)^{1/q}.$$

*We have also, when $p = \infty$,*

$$||f||_{\dot{F}_{\infty,w}^{\gamma,q}} \simeq \left|\left| N_{\gamma,q} f \right|\right|_{\infty,w} \simeq \left|\left| N_{\gamma,q}^\star f \right|\right|_{\infty,w}, \quad 0 < q < \infty,$$

*with*

$$N_{\gamma,q} f(x) = \sup_{R>0} \left( \frac{1}{w\left(B(x,R)\right)} \int_{B(x,R)} \int_0^R t^{-\gamma q} |v_t f(y)|^q w(y) \frac{dtdy}{t} \right)^{\frac{1}{q}}$$

*and*

$$N_{\gamma,q}^\star f(x) = \sup_{R>0} \left( \frac{1}{w\left(B(x,R)\right)} \int_{B(x,R)} \int_0^R t^{-\gamma q} (v_t^\star f(y))^q w(y) \frac{dtdy}{t} \right)^{\frac{1}{q}}.$$

## 4.5   Weighted Besov and Triebel-Lizorkin-type Spaces

Throughout this paper, $C$ denotes unspecified positive constant, possibly different at each occurrence; the symbol $A \preceq B$ means that $A \leq CB$. If $A \preceq B$ and

$B \preceq A$, then we write $A \simeq B$. The symbol $\lfloor s \rfloor$ denotes the maximal integer no more than $s$.

$w$ will be a fixed weight in $A_\infty$. We denote by $r_0$ the number $r_0 = \inf\{s : w \in A_s\}$ and by $r$ any number such that $0 < r < p/r_0$, $0 < p < \infty$, so that, in particular $w \in A_{\frac{p}{r}}$ and $w^{-\frac{r}{p-r}} \in A_{\frac{p}{p-r}}$. We also denote by $\delta$ the same reverse doubling constant of $w$ and $w^{-\frac{r}{p-r}}$. (See Lemma 4.2.1 and Remark 4.2.1). The Besov-type space $\dot{B}_{p,q}^{\gamma,\tau}$ and the Triebel-Lizorkin spaces $\dot{F}_{p,q}^{\gamma,\tau}$ were recently introduced and investigated in [28, 38, 39]. We define the weighted version of these spaces $\dot{B}_{p,q,w}^{\gamma,\tau}$ and $\dot{F}_{p,q,w}^{\gamma,\tau}$, with $w \in A_\infty$, as follows.

**Definition 4.5.1** *Let $w \in A_\infty$, $0 < p,q \le \infty$, $\gamma \in \mathbb{R}$ and $v \in \mathscr{S}(\mathbb{R}^n)$ and satisfies 4.4 and 4.5. The homogeneous Triebel-Lizorkin space $\dot{F}_{p,q,w}^{\gamma,\tau}$ is the set of all distribution $f \in \mathscr{S}_\infty'$ such that*

$$\|f\|_{\dot{F}_{p,q,w}^{\gamma,\tau}} = \|f\|_{\dot{F}_{p,q,w}^{\gamma,\tau,v}} = \sup_{Q \in \mathscr{Q}} \frac{1}{[w(Q)]^\tau} \left[ \int_Q \left( \sum_{j=j_Q}^{\infty} 2^{j\gamma q} |v_j f|^q \right)^{\frac{p}{q}} w(x)\,dx \right]^{\frac{1}{p}} < \infty.$$

*The homogeneous Besov space $\dot{B}_{p,q,w}^{\gamma,\tau}$ is the set of all distribution $f \in \mathscr{S}_\infty'$ such that*

$$\|f\|_{\dot{B}_{p,q,w}^{\gamma,\tau}} = \|f\|_{\dot{B}_{p,q,w}^{\gamma,\tau,v}} = \sup_{Q \in \mathscr{Q}} \frac{1}{[w(Q)]^\tau} \left[ \sum_{j=j_Q}^{\infty} \left( \int_Q 2^{j\gamma p} |v_j f|^p w(x)\,dx \right)^{\frac{q}{p}} \right]^{\frac{1}{q}} < \infty.$$

We note that the space $\dot{F}_{p,q,w}^{\gamma,\tau}$ is independent of the particular choice of the function $v$. The quasi-norms arising from different $v$ are equivalent. See [22].

In this work we study continuous characterizations of the these type weighted spaces. For comparison see for instance [7, 28, 29, 30, 34, 38, 39].

We need the following results. See [22].

**Theorem 4.5.1** *Let $w \in A_\infty$, $0 < p < \infty$, $0 < q \le \infty$, $r_0 = \inf\{s \ge 1 : w \in A_s\}$ and $\delta > 0$ is as in Lemma 4.2.1. If $f \in \dot{F}_{p,q,w}^{\gamma,\tau}$, then there exists a canonical way to find a representation of $f$ s.t $f \in S_L'$, where $L \equiv \max\left(-1, \lfloor \gamma + r_0 n(\tau - 1/p) \rfloor\right)$ if $\tau - 1/p \ge 0$ and $L \equiv \max\left(-1, \lfloor \gamma + n(r_0 - \delta)(\tau - 1/p) \rfloor\right)$ if $\tau - 1/p < 0$.*

*More precisely, assume for instance that $\tau - 1/p \ge 0$ and let $\varphi = \varphi_1$, $\psi = \psi_1 \in \mathscr{S}(\mathbb{R}^n)$ satisfying 4.4, 4.5 and 4.6. Then there exists a sequences of polynomials $\{P_N^1\}_{N=1}^{\infty}$, with degree of each $P_N^1$ no more than $L \equiv \lfloor \gamma + r_0 n(\tau - 1/p) \rfloor$ and $g_1 \in \mathscr{S}'(\mathbb{R}^n)$ s.t*

$$g_1 = \lim_{N \to \infty} \left( \sum_{j=-N \le j \le N} \tilde{\psi}_j \star \varphi_j \star f + P_N^1 \right) \quad in \quad \mathscr{S}'(\mathbb{R}^n). \tag{4.9}$$

*Moreover, if $g_2$ is the corresponding limit in 4.9 for some other $\varphi_2, \psi_2$ satisfying the same conditions as $\varphi_1, \psi_1$, then*

$$g_1 - g_2 \quad \in \quad \mathscr{P} \quad and \quad deg(g_1 - g_2) \leq L. \tag{4.10}$$

*We can take $g_1$ as a representation of the equivalent class $f + P(R^n)$ and we identify $f$ with its representative $g_1$. In the sense, $f \in S'_L$, with*

$$L \equiv \max(-1, \lfloor \gamma + r_0 n(\tau - 1/p) \rfloor).$$

*Similar conclusion holds whenever $\tau - 1/p < 0$ by taking*

$$L \equiv \max(-1, \lfloor \gamma + (r_0 - \delta)n(\tau - 1/p) \rfloor).$$

**Corollary 4.5.1** *Let $w \in A_\infty$ and $f \in \dot{F}^{\gamma,\tau}_{p,q,w}$ with $0 < p < \infty$, $0 < q \leq \infty$. Let $\varphi_1$, $\tau$ and $L$ is as in Theorem 4.5.1. Then there exists a sequence $\{P_N\}_{N=1}^\infty$ of polynomials with $degP_N \leq L$ and $g_1 \in S'(\mathbb{R}^n)$ s.t such that*

$$g_1 = \lim_{k \to \infty} \left( \sum_{2^{-nk} \leq l(Q) \leq 2^{nk}} \langle f, \varphi_Q \rangle \psi_Q + P_k \right) \quad in \quad \mathscr{S}'(\mathbb{R}^n). \tag{4.11}$$

*Moreover, if $g_2$ is the corresponding limit in 4.11 for some other $\varphi_2, \psi_2$ satisfying the same conditions as $\varphi_1, \psi_1$, then*

$$g_1 - g_2 \quad \in \quad \mathscr{P} \quad and \quad deg(g_1 - g_2) \leq L. \tag{4.12}$$

## 4.6 Continuous Characterizations

We begin this section by recalling the notions of some basic spaces of functions and their properties. Let $0 < p, q \leq \infty$, $0 \leq \tau < \infty$ and $w \in A_\infty$. The space $l^q\left(L^\tau_{p,w}\right)$ is defined to be the set of all sequences $g = \{g_j\}_{j \in \mathbb{Z}}$ of measurable functions on $\mathbb{R}^n$ s.t

$$\|g\|_{l^q(L^\tau_{p,w})} = \sup_{Q \in \mathscr{Q}} \frac{1}{[w(Q)]^\tau} \left[ \sum_{j=j_Q}^\infty \left( \int_Q |g_j(x)|^p w(x)dx \right)^{\frac{q}{p}} \right]^{\frac{1}{q}} < \infty.$$

Similarly, the space $L^\tau_{p,w}(l^q)$ with $0 < p < \infty$ is defined to be the set of all sequences $g = \{g_j\}_{j \in \mathbb{Z}}$ of measurable functions on $\mathbb{R}^n$ s.t

$$\|g\|_{L^\tau_{p,w}(l^q)} = \sup_{Q \in \mathscr{Q}} \frac{1}{[w(Q)]^\tau} \left[ \int_Q \left( \sum_{j=j_Q}^\infty |g_j(x)|^q \right)^{\frac{p}{q}} w(x)dx \right]^{\frac{1}{p}} < \infty$$

where $j_Q = -log_2 l(Q)$ and $l(Q)$ is the side length of the dyadic cube $Q$. We need the following Lemma which is a generalization of Lemma 2 in [25].

**Lemma 4.6.1** *Let $r_0 \geq 1$, $w \in A_{r_0}$, $0 < q$, $p < \infty$, $0 \leq \tau < \infty$, $\delta > n\tau r_0$ and $g = \{g_j\}_{j \in \mathbb{Z}}$ is a collection of measurable functions and a sequence of complex numbers $a = \{a_j\}_{j \in \mathbb{Z}}$ satisfying*

$$|a_j| \leq C \begin{cases} 2^j, & if \quad j \in \mathbb{Z}_- \\ 2^{-\delta j}, & if \quad j \in \mathbb{N}_0. \end{cases}$$

*Then there exists a positive constant $C$, independent of $g$ s.t*

$$\|G\|_{l^q(L^\tau_{p,w})} \leq C \|g\|_{l^q(L^\tau_{p,w})}$$

*and*

$$\|G\|_{L^\tau_{p,w}(l^q)} \leq C \|g\|_{L^\tau_{p,w}(l^q)}$$

*where $G_j(x) = \sum\limits_{m \in \mathbb{Z}} a_{j-m} g_m(x)$.*

The following corollary that generalizes the result of Rychkov, V. S. [25, Lemma 2] is a direct consequence of Lemma 4.6.1.

**Corollary 4.6.1** *Let $r_0 \geq 1$, $w \in A_{r_0}$, $0 < q$, $p < \infty$, $0 \leq \tau < \infty$, $\delta > n\tau r_0$ and $g = \{g_j\}_{j \in \mathbb{Z}}$ is sequences of measurable functions. Define $G_j(x) = \sum\limits_{m \in \mathbb{Z}} 2^{-|m-j|\delta} g_m(x)$. Then there exists a positive constant $C$, independent of $g$ s.t*

$$\|G\|_{l^q(L^\tau_{p,w})} \leq C \|g\|_{l^q(L^\tau_{p,w})}$$

*and*

$$\|G\|_{L^\tau_{p,w}(l^q)} \leq C \|g\|_{L^\tau_{p,w}(l^q)}.$$

**Lemma 4.6.2** *Let $0 \leq \tau < \infty$, $1 < p, q < \infty$ and $w \in A_p$. Define*

$$\|Mg\|_{L^\tau_{p,w}(l^q)} = \sup_Q \frac{1}{[w(Q)]^\tau} \left( \int_Q \left( \sum_{j=j_Q}^{\infty} |Mg_j(x)|^q \right)^{\frac{p}{q}} w(x)dx \right)^{\frac{1}{p}}.$$

*If $\tau - \frac{1}{p} < 0$ then we have the following general weighted Fefferman-Stein inequality*

$$\|Mg\|_{L^\tau_{p,w}(l^q)} \leq C \|g\|_{L^\tau_{p,w}(l^q)}.$$

We give the proof of Lemma 4.6.1 and Lemma 4.6.2 in the next section.

Let $0 < \varepsilon < \infty$, $R \in \mathbb{Z}_+ \cup \{-1\}$ and $v \in \mathscr{S}(R^n)$ such that:

$$\hat{v} > 0 \quad on \quad \{\xi \in \mathbb{R}^n : \varepsilon/2 \leq |\xi| \leq 2\varepsilon\} \tag{4.13}$$

and

$$\partial^\alpha \hat{v}(0) = 0 \quad for \quad all \quad |\alpha| \leq R. \tag{4.14}$$

Recall that $v_t \star f$ for $t \in \mathbb{R}$ are usually called the local means; see, for example, [35].

The following estimates, are widely used in this section.

**Lemma 4.6.3** *Let* $R \in \mathbb{Z}_+ \cup \{-1\}, v, \mu \in \mathscr{S}(\mathbb{R}^n)$ *satisfy 4.13 and 4.14 and* $f \in \mathscr{S}'_R(\mathbb{R}^n)$. *Then for all* $N \geq \lambda, j \in \mathbb{Z}$ *and* $x \in \mathbb{R}^n$

$$2^{j\gamma} v_j^\star f(x) \leq C \sum_{k \in \mathbb{Z}} 2^{k\gamma} \mu_k^\star f(x) \min\left(2^{(k-j)(R+1-\gamma)}, 2^{(k-j)(N-2\lambda+\gamma)}\right). \tag{4.15}$$

**Lemma 4.6.4** *Let* $R \in \mathbb{Z}_+ \cup \{-1\}, v \in \mathscr{S}(\mathbb{R}^n)$ *satisfy 4.13, 4.14 and* $f \in \mathscr{S}'_R(\mathbb{R}^n)$. *Then for all* $t \in [1, 2], N \geq \lambda, l \in \mathbb{Z}$ *and* $x \in \mathbb{R}^n$

$$|v_{2^{-j}t}^\star f(x)|^r \leq C \sum_{k=j}^{\infty} 2^{-(k-j)(Nr-n)} 2^{jn} \int_{\mathbb{R}^n} |v_{2^{-k}t} f(z)|^r (1 + |x-z|2^j)^{-\lambda r} dz \tag{4.16}$$

*where $r$ is an arbitrary fixed positive number and $C$ is a positive constant independent of $v, f, l, x$ and $t$, but may depend on $r$.*

*Proof.* The proof of Lemma 4.6.3 follows from Theorem 4.5.1 by using similar arguments in the proof of Theorem 1.2 in [39]. Using again Theorem 4.5.1 and arguing as in the proof of the estimate 2.29 in [37] and the proof of 2.10 in [39], to obtain Lemma 4.6.4. ■

We need also the following size estimates of Heidman type, see for instance Lemma 2.1 in [2].

**Lemma 4.6.5** *Let* $\mu, v \in \mathscr{S}(\mathbb{R}^n)$, *and* $\lambda \geq 0$. *Assume that $\mu$ has $R$ vanishing moments, then for any $0 < s \leq t < \infty$ we have*

$$\int_{\mathbb{R}^n} v_t \star \mu_s \left(1 + \frac{|z|}{t}\right)^\lambda dz \leq C \left(\frac{s}{t}\right)^{R+1}.$$

**Theorem 4.6.1** *Let* $w \in A_\infty, \gamma \in \mathbb{R}, \tau \geq 0, 0 < p, q \leq \infty$ *and* $R \in \mathbb{N} \cup \{-1\}$ *such that* $\gamma + n\tau r_0 < R+1$ *and* $v \in \mathscr{S}(\mathbb{R}^n)$ *satisfying 4.13 and 4.14. Assume $\lambda$ is large enough, then*

$$\|f\|_{\dot{F}_{p,q,w}^{\gamma,\tau}} \simeq \sup_{Q \in \mathscr{Q}} \frac{1}{[w(Q)]^\tau} \left[\int_Q \left(\sum_{j=j_Q}^{\infty} 2^{j\gamma q} |v_{j,\lambda}^\star f|^q\right)^{\frac{p}{q}} w(x) dx\right]^{\frac{1}{p}}.$$

*Proof.* To simplify notation, we put $v_{j,\lambda}^\star = v_j^\star$.
First, we prove that

$$\|\{2^{j\gamma} v_j^\star f\}_{j \in \mathbb{Z}}\|_{L_{p,w}^\tau(l^q)} = \sup_{Q \in \mathscr{Q}} \frac{1}{[w(Q)]^\tau} \left[\int_Q \left(\sum_{j=j_Q}^{\infty} 2^{j\gamma q} |v_j^\star f|^q\right)^{\frac{p}{q}} w(x) dx\right]^{\frac{1}{p}}$$

does not depend on the choose of $v$. To see this, we use Lemma 4.6.3 by choosing $N > 2\lambda - \gamma + n\tau, R > n\tau + \gamma - 1$ and then Lemma 4.6.1, to obtain

$$\|\{2^{j\gamma} v_j^\star f\}_{j \in \mathbb{Z}}\|_{L_{p,w}^\tau(l^q)} \precsim \|\{2^{j\gamma} \mu_j^\star f\}_{j \in \mathbb{Z}}\|_{L_{p,w}^\tau(l^q)}.$$

Where $\mu$ and $\nu$ are as in Lemma 4.6.3. In a similar manner we have

$$\left\|\{2^{j\gamma}\mu_j^{\star}f\}_{j\in\mathbb{Z}}\right\|_{L_{p,w}^{\tau}(l^q)} \preceq \left\|\{2^{j\gamma}\nu_j^{\star}f\}_{j\in\mathbb{Z}}\right\|_{L_{p,w}^{\tau}(l^q)}.$$

Thus

$$\left\|\{2^{j\gamma}\nu_j^{\star}f\}_{j\in\mathbb{Z}}\right\|_{L_{p,w}^{\tau}(l^q)} \simeq \left\|\{2^{j\gamma}\mu_j^{\star}f\}_{j\in\mathbb{Z}}\right\|_{L_{p,w}^{\tau}(l^q)}.$$

Secondly we prove that

$$\left\|\{2^{j\gamma}\nu_j^{\star}f\}_{j\in\mathbb{Z}}\right\|_{L_{p,w}^{\tau}(l^q)} \preceq \left\|\{2^{j\gamma}\nu_j f\}_{j\in\mathbb{Z}}\right\|_{L_{p,w}^{\tau}(l^q)}.$$

In fact, to fix a dyadic cube $Q$ with $l(Q) = 2^{-j_Q}$. For $j \geq j_Q$ and $x \in Q$, we have from Lemma 4.6.4 and Lemma 4.2.2

$$|\nu_j^{\star}f(x)|^r \leq C\sum_{k=j}^{\infty} 2^{-(k-j)(Nr-n)}2^{jn}\int_{\mathbb{R}^n}|\nu_k f(z)|^r(1+|x-z|2^j)^{-\lambda r}dz = I_{1j}(x)+I_{2j}(x)$$

with

$$I_{1j}(x) = \sum_{k=j}^{\infty} 2^{-(k-j)(Nr-n)}2^{jn}\int_{|x-z|\leq 2^{-j_Q}}|\nu_k f(z)|^r(1+|x-z|2^j)^{-\lambda r}dz$$

$$\leq C\sum_{k=j}^{\infty} 2^{-(k-j)(Nr-n)}2^{jn}\int_{Q^{\star}}|\nu_k f(z)|^r(1+|x-z|2^j)^{-\lambda r}dz$$

$$\leq C\sum_{k=j}^{\infty} 2^{-(k-j)(Nr-n)}M\left(\chi_{Q^{\star}}|\nu_k f|^r\right)(x).$$

Where $Q^{\star} = c_n Q$ is chosen so that $\{|x-z| \leq 2^{-j_Q}\} \subset Q^{\star}$ for all $x \in Q$. By Vector Fefferman-Stein inequality and a discrete version of the Hardy inequality,

$$\sum_j 2^{j\theta}\left(\sum_{k=j}^{\infty}|b_k|\right)^{\eta} \leq c\sum_j 2^{j\theta}|b_j|^{\eta}, \quad \eta,\theta > 0,$$

which we apply with $\theta = q\left(N+\gamma-\frac{n}{r}\right)$ and $\eta = \frac{q}{r}$, we have

$$\frac{1}{[w(Q)]^{\tau}}\left[\int_Q\left(\sum_{j=j_Q}^{\infty}2^{j\gamma q}|I_{1j}(x)|^{\frac{q}{r}}\right)^{\frac{p}{q}}w(x)dx\right]^{\frac{1}{p}}$$

$$\preceq \frac{1}{[w(Q)]^{\tau}}\left[\int_Q\left(\sum_{j=j_Q}^{\infty}\left[\sum_{k=j}^{\infty}2^{-(k-j)(Nr-n+\gamma r)}2^{k\gamma r}M\left(\chi_{Q^{\star}}|\nu_k f|^r\right)(x)\right]^{\frac{q}{r}}\right)^{\frac{p}{q}}w(x)dx\right]^{\frac{1}{p}}$$

$$\preceq \frac{1}{[w(Q)]^{\tau}}\left[\int_Q\left(\left[\sum_{k=j_Q}^{\infty}2^{k\gamma r}M\left(\chi_{Q^{\star}}|\nu_k f|^r\right)(x)\right]^{\frac{q}{r}}\right)^{\frac{p}{q}}w(x)dx\right]^{\frac{1}{p}}$$

$$\preceq \|f\|_{\dot{F}_{p,q,w}^{\gamma,\tau}}.$$

On the other hand, we have by Hölder's inquality and Lemma 4.2.3

$$I_{2j}(x) = \sum_{k=j}^{\infty} 2^{-(k-j)(Nr-n)} 2^{jn} \int_{|x-z|>2^{-jQ}} |v_k f(z)|^r (1+|x-z|2^j)^{-\lambda r} dz$$

$$\preceq \sum_{k=j}^{\infty} 2^{-(k-j)(Nr-n)} 2^{jn} \left( \int_{|x-z|>2^{-jQ}} |v_k f(z)|^p (1+|x-z|2^j)^{(-\lambda r+n)\frac{r}{p}} w(z) dz \right)^{\frac{r}{p}}$$

$$\times \left( \int_{\mathbb{R}^n} (1+|x-z|2^j)^{-\frac{np}{p-r}} w^{\frac{-r}{p-r}}(z) dz \right)^{1-\frac{r}{p}}$$

$$\preceq \sum_{k=j}^{\infty} 2^{-(k-j)(Nr-n)} 2^{jn} \left( \int_{|x-z|>2^{-jQ}} |v_k f(z)|^p (1+|x-z|2^j)^{(-\lambda r+n)\frac{r}{p}} w(z) dz \right)^{\frac{r}{p}}$$

$$\times 2^{-nj} \left( \int_Q w dz \right)^{\frac{-r}{p}}$$

$$\preceq \sum_{k=j}^{\infty} 2^{-(k-j)(Nr-n)} II_{jk} \left( \int_Q w dz \right)^{-\frac{r}{p}},$$

where

$$II_{jk} = \left( \int_{|x-z|>2^{-jQ}} |v_k f(z)|^p (1+|x-z|2^j)^{(-\lambda r+n)\frac{r}{p}} w(z) dz \right)^{\frac{r}{p}}$$

$$\preceq 2^{(-\lambda r+n)(j-jQ)} \left( \sum_{m=1}^{\infty} 2^{-mr(\lambda r-n)\frac{p}{r}} \int_{|z-x|\leq 2^{-jQ+m+1}} |v_k f(z)|^p w(z) dz \right)^{\frac{r}{p}}$$

$$\preceq 2^{(-\lambda r+n)(j-jQ)} \sum_{m=1}^{\infty} 2^{-mr(\lambda r-n)} \left( \int_{|z-x|\leq 2^{-jQ+m+1}} |v_k f(z)|^p w(z) dz \right)^{\frac{r}{p}}$$

$$\preceq 2^{(-\lambda r+n)(j-jQ)} \sum_{m=1}^{\infty} 2^{-mr(\lambda r-n)} \left( \int_{|z-x|\leq 2^{-jQ+m+1}} |v_k f(z)|^p w(z) dz \right)^{\frac{r}{p}}$$

$$\preceq 2^{-\gamma jr} 2^{(-\lambda r+n)(j-jQ)} \left( \|f\|_{\dot{F}_{p,q,w}^{\gamma,\tau}} \right)^r \sum_{m=1}^{\infty} 2^{-mr(\lambda r-n)} \left( \int_{|z-x|\leq 2^{-jQ+m+1}} w(z) dz \right)^{r\tau}$$

$$\preceq 2^{-\gamma kr} \left( \|f\|_{\dot{F}_{p,q,w}^{\gamma,\tau}} \right)^r \sum_{m=1}^{\infty} 2^{-m(\lambda r-n)} \left( \int_{|z-x|\leq 2^{-jQ+m+1}} w(z) dz \right)^{r\tau}$$

$$\preceq 2^{-\gamma kr} 2^{-(\lambda r+n)(j-j_Q)} \left(||f||_{\dot{F}^{\gamma,\tau}_{p,q,w}}\right)^r \sum_{m=1}^{\infty} 2^{-mr(\lambda r-n)} 2^{r\tau r_0 nm} \left(\int\limits_{|z-x|\leq 2^{-j_Q}} w(z)dz\right)^{r\tau}$$

$$\preceq 2^{-\gamma kr} 2^{(-\lambda r+n)(j-j_Q)} \left(||f||_{\dot{F}^{\gamma,\tau}_{p,q,w}}\right)^r \sum_{m=1}^{\infty} 2^{-mr(\lambda r-n(1+\tau r_0))} \left(\int\limits_{|z-x|\leq 2^{-j_Q}} w(z)dz\right)^{r\tau}.$$

It follows that

$$\sum_{k=j}^{\infty} 2^{-(k-j)(Nr-n)} 2^{-\gamma kr} \left(\int_Q wdz\right)^{-\frac{r}{p}} II_{jk}$$

$$\preceq 2^{-\gamma jr} 2^{(-\lambda r+n)(j-j_Q)} \sum_{k=j}^{\infty} 2^{-(k-j)(Nr+\gamma r-n)} \left(\int_Q wdz\right)^{-\frac{r}{p}} I_{2k}$$

$$\preceq 2^{-\gamma jr} 2^{(-\lambda r+n)(j-j_Q)} \left(||f||_{\dot{F}^{\gamma,\tau}_{p,q,w}}\right)^r \left(\int_Q wdz\right)^{r\tau} \left(\int_Q wdz\right)^{-\frac{r}{p}}$$

$$\preceq 2^{-\gamma jr} 2^{(-\lambda r+n+nr_0\frac{r}{p})(j-j_Q)} \left(||f||_{\dot{F}^{\gamma,\tau}_{p,q,w}}\right)^r \left(\int_Q wdz\right)^{r\tau-\frac{r}{p}}.$$

Choose $-\lambda r + n + nr_0\frac{r}{p} < 0$ to get

$$\frac{1}{[w(Q)]^\tau} \left[\int_Q \left(\sum_{j=j_Q}^{\infty} 2^{j\gamma q}|I_{2j}(x)|^{\frac{q}{r}}\right)^{\frac{p}{q}} w(x)dx\right]^{\frac{1}{p}} \preceq \sum_{j=j_Q}^{\infty} 2^{(-\lambda r+n+nr_0\frac{r}{p})(\frac{j-j_Q}{r})} ||f||_{\dot{F}^{\gamma,\tau}_{p,q,w}} \preceq ||f||_{\dot{F}^{\gamma,\tau}_{p,q,w}}.$$

◼

**Theorem 4.6.2** *Let $w \in A_\infty$, $\gamma \in \mathbb{R}$, $\tau \geq 0$, $0 < p,q \leq \infty$ and $R \in \mathbb{N} \cup \{-1\}$ such that $\gamma + n\tau r_0 < R+1$ and $v \in \mathscr{S}(\mathbb{R}^n)$ satisfying 4.13 and 4.14. Assume $\lambda$ is large enough, then the space $\dot{F}^{\gamma,\tau}_{p,q,w}$ is characterized by*

$$\dot{F}^{\gamma,\tau}_{p,q,w} = \{f \in S'_R(\mathbb{R}^n) : ||f||^{(i)}_{\dot{F}^{\gamma,\tau}_{p,q,w}} < \infty\}, \ i = 1,\ldots,6,$$

*where*

$$\|f\|^{(1)}_{\dot{F}^{\gamma,\tau}_{p,q,w}} = \sup_{Q\in\mathscr{Q}} \frac{1}{[w(Q)]^\tau} \left[\int_Q \left(\int_0^{l(Q)} t^{-\gamma q} |v_t f(x)|^q \frac{dt}{t}\right)^{\frac{p}{q}} w(x)dx\right]^{\frac{1}{p}}$$

$$\|f\|^{(2)}_{\dot{F}^{\gamma,\tau}_{p,q,w}} = \sup_{Q\in\mathscr{Q}} \frac{1}{[w(Q)]^\tau} \left[\int_Q \left(\int_0^{l(Q)} t^{-\gamma q} |v^\star_{t,\lambda} f(x)|^q \frac{dt}{t}\right)^{\frac{p}{q}} w(x)dx\right]^{\frac{1}{p}}$$

$$\|f\|^{(3)}_{\dot{F}^{\gamma,\tau}_{p,q,w}} = \sup_{Q\in\mathscr{Q}} \frac{1}{[w(Q)]^\tau} \left[\int_Q \left(\int_0^{l(Q)} t^{-\gamma q} \int_{|z|<t} |v_t f(x+z)|^q \frac{dt}{t^{n+1}}\right)^{\frac{p}{q}} w(x)dx\right]^{\frac{1}{p}}$$

$$\|f\|^{(4)}_{\dot{F}^{\gamma,\tau}_{p,q,w}} = \sup_{Q\in\mathscr{Q}} \frac{1}{[w(Q)]^\tau} \left[\int_Q \left(\sum_{j=j_Q}^{\infty} 2^{j\gamma q} |v^\star_{2^{-j},\lambda} f|^q\right)^{\frac{p}{q}} w(x)dx\right]^{\frac{1}{p}}$$

$$\|f\|^{(5)}_{\dot{F}^{\gamma,\tau}_{p,q,w}} = \sup_{Q\in\mathscr{Q}} \frac{1}{[w(Q)]^\tau} \left[\int_Q \left(\sum_{j=j_Q}^{\infty} 2^{j\gamma q} |v_{2^{-j}} f|^q\right)^{\frac{p}{q}} w(x)dx\right]^{\frac{1}{p}},$$

*and*

$$\|f\|^{(6)}_{\dot{F}^{\gamma,\tau}_{p,q,w}} = \sup_{Q\in\mathscr{Q}} \frac{1}{[w(Q)]^\tau} \times$$

$$\left[\int_Q \left(\sum_{j=j_Q}^{\infty} 2^{j\gamma q} \left[\int_{\mathbb{R}^n} |v_{2^{-j}} f(y+z)|^r (1+2^j|x-z|)^{-\lambda r} dz|\right]^{\frac{q}{r}}\right)^{\frac{p}{q}} w(x)dx\right]^{\frac{1}{p}},$$

*with $0 < r < min(p/r_0, q)$, where $r_0 = inf\{s : w \in A_s\}$, s.t, $w \in A_{r_0}$. Furthermore, all $\|.\|^{(i)}_{\dot{F}^{\gamma,\tau}_{p,q,w}}$, $i = 1,\ldots,6$, are equivalent (quasi-)norms in $\dot{F}^{\gamma,\tau}_{p,q,w}$.*

Proof. We adapt here the proof of Theorem 3.1 in [20]. We prove Theorem 4.6.2 in three steps.

*Step 1*. In this step, for same $v$, we prove that

$$\|f\|^{(1)}_{\dot{F}^{\gamma,\tau}_{p,q,w}} \simeq \|f\|^{(2)}_{\dot{F}^{\gamma,\tau}_{p,q,w}} \simeq \|f\|^{(4)}_{\dot{F}^{\gamma,\tau}_{p,q,w}} \simeq \|f\|^{(5)}_{\dot{F}^{\gamma,\tau}_{p,q,w}}. \tag{4.17}$$

It is clear from the above Definitions that

$$\|f\|^{(1)}_{\dot{F}^{\gamma,\tau}_{p,q,w}} \leq \|f\|^{(2)}_{\dot{F}^{\gamma,\tau}_{p,q,w}} \text{ and} \|f\|^{(5)}_{\dot{F}^{\gamma,\tau}_{p,q,w}} \leq \|f\|^{(4)}_{\dot{F}^{\gamma,\tau}_{p,q,w}}. \tag{4.18}$$

Let $0 < r < min(p/r_0, q)$, where $r_0 = inf\{s : w \in A_s\}$, s.t, $w \in A_{r_0}$, in particular, $w \in A_{p/r}$. Fix a dyadic cube $Q(x_0, l(Q))$ centred at some point $x_0$ with side length $l(Q) = 2^{-l}$. Let $x \in Q$, $1 \leq t \leq 2$ and set $v_{t,\lambda}^\star = v_t^\star$. Then we have by Lemma 4.6.4

$$\left( \int_0^{jQ} t^{-\gamma q} |v_t^\star f(x)|^q \frac{dt}{t} \right)^{\frac{p}{q}} = \left( \sum_{j=l}^\infty \int_{2^{-j}}^{2^{-j+1}} t^{-\gamma q} |v_t^\star f(x)|^q \frac{dt}{t} \right)^{\frac{p}{q}} = \left( \sum_{j=l}^\infty 2^{j\gamma q} \int_1^2 t^{-\gamma q} |v_{2^{-j}t}^\star f(x)|^q \frac{dt}{t} \right)^{\frac{p}{q}},$$

hence

$$\left( \int_0^{jQ} t^{-\gamma q} |v_t^\star f(x)|^q \frac{dt}{t} \right)^{\frac{p}{q}} \preceq$$

$$\left( \sum_{j=l}^\infty 2^{j\gamma q} \int_1^2 t^{-\gamma q} \left( \sum_{k=0}^\infty 2^{-kNr} 2^{(k+j)n} \int_{\mathbb{R}^n} |v_{2^{-k-j}t} f(z)|^r (1 + |x - z|2^j)^{-\lambda r} dz \right)^{\frac{q}{r}} \frac{dt}{t} \right)^{\frac{p}{q}}.$$

Minkowski's inequality for integrals yields to

$$\int_1^2 t^{-\gamma q} \left( \sum_{k=0}^\infty 2^{-kNr} 2^{(k+j)n} \int_{\mathbb{R}^n} |v_{2^{-k-j}t} f(z)|^r (1 + |x - z|2^j)^{-\lambda r} dz \right)^{\frac{q}{r}} \frac{dt}{t}$$

$$\leq C \left[ \sum_{k=0}^\infty 2^{-kNr} 2^{(k+j)n} \int_{\mathbb{R}^n} \left( \int_1^2 t^{-\gamma q} |v_{2^{-k-j}t} f(z)|^q \frac{dt}{t} \right)^{\frac{r}{q}} (1 + 2^j |x - z|)^{-\lambda r} dz \right]^{\frac{q}{r}}$$

$$\tag{4.19}$$

$$\leq C \left[ \sum_{k=0}^\infty 2^{-kNr} 2^{kn} I_{jk,1} + \sum_{k=0}^\infty 2^{-kNr} 2^{kn} I_{jk,2} \right]^{\frac{q}{r}},$$

where

$$I_{jk,1} = 2^{jn} \int_{Q^*} (1 + 2^j |x - z|)^{-\lambda r} \left( \int_1^2 t^{-\gamma q} |v_{2^{-k-j}t} f(x)|^q \frac{dt}{t} \right)^{\frac{r}{q}} dz$$

$$\leq M \left[ \left( \int_1^2 |\chi_{Q^*} t^{-\gamma q} v_{2^{-k-j}t} f(.) \frac{dt}{t}|^q \right)^{\frac{r}{q}} \right] (x)$$

and $Q^* = c_n Q$ be such that $\{|x - z| \leq 2^{-l}\} \subseteq Q^*$ for every $x \in Q$. Hence

$$\sum_{k=0}^\infty 2^{-kNr} 2^{kn} I_{jk,1} \leq \sum_{k=0}^\infty 2^{j\gamma r} 2^{-kNq} 2^{kn} M \left[ \left( \int_1^2 |\chi_{Q^*} t^{-\gamma q} v_{2^{-k-j}t} f(.) \frac{dt}{t}|^q \right)^{\frac{r}{q}} \right] (x).$$

Letting $\varepsilon > 0$ and $\lambda > \max(\varepsilon, \varepsilon + \frac{n}{r} - \gamma)$, from Hölder's inequality we deduce that

$$\sum_{k=0}^{\infty} 2^{-kNr} 2^{kn} I_{jk,1} \leq C \left( \sum_{k=0}^{\infty} 2^{-k(N-\varepsilon)q} 2^{kn\frac{q}{r}+j\gamma q} \left( M\left[ \left( \int_{1}^{2} |\chi_{Q^*} t^{-\gamma q} v_{2^{-k-j}t} f(.)\frac{dt}{t}|^q\right)^{\frac{r}{q}} \right](x) \right)^{\frac{q}{r}} \right)^{\frac{r}{q}}.$$

It follows

$$\left( \int_{Q} \left( \sum_{j=l}^{\infty} \left( \sum_{k=0}^{\infty} 2^{j\gamma r} 2^{-kNr} 2^{kn} I_{jk,1} \right)^{\frac{q}{r}} \right)^{\frac{p}{q}} w(x)dx \right)^{\frac{1}{p}}$$

$$\leq \left( \int_{Q} \left( \sum_{j=l,k=0}^{\infty} 2^{j\gamma r} 2^{-k(N-\varepsilon)q} 2^{kn\frac{q}{r}} \left( M\left[ \left( \int_{1}^{2} |\chi_{Q^*} t^{-\gamma q} v_{2^{-k-j}t} f(.)\frac{dt}{t}|^q\right)^{\frac{r}{q}} \right](x) \right)^{\frac{q}{r}} \right)^{\frac{p}{q}} w(x)dx \right)^{\frac{1}{p}}$$

$$\leq \left( \int_{Q} \left( \sum_{j=l,k=0}^{\infty} 2^{-k(N-\varepsilon)q+kn\frac{q}{r}-k\gamma q} \left( M\left[ \left( \int_{2^{-k-j}}^{2^{-k-j+1}} |\chi_{Q^*} t^{-\gamma q} v_t f(.)\frac{dt}{t}|^q\right)^{\frac{r}{q}} \right](x) \right)^{\frac{q}{r}} \right)^{\frac{p}{q}} w(x)dx \right)^{\frac{1}{p}}$$

using the Fefferman-Stein vector inequality to get

$$\left( \int_{Q} \left( \sum_{j=l}^{\infty} \left( \sum_{k=0}^{\infty} 2^{j\gamma r} 2^{-kNr} 2^{kn} I_{jk,1} \right)^{\frac{q}{r}} \right)^{\frac{p}{q}} w(x)dx \right)^{\frac{1}{p}}$$

$$\leq \left( \int_{Q^*} \left( \sum_{j=l,k=0}^{\infty} 2^{-k(N-\varepsilon)q+kn\frac{q}{r}-k\gamma q} \int_{2^{-k-j}}^{2^{-k-j+1}} |t^{-\gamma q} v_t f(x)\frac{dt}{t}| \right)^{\frac{p}{q}} w(x)dx \right)^{\frac{1}{p}}$$

$$\leq \left( \int_{Q^*} \left( \int_{0}^{2^{-l+1}} t^{-\gamma q} |v_t f(x)|\frac{dt}{t} \right)^{\frac{p}{q}} w(x)dx \right)^{\frac{1}{p}}.$$

We conclude that

$$\sup_{Q\in\mathscr{Q}}\frac{1}{[w(Q)]^{\tau}}\left(\int_{Q}\left(\sum_{j=l}^{\infty}\left(\sum_{k=0}^{\infty}2^{j\gamma r}2^{-kNr}2^{kn}I_{jk,1}\right)^{\frac{q}{r}}\right)^{\frac{p}{q}}w(x)dx\right)^{\frac{1}{p}}$$

$$\leq C\sup_{Q\in\mathscr{Q}}\frac{1}{[w(Q)]^{\tau}}\left(\int_{Q^{*}}\left(\int_{0}^{2^{-l+1}}t^{-\gamma q}|v_{t}f(x)|^{q}\frac{dt}{t}\right)^{\frac{p}{q}}w(x)dx\right)^{\frac{1}{p}}. \qquad (4.20)$$

Let $\delta > 0$, and $\lambda > max\left(\frac{n}{r},\delta,\delta+\frac{n}{r}-\tau\frac{np}{r}\right)$, then

$$I_{jk,2}=2^{jn}\int_{|x-z|>2^{-l}}(1+2^{j}|x-z|)^{-\lambda r}\left(\int_{1}^{2}t^{-\gamma q}|v_{2^{-k-j}t}f(x)|^{q}\frac{dt}{t}\right)^{\frac{r}{q}}dz$$

$$=2^{jn}\sum_{m=1}^{\infty}\int_{2^{m-l}<|x-z|<2^{m-l+1}}(1+2^{j}|x-z|)^{-\lambda r}\left(\int_{1}^{2}t^{-\gamma q}|v_{2^{-k-j}t}f(x)|^{q}\frac{dt}{t}\right)^{\frac{r}{q}}dz$$

$$\leq 2^{j(n-\lambda r)}\sum_{m=1}^{\infty}2^{-(m-l)\lambda r}\int_{|x_{0}-z|<2^{m-l+1}}\left(\int_{1}^{2}t^{-\gamma q}|v_{2^{-k-j}t}f(x)|^{q}\frac{dt}{t}\right)^{\frac{r}{q}}dz$$

$$\leq C2^{j(n-\lambda r)}\sum_{m=1}^{\infty}2^{-(m-l)\lambda r}2^{(m-l)n}M\left[\left(\int_{1}^{2}\chi_{Q(x_{0},2^{m-l+1})}(.)t^{-\gamma q}|v_{2^{-k-j}t}f(.)|^{q}\frac{dt}{t}\right)^{\frac{r}{q}}\right](x)$$

$$\leq C2^{(j-l)(n-\lambda r)}$$

$$\times\left(\sum_{m=1}^{\infty}2^{-m(\lambda-\delta)q}2^{mn\frac{q}{r}}\left(M\left[\left(\int_{1}^{2}\chi_{Q(x_{0},2^{m-l+1})}(.)t^{-\gamma q}|v_{2^{-k-j}t}f(.)|^{q}\frac{dt}{t}\right)^{\frac{r}{q}}\right]\right)^{\frac{r}{q}}(x)\right)^{\frac{r}{q}}.$$

Using again Hölder's inequality and then by $\lambda r > n$ we have

$$\left(\sum_{k=0}^{\infty}2^{-kNr}2^{kn}I_{jk,2}\right)^{\frac{q}{r}}\leq C\sum_{m=1}^{\infty}2^{-m(\lambda-\delta)q}2^{mn\frac{q}{r}}\sum_{k=0}^{\infty}2^{-k(N-\delta_{1})q}2^{kn\frac{q}{r}}$$

$$\times\left(M\left[\left(\int_{1}^{2}\chi_{Q(x_{0},2^{m-l+1})}(.)t^{-\gamma q}|v_{2^{-k-j}t}f(.)|^{q}\frac{dt}{t}\right)^{\frac{r}{q}}\right]\right)^{\frac{q}{r}}(x).$$

From Fefferman-Stein vector inequality we obtain

$$\left( \int_Q \left( \sum_{j=l}^{\infty} \left( \sum_{k=0}^{\infty} 2^{j\gamma r} 2^{-kNr} 2^{kn} I_{jk,2} \right)^{\frac{q}{r}} \right)^{\frac{p}{q}} w(x)\,dx \right)^{\frac{1}{p}}$$

$$\leq C \sum_{m=1}^{\infty} 2^{-m(\lambda-\delta)} 2^{\frac{mn}{r}} \left( \int_{Q(x_0,2^{m-l+1})} \left( \int_0^{2^{-l+1}} |t^{-\gamma q} v_t f(x) \frac{dt}{t} \right)^{\frac{p}{q}} w(x)\,dx \right)^{\frac{1}{p}}$$

$$\leq C \sum_{m=1}^{\infty} 2^{-m(\lambda-\delta)} 2^{\frac{mn}{r}} [w(Q(x_0,2^{m-l+1}))]^{\tau}$$

$$\times \frac{1}{[w(Q(x_0,2^{m-l+1}))]^{\tau}} \left( \int_{Q(x_0,2^{m-l+1})} \left( \int_0^{2^{m-l+1}} |t^{-\gamma q} v_t f(x) \frac{dt}{t} \right)^{\frac{p}{q}} w(x)\,dx \right)^{\frac{1}{p}}$$

$$\leq C [w(Q(x_0,2^{-l}))]^{\tau} \sup_{P \in \mathscr{Q}} \frac{1}{[w(P)]^{\tau}} \left[ \int_P \left( \int_0^{l(P)} t^{-\gamma q} |v_t f(x)|^q \frac{dt}{t} \right)^{\frac{p}{q}} w(x) \right]^{\frac{1}{p}}$$

$$\times \sum_{m=1}^{\infty} 2^{-m(\lambda-\delta)} 2^{\frac{mn}{r}} 2^{-mn\tau\frac{p}{r}}$$

$$\leq C [w(Q(x_0,2^{-l}))]^{\tau} \sup_{P \in \mathscr{Q}} \frac{1}{[w(P)]^{\tau}} \left[ \int_P \left( \int_0^{l(P)} t^{-\gamma q} |v_t f(x)|^q \frac{dt}{t} \right)^{\frac{p}{q}} w(x) \right]^{\frac{1}{p}} .$$

It follows that

$$\sup_{Q \in \mathscr{Q}} \frac{1}{[w(Q)]^{\tau}} \left( \int_Q \left( \sum_{j=l}^{\infty} \left( \sum_{k=0}^{\infty} 2^{j\gamma r} 2^{-kNr} 2^{kn} I_{jk,2} \right)^{\frac{q}{r}} \right)^{\frac{p}{q}} w(x)\,dx \right)^{\frac{1}{p}}$$

$$\leq \sup_{Q \in \mathscr{Q}} \frac{1}{[w(Q)]^{\tau}} \left[ \int_Q \left( \int_0^{l(Q)} t^{-\gamma q} |v_t f(x)|^q \frac{dt}{t} \right)^{\frac{p}{q}} w(x) \right]^{\frac{1}{p}} . \tag{4.21}$$

Combining 4.20 and 4.21 to conclude that

$$\sup_{Q \in \mathscr{Q}} \frac{1}{[w(Q)]^{\tau}} \left[ \int_Q \left( \int_0^{l(Q)} t^{-\gamma q} |v_t^{\star} f(x)|^q \frac{dt}{t} \right)^{\frac{p}{q}} w(x) \right]^{\frac{1}{p}}$$

$$\leq C \sup_{Q \in \mathscr{Q}} \frac{1}{[w(Q)]^{\tau}} \left[ \int_Q \left( \int_0^{l(Q)} t^{-\gamma q} |v_t f(x)|^q \frac{dt}{t} \right)^{\frac{p}{q}} w(x) \right]^{\frac{1}{p}} .$$

Thus

$$||f||^{(1)}_{\dot{F}^{\gamma,\tau}_{p,q,w}} \simeq ||f||^{(2)}_{\dot{F}^{\gamma,\tau}_{p,q,w}}. \tag{4.22}$$

To prove that

$$||f||^{(1)}_{\dot{F}^{\gamma,\tau}_{p,q,w}} \preceq ||f||^{(4)}_{\dot{F}^{\gamma,\tau}_{p,q,w}}$$

we write

$$\left( \int_0^{l(Q)} t^{-\gamma q} |v_t f(x)|^q \frac{dt}{t} \right)^{\frac{p}{q}} = \left( \sum_{k=j_Q}^{\infty} 2^{k\gamma q} \int_1^2 t^{-\gamma q} |v_{2^{-k}t} f(x)|^q \frac{dt}{t} \right)^{\frac{p}{q}}$$

and using Theorem 4.5.1 to obtain

$$v_{2^{-k}t} f(x) = \sum_{j\in\mathbb{Z}} v_{2^{-k}t} \star \mu_j \star \varphi_j \star f(x) = \sum_{j\leq k} v_{2^{-k}t} \star \mu_j \star \varphi_j f(x) + \sum_{j\geq k+1} v_{2^{-k}t} \star \mu_j \star \varphi_j f(x).$$

Fix $1 \leq t \leq 2$ and $\lambda \geq 0$. If $j \geq k$ then $2^{-j} \leq 2^{-k} \leq 2^{-k}t$. Since $\mu$ has all vanishing moments, we have by Lemma 4.6.5

$$|v_{2^{-k}t} \star \mu_j \star \varphi_j f(x)| = \left| \int_{\mathbb{R}^n} v_{2^{-k}t} \star \mu_{2^{-j}}(x-z) \varphi_j f(z) dz \right|$$

$$\leq \varphi_j^\star f(x) \int_{\mathbb{R}^n} |v_{2^{-k}t} \star \mu_{2^{-j}}(x-z)| \left(1 + 2^j|x-z|\right)^{\lambda} dz$$

$$\preceq \varphi_j^\star f(x) \int_{\mathbb{R}^n} |v_{2^{-k}t} \star \mu_{2^{-j}}(x-z)| \left(1 + 2^j|x-z|\right)^{\lambda} dz$$

$$\preceq 2^{-(j-k)(N+\gamma)} 2^{(j-k)\gamma} \varphi_j^\star f(x), \ \forall N > 0.$$

If $j \leq k+1$ then $2^{-k}t \leq c2^{-j}$ and since $v$ has $R$ vanishing moments, we have by Lemma 4.6.5

$$|v_{2^{-k}t} \star \mu_j \star \varphi_j f(x)| \leq \varphi_j^\star f(x) \left| \int_{\mathbb{R}^n} |v_{2^{-k}t} \star \mu_{2^{-j}}(x-z)| \left(1 + 2^j|x-z|\right)^{\lambda} dz \right|$$

$$\preceq \varphi_j^\star f(x) \left| \int_{\mathbb{R}^n} |v_{2^{-j}t} \star \mu_{2^{-j}}(x-z)| \left(1 + 2^k t^{-1}|x-z|\right)^{\lambda} dz \right|$$

$$\preceq 2^{-(k-j)(R+1)} t^{R+1} \varphi_j^\star f(x) \leq 2^{-(k-j)(R+1-\gamma)} 2^{(j-k)\gamma} \varphi_j^\star f(x).$$

Choose $N > R + 1 - 2\gamma$ to obtain

$$v_{2^{-k}t} f(x) \preceq 2^{-k\gamma} \sum_{j\in\mathbb{Z}} 2^{-|j-k|(R+1-\gamma)} 2^{j\gamma} \varphi_j^\star f(x) = 2^{-k\gamma} G_k(x),$$

with

$$G_k(x) = \sum_{j\in\mathbb{Z}} 2^{-|j-k|(R+1-\gamma)} 2^{j\gamma} \varphi_j^\star f(x).$$

Using Lemma 4.6.1 by taking $R > \gamma + n\tau r_0 - 1$ to conclude

$$\|f\|^{(1)}_{\dot{F}^{\gamma,\tau}_{p,q,w}} \precsim \sup_{Q \in \mathscr{Q}} \frac{1}{[w(Q)]^{\tau}} \left[ \int_Q \left( \sum_{k=j_Q}^{\infty} |G_k(x)|^q \right)^{\frac{q}{p}} w(x)dx \right]^{\frac{1}{p}}$$

$$\precsim \|f\|^{(4)}_{\dot{F}^{\gamma,\tau}_{p,q,w}}.$$

With slight modifications of the argument given in [2] we have also

$$\|f\|^{(4)}_{\dot{F}^{\gamma,\tau}_{p,q,w}} \precsim \|f\|^{(2)}_{\dot{F}^{\gamma,\tau}_{p,q,w}}. \tag{4.23}$$

The above estimates and Theorem 4.6.1 lead to 4.17.

*Step 2.* We now prove

$$\|f\|^{(2)}_{\dot{F}^{\gamma,\tau}_{p,q,w}} \precsim \|f\|^{(3)}_{\dot{F}^{\gamma,\tau}_{p,q,w}} \tag{4.24}$$

We have from 4.19

$$\left( \int_0^{l(Q)} t^{-\gamma q} |v_t^{\star} f(x)|^q \frac{dt}{t} \right)^{\frac{p}{q}} \tag{4.25}$$

$$\precsim \left[ \sum_{k=0}^{\infty} 2^{-kNr} 2^{(k+j)n} \int_{\mathbb{R}^n} \left( \int_1^2 \int_{|y|<t} t^{-\gamma q} |v_{2^{-k-j}t} f(z)|^q \frac{dydt}{t^{n+1}} \right)^{\frac{r}{q}} (1 + 2^j|x-z|)^{-\lambda r} dz \right]^{\frac{q}{r}} \tag{4.26}$$

$$\precsim \left[ \sum_{k=0}^{\infty} 2^{-kNr} 2^{(k+j)n} \int_{\mathbb{R}^n} \left( \int_1^2 \int_{|y|<t} t^{-\gamma q} |v_{2^{-k-j}t} f(y+z)|^q \frac{dydt}{t^{n+1}} \right)^{\frac{r}{q}} (1 + 2^j|x-z|)^{-\lambda r} dz \right]^{\frac{q}{r}}. \tag{4.27}$$

We continue with analogous arguments as after 4.19 and end up with 4.24.

*Step 3.* Using Lemma 4.6.4, Lemma 4.6.1 and arguing as in *Step 1* to obtain

$$\|f\|^{(5)}_{\dot{F}^{\gamma,\tau}_{p,q,w}} \precsim \|f\|^{(6)}_{\dot{F}^{\gamma,\tau}_{p,q,w}}.$$

The estimate

$$\|f\|^{(6)}_{\dot{F}^{\gamma,\tau}_{p,q,w}} \precsim \|f\|^{(5)}_{\dot{F}^{\gamma,\tau}_{p,q,w}}$$

is immediate from the following easy inequality

$$\int_{\mathbb{R}^n} |v_{2^{-j}} f(y+z)|^r (1 + 2^j|x-z|)^{-\lambda r} dz \precsim (v_{2^{-j}}^{\star})^r_{\lambda/2}(x)$$

Thus

$$\|f\|^{(6)}_{\dot{F}^{\gamma,\tau}_{p,q,w}} \simeq \|f\|^{(5)}_{\dot{F}^{\gamma,\tau}_{p,q,w}}.$$

The proof Theorem 4.6.2 is complete. ■

The Besov-type spaces also have the characterizations as in Theorem 4.6.2, whose proofs are similar to that of Theorem 4.6.2.

**Theorem 6.3.** Let $\gamma \in \mathbb{R}$, $\tau \geq 0$, $0 < p,q \leq \infty$ and $R \in \mathbb{N} \cup \{-1\}$ such that $\gamma + n\tau r_0 < R+1$ and $v \in \mathscr{S}(\mathbb{R}^n)$ satisfying 4.13 and 4.14. Assume $\lambda$ is large enough, then the space $\dot{B}^{\gamma,\tau}_{p,q,w}$ is characterized by

$$\dot{B}^{\gamma,\tau}_{p,q,w} = \{f \in S'_R(\mathbb{R}^n) : \|f\|^{(i)}_{\dot{B}^{\gamma,\tau}_{p,q,w}} < \infty\}, \ i = 1,\ldots,5,$$

where

$$\|f\|^{(1)}_{\dot{B}^{\gamma,\tau}_{p,q,w}} = \sup_{Q \in \mathscr{Q}} \frac{1}{[w(Q)]^\tau} \left[ \int_0^{l(Q)} t^{-\gamma q} \int_Q \left( |v_t f(x)|^p \frac{dt}{t} \right)^{\frac{q}{p}} w(x)dx \right]^{\frac{1}{q}}$$

$$\|f\|^{(2)}_{\dot{B}^{\gamma,\tau}_{p,q,w}} = \sup_{Q \in \mathscr{Q}} \frac{1}{[w(Q)]^\tau} \left[ \int_0^{l(Q)} t^{-\gamma q} \int_Q \left( |v^\star_{t,\lambda} f(x)|^p \frac{dt}{t} \right)^{\frac{q}{p}} w(x)dx \right]^{\frac{1}{q}}$$

$$\|f\|^{(3)}_{\dot{B}^{\gamma,\tau}_{p,q,w}} = \sup_{Q \in \mathscr{Q}} \frac{1}{[w(Q)]^\tau} \left[ \sum_{j=j_Q}^\infty 2^{j\gamma q} \int_Q \left( |v^\star_{2^{-j},\lambda} f|^p \right)^{\frac{q}{p}} w(x)dx \right]^{\frac{1}{q}}$$

$$\|f\|^{(4)}_{\dot{B}^{\gamma,\tau}_{p,q,w}} = \sup_{Q \in \mathscr{Q}} \frac{1}{[w(Q)]^\tau} \left[ \sum_{j=j_Q}^\infty 2^{j\gamma q} \int_Q \left( |v_{2^{-j}} f|^p \right)^{\frac{q}{p}} w(x)dx \right]^{\frac{1}{q}},$$

and

$$\|f\|^{(5)}_{\dot{B}^{\gamma,\tau}_{p,q,w}} = \sup_{Q \in \mathscr{Q}} \frac{1}{[w(Q)]^\tau} \times$$

$$\left[ \sum_{j=j_Q}^\infty 2^{j\gamma q} \left[ \int_Q \left( \int_{\mathbb{R}^n} |v_{2^{-j}} f(y+z)|^r (1 + 2^j |x - z|)^{-\lambda r} dz \right)^{\frac{p}{r}} w(x)dx \right]^{\frac{q}{p}} \right]^{\frac{1}{q}},$$

with $0 < r < p/r_0$, where $r_0 = \inf\{s : w \in A_s\}$, s.t, $w \in A_{r_0}$. Furthermore, all $\|\cdot\|^{(i)}_{\dot{B}^{\gamma,\tau}_{p,q,w}}$, $i = 1,\ldots,6$, are equivalent (quasi-)norms in $\dot{B}^{\gamma,\tau}_{p,q,w}$.

## 4.7  Proofs of Auxiliary Results

Proof of Lemma 4.6.1. By similarity, we only prove the first inequality.
First we note that if $k \geq j_Q$ then

$$\frac{1}{[w(Q)]^\tau} \left( \int_Q |g_k(x)|^q w(x)dx \right)^{\frac{1}{q}} \leq \frac{1}{[w(Q)]^\tau} \left[ \sum_{j=j_Q}^{\infty} \left( \int_Q |g_j(x)|^p w(x)dx \right)^{\frac{q}{p}} \right]^{\frac{1}{q}}$$

$$\leq \|g\|_{l^q\left(L_{p,w}^\tau\right)}$$

and

$$\frac{1}{[w(Q)]^\tau} \left( \int_Q |g_k(x)|^p w(x)dx \right)^{\frac{1}{p}} \leq \frac{1}{[w(Q)]^\tau} \left[ \int_Q \left( \sum_{j=j_Q}^{\infty} |g_j|^q \right)^{\frac{p}{q}} w(x)dx \right]^{\frac{1}{p}}$$

$$\leq \|g\|_{L_{p,w}^\tau(l^q)}.$$

Through the proof we take into the account that $a \in l^r(\mathbb{Z})$, $\forall\, r > 0$. We begin by considering the case $0 < p \leq 1$.

Fix a dyadic cube $Q$ and using the Young's inequality

$$\forall\ 0 < \varepsilon < 1, \quad \forall\ z_i \in \mathbb{C} : \left( \sum_{m \in \mathbb{Z}} |z_m| \right)^\varepsilon \leq \sum_{m \in \mathbb{Z}} |z_m|^\varepsilon \tag{4.28}$$

to obtain

$$I_Q = \frac{1}{[w(Q)]^\tau} \left[ \sum_{j=j_Q}^{\infty} \left( \int_Q \left| \sum_{m \in \mathbb{Z}} a_{j-m} g_m(x) \right|^p w(x)dx \right)^{\frac{q}{p}} \right]^{\frac{1}{q}}$$

$$\leq \frac{1}{[w(Q)]^\tau} \left[ \sum_{j=j_Q}^{\infty} \left( \sum_{m \in \mathbb{Z}} |a_{j-m}|^p \int_Q |g_m(x)|^p w(x)dx \right)^{\frac{q}{p}} \right]^{\frac{1}{q}}.$$

Assume $0 < q < p$. Then 4.28 implies

$$I_Q \leq \frac{1}{[w(Q)]^\tau} \left[ \sum_{j=j_Q}^{\infty} \sum_{m \in \mathbb{Z}} |a_{j-m}|^q \left( \int_Q |g_m(x)|^p w(x)dx \right)^{\frac{q}{p}} \right]^{\frac{1}{q}} = J_Q + H_Q$$

with

$$J_Q = \frac{1}{[w(Q)]^\tau} \left[ \sum_{j=j_Q}^{\infty} \sum_{m=j_Q}^{\infty} |a_{j-m}|^q \left( \int_Q |g_m(x)|^p w(x)dx \right)^{\frac{q}{p}} \right]^{\frac{1}{q}}$$

$$= \frac{1}{[w(Q)]^\tau} \left[ \sum_{m=j_Q}^{\infty} \sum_{j=j_Q}^{\infty} |a_{j-m}|^q \left( \int_Q |g_m(x)|^p w(x)dx \right)^{\frac{q}{p}} \right]^{\frac{1}{q}}$$

$$\leq C|g||_{l^q\left(L_{p,w}^\tau\right)}$$

and

$$H_Q = \frac{1}{[w(Q)]^\tau} \left[ \sum_{j=j_Q}^{\infty} \sum_{m=-\infty}^{j_Q-1} |a_{j-m}|^q \left( \int_Q |g_m(x)|^p w(x)dx \right)^{\frac{q}{p}} \right]^{\frac{1}{q}}$$

$$\leq C \frac{1}{[w(Q)]^\tau} \left[ \sum_{j=j_Q}^{\infty} \sum_{m=-\infty}^{j_Q-1} 2^{-|m-j|\delta q} \left( \int_Q |g_m(x)|^p w(x)dx \right)^{\frac{q}{p}} \right]^{\frac{1}{q}}$$

$$\leq C \left[ \sum_{j=j_Q}^{\infty} \sum_{m=-\infty}^{j_Q-1} 2^{(m-j)\delta q} \left( \frac{w(Q_m)}{w(Q)} \right)^{q\tau} \frac{1}{[w(Q_m)]^{q\tau}} \left( \int_{Q_m} |g_m(x)|^p w(x)dx \right)^{\frac{q}{p}} \right]^{\frac{1}{q}},$$

where $Q_m$ is a dyadic cube containing $Q$ with side length $l(Q_m) = 2^{-m}, m \leq j_Q - 1$. From the $A_{r_0}$ property, we have

$$\left( \frac{w(Q_m)}{w(Q)} \right)^{q\tau} \leq C \left( \frac{|Q_m|}{|Q|} \right)^{q\tau r_0} \simeq 2^{nq\tau r_0(j_Q-m)}.$$

It follows that

$$H_Q \leq C \left[ \sum_{j=j_Q}^{\infty} 2^{(m-j)\delta q} \sum_{m=-\infty}^{j_Q-1} 2^{nq\tau r_0(j_Q-m)} \right]^{\frac{1}{q}} ||g||_{l^q\left(L_{p,w}^\tau\right)}$$

$$\leq C \left[ \sum_{j=j_Q}^{\infty} 2^{(j_Q-j)\delta q} \sum_{m=-\infty}^{j_Q-1} 2^{q(\delta-n\tau r_0)(m-j_Q)} \right]^{\frac{1}{q}} ||g||_{l^q\left(L_{p,w}^\tau\right)}$$

$$\leq C||g||_{l^q\left(L_{p,w}^\tau\right)}.$$

Now assume that $q > p$ and write

$$\frac{1}{[w(Q)]^{\tau}} \left[ \sum_{j=j_Q}^{\infty} \left( \sum_{m \in \mathbb{Z}} |a_{m-j}|^p \int_Q |g_m(x)|^p w(x) dx \right)^{\frac{q}{p}} \right]^{\frac{1}{q}}$$

$$= \frac{1}{[w(Q)]^{\tau}} \left[ \sum_{j=j_Q}^{\infty} \left( \sum_{m \in \mathbb{Z}} |a_{m-j}|^{\frac{(\delta-\varepsilon)p}{\delta}} |a_{m-j}|^{\frac{\varepsilon p}{\delta}} \int_Q |g_m(x)|^p w(x) dx \right)^{\frac{q}{p}} \right]^{\frac{1}{q}}.$$

Choose $0 < \varepsilon < \delta - n\tau r_0$, arguing as before and using the following Hölder's inequality

$$\sum_{m \in \mathbb{Z}} |x_m y_m| \leq \left( \sum_{m \in \mathbb{Z}} |x_m|^r \right)^{1/r} \left( \sum_{m \in \mathbb{Z}} |y_m|^{r'} \right)^{1/r'}$$

where $\frac{1}{r} + \frac{1}{r'} = 1$, $x_m, y_m$ are in $\mathbb{C}$ and $r = \frac{q}{p}$, we obtain

$$I_Q \leq \frac{1}{[w(Q)]^{\tau}} \left[ \sum_{j=j_Q}^{\infty} \sum_{m \in \mathbb{Z}} |a_{m-j}|^{\frac{(\delta-\varepsilon)q}{\delta}} \left( \int_Q |g_m(x)|^p w(x) dx \right)^{\frac{q}{p}} \right]^{\frac{1}{q}}$$

$$\leq \frac{1}{[w(Q)]^{\tau}}$$

$$\times \left[ \sum_{j=j_Q}^{\infty} \sum_{m=j_Q}^{\infty} |a_{m-j}|^{\frac{(\delta-\varepsilon)q}{\delta}} \left( \int_Q |g_m(x)|^p w(x) dx \right)^{\frac{q}{p}} + \sum_{j=j_Q}^{\infty} \sum_{m=-\infty}^{j_Q-1} 2^{-|m-j|(\delta-\varepsilon)q} \ldots \right]^{\frac{1}{q}}$$

$$\leq C \|g\|_{l^q(L_{p,w}^{\tau})}.$$

If $1 < p \leq \infty$, we use Minkowski's inequality to get

$$I_Q \leq \frac{1}{[w(Q)]^{\tau}} \left[ \sum_{j=j_Q}^{\infty} \left( \sum_{m \in \mathbb{Z}} |a_{m-j}| \left( \int_Q |g_m(x)|^p w(x) dx \right)^{\frac{1}{p}} \right)^q \right]^{\frac{1}{q}}.$$

Applying Hölder's inequality if $1 < q \leq \infty$ or 4.28 if $0 < q \leq 1$ to conclude. ■

*Proof of Lemma 4.6.2.* We adapt here the proof of [34, Lemma 2.5]. Assume $\tau - 1/p = -\varepsilon < 0$ and denote by $\delta > 0$ the reverse-doubling constant of the weight $w \in A_p$. Pick any $x_0 \in \mathbb{R}^n$, and let $Q$ be a cube containing $x_0$ with a side length $l(Q) = r$. Write

$$g_j = g_{0j} + \sum_{j=1}^{\infty} g_{ij}$$

*with*

$$g_{0j} = \chi_{B(x_0, 2r)} g_j \quad \text{and} \quad g_{ij} = \chi_{B(x_0, 2^{i+1}r) \setminus B(x_0, 2^i r)} g_j \quad \text{for} \quad i \geq 1.$$

The Stein-Fefferman inequality implies

$$\left(\int_Q \left(\sum_{j=j_Q}^{\infty} |Mg_{0j}(x)|^q\right)^{\frac{p}{q}} w(x)dx\right)^{\frac{1}{p}} \leq \left(\int_{\mathbb{R}^n} \left(\sum_{j=j_Q}^{\infty} |g_{0j}(x)|^q\right)^{\frac{p}{q}} w(x)dx\right)^{\frac{1}{p}}$$

$$\leq C\|g\|_{L_{p,w}^{\tau}(l^q)} \left(\int_{B(x_0,r)} w(y)dy\right)^{\tau}.$$

On the other hand for $i \geq 1$ and $x \in B(x_0,r)$, we have

$$Mg_{ij}(x) = \sup_{R>0} \frac{1}{|B(x,R)|} \int_{B(x,R)\cap\{2^i r<|y-x_0|<2^{i+1}r\}} |g_j(y)|dx \leq C(2^i r)^{-n} \int_{\mathbb{R}^n} |g_{ij}(y)|dy.$$

Now, the generalized Minkowski's inequality leads to

$$\left(\sum_{j=j_Q}^{\infty} |Mg_{ij}(x)|^q\right)^{\frac{1}{q}} \leq C(2^i r)^{-n} \left(\sum_{j=j_Q}^{\infty} \left(\int_{\mathbb{R}^n} |g_{ij}(y)|dy\right)^q\right)^{\frac{1}{q}}$$

$$\leq C(2^i r)^{-n} \int_{B(x_0,2^{i+1}r)} \left(\sum_{j=j_Q}^{\infty} |g_{ij}|^q\right)^{\frac{1}{q}} dy$$

$$\leq C(2^i r)^{-n} \left(\int_{B(x_0,2^{i+1}r)} \left(\sum_{j=j_Q}^{\infty} |g_{ij}|^q\right)^{\frac{p}{q}} w(y)dy\right)^{\frac{1}{p}} \left(\int_{B(x_0,2^{i+1}r)} w^{\frac{-p'}{p}}\right)^{\frac{1}{p'}}$$

$$\leq C(2^i r)^{-n}\|g\|_{L_{p,w}^{\tau}(l^q)} \left(\int_{B(x_0,2^{i+1}r)} w^{\frac{-p'}{p}}\right)^{\frac{1}{p'}} \left(\int_{B(x_0,2^{i+1}r)} w(y)dy\right)^{\tau}$$

$$\leq C\|g\|_{L_{p,w}^{\tau}(l^q)} (2^i r)^{-n} \left(\int_{B(x_0,2^{i+1}r)} w^{\frac{-p'}{p}}\right)^{\frac{1}{p'}} \left(\int_{B(x_0,2^{i+1}r)} w(y)dy\right)^{\frac{1}{p}-\varepsilon}$$

$$\leq C\|g\|_{L_{p,w}^{\tau}(l^q)} \left(\int_{B(x_0,2^{i+1}r)} w(y)dy\right)^{-\varepsilon}$$

$$\leq C\|g\|_{L_{p,w}^{\tau}(l^q)} \left(\int_{B(x_0,r)} w(y)dy\right)^{-\varepsilon} \left(\frac{\int_{B(x_0,r)} w(y)dy}{\int_{B(x_0,2^{i+1}r)} w(y)dy}\right)^{\varepsilon}$$

$$\leq C2^{-in\varepsilon\delta}\|g\|_{L_{p,w}^{\tau}(l^q)} \left(\int_{B(x_0,r)} w(y)dy\right)^{\tau-\frac{1}{p}}.$$

Hence

$$\left(\sum_{j=j_Q}^{\infty} M^q\left(\sum_{i=1}^{\infty} g_{ij}\right)(x)\right)^{\frac{1}{q}} \leq C\left(\sum_{j=j_Q}^{\infty}\left(\sum_{i=1}^{\infty} Mg_{ij}(x)\right)^q\right)^{\frac{1}{q}}$$

$$\leq C\sum_{i=1}^{\infty}\left(\sum_{j=j_Q}^{\infty} M^q g_{ij}(x)\right)^{\frac{1}{q}}$$

$$\leq C\sum_{i=1}^{\infty} 2^{-in\varepsilon\delta}\|g\|_{L_{p,w}^{\tau}(l^q)}\left(\int_{B(x_0,r)} w(y)dy\right)^{\tau-\frac{1}{p}}$$

$$\leq C\|g\|_{L_{p,w}^{\tau}(l^q)}\left(\int_{B(x_0,r)} w(y)dy\right)^{\tau-\frac{1}{p}}.$$

It follows that

$$\left(\int_Q \left(\sum_{j=j_Q}^{\infty} |Mg_{ij}(x)|^q\right)^{\frac{p}{q}} w(x)dx\right)^{\frac{1}{p}} \leq C\|g\|_{L_{p,w}^{\tau}(l^q)}\left(\int_{B(x_0,r)} w(y)dy\right)^{\tau}.$$

We conclude that

$$\|Mg\|_{L_{p,w}^{\tau}(l^q)} \leq C\|g\|_{L_{p,w}^{\tau}(l^q)}.$$

■

# Acknowledgements

The author is grateful to the reviewers for their helpful remarks and careful corrections.

# References


[1] Bui, H.-Q. 1996. Weighted Besov and Triebel spaces, interpolation by the real method. Hiroshima Mathematical Journal 12(3): 581–605.

[2] Bui, H.-Q., Palusznsky, M. and Taibleson, M. 1996. A maximal function characterization of weighted Besov-Lipschitz and Triebel-Lizorkin spaces. Studia Mathematica 119(3): 219–246.

[3] Bui, H.-Q., Palusznsky, M. and Taibleson, M. 1997. Characterization of the Besov-Lipschitz and Triebel-Lizorkin spaces, the case $q < 1$. The Journal of Fourier Analysis and Applications 3: 837–846.

[4] Duoandikoetxea, J. 2001. Fourier analysis, Graduate Studies in Mathematics, 29. American Mathematical Society, Providence, RI.

[5] Fefferman, C. and Stein, E. 1971. Hp spaces of several variables. Acta Math. 129(1972): 137–193.

[6] Fefferman, C. and Stein, E. 1989. Some maximal inequalities. Amer. J. Math. 93: 107–115.

[7] Frazier, M. and Jawerth, B. 1990. A discrete transform and decomposition of distribution spaces. J. Funct. Anal. 93: 34–170. MR1070037 (92a:46042).

[8] Garcia-Cuerva, J. and Rubio de Francia, J.L. 1974. Weighted Norm Inequalities and Related Topics, North-Holland, Amsterdam.

[9] Grafakos, L. 2004. Classical and Modern Fourier Analysis. Hall. Pearson/Prentice-Hall

[10] Grant, W. and Shiying, Z. 1995. $\varepsilon$-families of operators in Triebel-Lizorkin and tent spaces. Can. J. Math. 47(5): 1095–1120.

[11] Gürbüz, F. 2018. Marcinkiewicz integrals with rough kernel associated with Schrödinger operators and commutators on generalized vanishing local Morrey spaces. Tbilisi Mathematical Journal 11(3): 133–156.

[12] Gürbüz, F. 2020. On the behaviors of rough multilinear fractional integral and multi-sublinear fractional maximal operators both on product Lp and weighted Lp spaces. Int. J. Nonlinear Sci. Numer. Simul. 21 (7-8): 715–726.

[13] Gürbüz, F. 2020. Some inequalities for the multilinear singular integrals with Lipschitz functions on weighted Morrey spaces. J. Inequal. Appl. Paper No. 134: 1–10.

[14] Gürbüz, F. 2018. Generalized weighted Morrey estimates for Marcinkiewicz integrals with rough kernel associated with Schrödinger operator and their commutators. Chinese Annals of Mathematics, Series B 41(1): 77-98.

[15] Gürbüz, F. 2020. On the behaviors of rough multilinear fractional integral and multi-sublinear fractional maximal operators both on product $L_p$ and weighted $L_p$ spaces. International Journal of Nonlinear Sciences and Numerical Simulation 21(7-8): 715–726.

[16] Gürbüz, F. 2021. Product generalized local Morrey spaces and commutators of multi-sublinear operators generated by multilinear Calderón-Zygmund operators and local Campanato functions. Filomat, 35(9): 2849–2868.

[17] Gürbüz, F. 2017. Some estimates for generalized commutators of rough fractional maximal and integral operators on generalized weighted Morrey spaces. Canad. Math. Bull. 60(1): 131–145.

[18] Komori, Y. and Shirai, S. 2009. Weighted Morrey spaces and a singular integral operator. Math. Nachr. 282(2): 219–231.

[19] Liang, Y., Sawano, Y., Ullrich, T., Yang, D. and Yuan, W. 2013. A new framework for generalized Besov-type and Triebel-Lizorkin-type spaces. Dissertationes Math. (Rozprawy Mat.) 489: 114.

[20] Liang, Y., Sawano, Y., Ullrich, T., Yang, D. and Yuan, W. 2012. New characterizations of Besov-Triebel-Lizorkin-Hausdorff spaces including Coorbits and wavelets. Journal of Fourier Analysis and Applications 18: 1067–1111.

[21] Loulit, A. 2020. Calderon-Zygmund singular estimates on weighted function spaces. Pacific Journal of Mathematics 307(1): 197–220. https://doi.org/10.2140/pjm.2020.307.197.

[22] Loulit, A. 2021. Atomic and molecular decompositions of weighted Triebel-Lizorkin-type spaces. Romanian Journal of Mathematics and Computer Science, Special Issue 11(1).

[23] Morrey, C.B. 1938. On the solutions of quasi-linear elliptic partial differential equations. Trans. Amer. Math. Soc. 43: 126–166.

[24] Muckenhoupt, B. 1972. Weighted norm inequalities for the Hardy maximal function. Trans. Amer. Math. Soc. 165: 207–226.

[25] Rychkov, S.V. 1999. On a theorem of Bui, Paluszynski and Taibleson. Proc. Steklov Inst., 227: 280–292.

[26] Sawano, Y. 2009. A note on Besov-Morrey spaces and Triebel-Lizorkin-Morrey spaces. Acta Math. Sin. (Engl. Ser.) 25: 1223–1242.

[27] Samko, N. 2009. Weighted Hardy and singular operators in Morrey spaces. J. Math. Anal. Appl. 350(1): 56–72.

[28] Sawano, Y., Yang, D. and Yuan, D. 2010. New applications of Besov-type and Triebel-Lizorkin-type spaces. J. Math. Anal. Appl. 363: 73–85.

[29] Sawano, Y. and Tanaka, H. 2007. Decompositions of Besov-Morrey spaces and Triebel-Lizorkin-Morrey spaces. Math. Z. 257: 871–905.

[30] Sawano, Y. 2009. A note on Besov-Morrey spaces and Triebel-Lizorkin-Morrey spaces. Acta Math. Sin. (Engl. Ser.) 25: 1223–1242.

[31] Yuan, W., Sickel, Y. and Yang, D. 2010. Morrey and Campanato Meet Besov, Lizorkin and Triebel. Lecture Notes in Mathematics vol 2005, Springer-Verlag, Berlin.

[32] Stein, E. 1993. Harmonic Analysis. Princeton University Press, Princeton, NJ.

[33] Strömberg, J.O. and Torchinsky, A. 1989. Weighted Hardy Spaces. Lecture Notes in Mathematics, vol. 1381, Springer-Verlag, Berlin.

[34] Tang, L. and Xu, J. 2005. Some properties of Morrey type Besov-Triebel spaces. Math. Nachr. 278(2005): 904–914.

[35] Triebel, H. 1992. Theory of Function Spaces II. Birkhïauser Verlag, Basel.

[36] Triebel, H. 2006. Theory of Function Spaces III. Birkhïauser Verlag, Basel.

[37] Ullrich, T. 2012. Continuous characterization of Besov-Lizorkin-Triebel spaces and new interpretations as coorbits. J. Funct. Space Appl. Article ID 163213.

[38] Yang, D. and Yuan, W. 2008. A new class of function spaces connecting Triebel-Lizorkin spaces and Q spaces. Journal of Functional Analysis 255: 2760–2809.

[39] Yang, D. and Yuan, W. 2010. Characterizations of Besov-type and Triebel-Lizorkin-type spaces via maximal functions and local means. Nonlinear Analysis 73: 3805–3820.

[40] Wu, S., Yang, D. and Yuan, W. 2017. Equivalent quasi-norms of Besov–Triebel–Lizorkin-type spaces via derivatives. Results Math. 72: 813–841.

[41] Wu, S., Yang, D. and Zhuo, C. 2018. Variable 2-microlocal Besov-Triebel-Lizorkin-type spaces. Acta Math. Sin. (Engl. Ser.) 34: 699–748.

[42] Yuan, W. Haroske, D., Moura, S., Skrzypczak, D.L. and Yang, D. 2015. Limiting embeddings in smoothness Morrey spaces, continuity envelopes and applications. J. Approx. Theory 192: 306–335.

[43] Yuan, W., Sickel, W. and Yang, D. 2020. The Haar system in Besov-type spaces. Studia Math. 253: 129–162.

[44] Yuan, W., Sickel, W. and Yang, D. 2005. Morrey and Campanato Meet Besov, Lizorkin and Triebel. Lecture Notes in Mathematics, Springer-Verlag, Berlin, 2010, xi+281pp.

[45] Zhuo, C., Chang, D.-C. and Yang, D. 2019. Ball average characterizations of variable Besov-type spaces. Taiwanese J. Math. 23; 427–452.

[46] Zhuo, C., Yang, D. and Yuan, W. 2014. Hausdorff Besov-type and Triebel–Lizorkin-type spaces and their applications. J. Math. Anal. Appl. 412: 998–1018.

# Chapter 5

# Variable Exponent Vanishing Morrey Type Spaces on Unbounded Domains

*Ferit Gürbüz*

## 5.1 Preliminaries and Introduction

Before giving the main results, we will give some background material that is required for later chapters. This section is a summary of the basic definitions and results about measure theory which will be used throughout the book. Most of the materials are familiar to the reader and may be omitted. This section, included for the sake of completeness, should serve to settle questions of notation and such, sometimes we will also refer the interested readers to some papers and references. We will set forth the basic concepts of measure theory and develop the theory of integration on abstract measure spaces, paying particular attention to the Lebesgue integral on the Euclidean space $\mathbb{R}^n$. In particular, we prove Hölder's

Department of Mathematics, Kırklareli University, Kırklareli, 39100, Turkey.
Email: feritgurbuz@klu.edu.tr

inequality for integrals and embeddings between Lebesgue and Morrey spaces are characterized.

Now, we give the precise definition of a normed linear space.

Let $X$ be a linear space over the real number field $K$. A real-valued function $\|\cdot\| : X \to \mathbb{R}$, $x \to \|x\|$ defined on $X$ is called a norm on $X$ if it satisfies the following three conditions $(N1)$, $(N2)$ and $(N3)$:

$$(N1)\ \|x\| \geq 0 \text{ and } \|x\| = 0 \Leftrightarrow x = \theta$$
$$(N2)\ \|ax\| = |a|\,\|x\|,\ \forall x \in X \text{ and } \forall a \in K$$
$$(N3)\ \|x+y\| \leq \|x\| + \|y\|,\ \forall x,y \in X \text{ (the triangle inequality).}$$

A linear space $X$ equipped with a norm $\|\cdot\|$ is called a normed linear space. The topology on $X$ is defined by the metric

$$d(x,y) = \|x - y\|.$$

The convergence

$$\lim_{n \to \infty} \|x_n - x\| = 0$$

in $X$ is denoted by $x_n \to x$ and and we say that the sequence $(x_n)$ converges strongly to $x$. A sequence $(x_n)$ in $X$ is called a Cauchy sequence if it satisfies the condition

$$\lim_{n,m \to \infty} \|x_n - x_m\| = 0.$$

A normed linear space $X$ is called a Banach space if it is complete, that is, if every Cauchy sequence in $X$ converges strongly to a point in $X$.

Two norms $\|\cdot\|_1$ and $\|\cdot\|_2$ defined on the same linear space $X$ are said to be equivalent if there exist constants $c > 0$ and $C > 0$ such that

$$c\|\cdot\|_1 \leq \|\cdot\|_2 \leq C\|\cdot\|_1$$

for all $x \in X$. Also, equivalent norms induce the same topology.

**Example 1** *If $\|x+y\| = \|x\| + \|y\|$ for two vectors $x$ and $y$ in a normed space, then show that $\|\lambda x + \mu y\| = \lambda\,\|x\| + \mu\,\|y\|$ for all scalars $\lambda,\ \mu \geq 0$.*

**Solution**  If $\lambda \geq \mu$, then

$$\lambda\,\|x\| + \mu\,\|y\| \geq \|\lambda x + \mu y\| = \|\lambda(x+y) + (\mu - \lambda)y\|$$
$$\geq \lambda\,\|x+y\| - (\lambda - \mu)\,\|y\| = \lambda\,\|x\| + \mu\,\|y\|.$$

Hence, $\|\lambda x + \mu y\| = \lambda\,\|x\| + \mu\,\|y\|$ holds for all $\lambda,\ \mu \geq 0$.  ■

Let $X, Y$ be linear spaces over the same scalar field $K$. A mapping $T$ defined on a linear subspace $D(T)$ of $X$ and taking values in $Y$ is said to be linear if it preserves the operations of addition and scalar multiplication:

$$(L1) \ T(x+y) = Tx + Ty \text{ for all } x, y \in D(T)$$
$$(L2) \ T(\alpha x) = \alpha Tx \text{ for all } x \in D(T) \text{ and } \alpha \in K.$$

We often write $Tx$, rather than $T(x)$, if $T$ is linear. We let

$$R(T) = \{Tx : x \in D(T)\}$$
$$N(T) = \{x \in D(T) : Tx = 0\},$$

and call them the domain, the range and the null space of $T$, respectively. The mapping $T$ is called a linear operator from $D(T) \subset X$ into $Y$. We also say that $T$ is a linear operator from $X$ into $Y$ with domain $D(T)$. In the particular case when $Y = K$, the mapping $T$ is called a linear functional on $D(T)$. In other words, a linear functional is a $K$-valued function on $D(T)$ that satisfies conditions $(L1)$ and $(L2)$. If a linear operator $T$ is a one-to-one map of $D(T)$ onto $R(T)$, then it is easy to see that the inverse mapping $T^{-1}$ is a linear operator on $R(T)$ onto $D(T)$. The mapping $T^{-1}$ is called the inverse operator or simply the inverse of $T$. A linear operator $T$ admits the inverse $T^{-1}$ if and only if $Tx = 0$ implies that $x = 0$. Let $T_1$ and $T_2$ be two linear operators from a linear space $X$ into a linear space $Y$ with domains $D(T_1)$ and $D(T_2)$, respectively. Then we say that $T_1 = T_2$ if and only if $D(T_1) = D(T_2)$ and $T_1 x = T_2 x$ for all $x \in D(T_1) = D(T_2)$. If $D(T_1) \subset D(T_2)$ and $T_1 x = T_2 x$ for all $x \in D(T_1)$, then we say that $T_2$ is an extension of $T_1$ and also that $T_1$ is a restriction of $T_2$, and we write $T_1 \subset T_2$. Moreover, we call a linear operator $T$ bounded if there is a constant $C$ such that for all $x \in X$,

$$\|Tx\| \leq C\|x\|.$$

We call the least such $C$ the norm of $T$ and denote it by $\|T\|$. Thus,

$$\|T\| = \sup_{x \in X, x \neq \theta} \frac{\|Tx\|}{\|x\|}.$$

$T$ linear functional on a linear space $X$ is a linear operator from $X$ to the space $\mathbb{R}$ of real numbers. Let $a, b \in Y$ and let $\gamma$ be a scalar.

**Definition 5.1** **(Homogeneous Functions)** A function $f(x_1, x_2, \ldots, x_n)$ of several variables is called a homogeneous function if the relation

$$f(\lambda x_1, \lambda x_2, \ldots, \lambda x_n) = \lambda^m f(x_1, x_2, \ldots, x_n)$$

holds for arbitrary $\lambda$. The number $m$ is the degree of homogeneity.

**Example 2** *For $f(x,y) = x^2 - 3xy + y^2 + x\sqrt{xy + \frac{x^3}{y}}$, the degree of homogeneity is $m = 2$ and for $f(x,y) = \frac{x+z}{2x-3y}$, the degree of homogeneity is $m = 0$.*

**Example 3** *Let $T : X \to Y$ be a surjective one-to-one operator between two vector spaces. Show that $T^{-1}$ (the inverse of $T$) is a linear operator.*

**Solution**   Let $G : Y \to X$ denote the inverse of $T$. That is, we have $TG = I_Y$ and $GT = I_X$. Let $a, b \in Y$ and let $\xi$ be a scalar. For the additivity of $G$ note that

$$T\left[(Ga + Gb)\right] = TGa + TGb = a + b = T\left[G(a+b)\right].$$

Since $T$ is one-to-one, it follows that $G(a+b) = G(a) + G(b)$. That is, $G$ is additive. Similarly, for the homogeneity of $G$ observe that

$$T\left[G(\alpha a)\right] = \alpha a = \alpha TGa = T\left[\alpha Ga\right]$$

and so $G(\alpha a) = \alpha G(a)$. Thus, $G = T^{-1}$ is a linear operator.   ■

**Example 4** *If $T : X \to Y$ is bounded operator between normed spaces. Then show that*
$$\|T\| = \min\left\{C \geq 0 : \|Tx\| \leq C\|x\| \text{ for all } x \in X\right\}.$$

**Solution**   Let

$$C_0 = \min\left\{C \geq 0 : \|Tx\| \leq C\|x\| \text{ for all } x \in X\right\}.$$

In fact, note that the infimum is a minimum. Now if $C \geq 0$ satisfies

$$\|Tx\| \leq C\|x\|$$

for all $x \in X$, then
$$\|T\| = \sup_{\|x\|=1} \|Tx\| \leq C,$$

and so
$$\|T\| \leq C_0.$$

On the other hand,
$$\|Tx\| \leq C\|x\|$$

for each $x \in X$ implies
$$C_0 \leq \|T\|.$$

Thus,
$$\|T\| = C_0.$$

■

Before dealing with integration, let us review some elementary facts and notation that will be needed. The real numbers are denoted by $\mathbb{R}$, while the complex numbers are denoted by $\mathbb{C}$ and $\bar{z}$ is the complex conjugate of $z$. It will be assumed that the reader is equipped with a knowledge of the fundamentals of the calculus on $n$-**dimensional Euclidean space** $\mathbb{R}^n = \{(x_1, x_2, \ldots, x_n) : \text{each } x_i \text{ is in } \mathbb{R}\}$. An important class of functions consists of the **characteristic functions** of sets. If $E$ is a set we define: $\chi_E(x) = \begin{cases} 1 & \text{if } x \in E \\ 0 & \text{if } x \notin E \end{cases}$. Denote by $|E| = \int_E dx$ the Lebesgue measure of the set $E \subset \mathbb{R}^n$ and the characteristic function of a set $E$ is denoted by $\chi_E$, hence $\chi_E(x) = 1$ if $x \in E$ and zero otherwise. $E^C$ will always denote the complement of $E$.

Let $\mathbb{R}^n$, $n \geq 1$, denote the $n$-dimensional Euclidean space and $x = (x_1, x_2, \ldots, x_n)$, $\xi = (\xi_1, \xi_2, \ldots, \xi_n) \ldots$ etc. be points of $\mathbb{R}^n$. The scalar product of elements in $\mathbb{R}^n$ is denoted by $x.\xi = \sum_{i=1}^{n} x_i \xi_i$ and, in particular, $|x| = (x.x)^{1/2} = \left( \sum_{i=1}^{n} x_i^2 \right)^{1/2}$ is the distance of $x$ to the origin. A multiindex $\alpha = (\alpha_1, \alpha_2, \ldots, \alpha_n)$ is an n-tuple of nonnegative integers and $|\alpha| = \alpha_1 + \alpha_2 \ldots + \alpha_n$, $x^\alpha = x_1^{\alpha_1} \ldots x_n^{\alpha_n}$. We expect the reader to know some elementary inequalities such as the triangle inequality, $|x| + |y| \geq |x - y|$.

$B(x, r)$ is the open ball with center $x$ and radius $r$, $\overline{B(x, r)}$ is the closed ball and the sphere $S(x, r)$ is the boundary of $B(x, r)$. $d(x, B) = \inf_{y \in B} |x - y|$ denotes the Euclidean distance of the point $x \in \mathbb{R}^n$ from the non-empty subset $B$ of $\mathbb{R}^n$ and $d(A, B) = \inf_{x \in A, y \in B} |x - y|$ denotes the Euclidean distance between the non-empty sets $A$ and $B$. If $B$ is closed and $x \notin B$, then $d(x, B) > 0$ and, if $A$ is compact, $B$ is closed and the two sets are disjoint, then $d(A, B) > 0$.

Suppose $E$ is an open subset of $\mathbb{R}^n$ and consider the open sets

$$E_{(m)} = \left\{ x \in E : d(x, \partial E) > \frac{1}{m}, \ |x| < m \right\}, \ m \in \mathbb{N}.$$

It is easy to check the following four properties:
1. Every $\overline{E_{(m)}}$ is a compact subset of $E$,
2. $\overline{E_{(m)}} \subset E_{(m+1)}$ for all $m$,
3. $\bigcup_{m=1}^{\infty} E_{(m)} = E$ and
4. Every compact subset of $E$ is contained in $E_{(m)}$ for a sufficiently large $m$.

This increasing sequence $\{E_{(m)}\}$ of open sets, or any other with the same four properties, is called an **open exhaustion** of $E$. The increasing sequence $\{F_{(m)}\}$, with $F_{(m)} = \overline{E_{(m)}}$, where $\{E_{(m)}\}$ is an open exhaustion of $E$, is called a **compact exhaustion** of $E$.

In general, if $f$ is a function from some set $A$ (e.g., some subset of $\mathbb{R}^n$) with values in some set $B$ (e.g., the real numbers), we denote this fact by $f : A \to B$. If $x \in A$, we write $x \to f(x)$, the bar on the arrow serving to distinguish the image of a single point $x$ from the image of the whole set $A$.

In the space $\mathbb{R}^n$, $n \geq 1$, the Lebesgue measure is denoted by $dx = dx_1...dx_n$. In what follows, all sets and functions considered will be Lebesgue measurable and $\int_{\mathbb{R}^n} f\, dx$ stands for $\int_{\mathbb{R}^n} f(x_1,...,x_n)\, dx_1...dx_n$, that is,

$$\int_{\mathbb{R}^n} f(x)\, dx = \int \cdots \int f(x_1,...,x_n)\, dx_1...dx_n$$

will denote the (Lebesgue) integral of the function $f$ over the entire space $\mathbb{R}^n$.

$S^{n-1} = \{x \in \mathbb{R}^n : |x| = 1\}$ represents the unit sphere in Euclidean $n$-dimensional space $\mathbb{R}^n$ $(n \geq 2)$ and $dx'$ is its surface measure. If $x \neq 0 = (0,...,0)$, then $x = |x|\frac{x}{|x|} = rx'$, where $x' = \frac{x}{|x|}$ is unit vector corresponding to $x$ on $S^{n-1}$.

We shall frequently use polar coordinates in $\mathbb{R}^n$ and provide the details now.

Let $r = |x|$. For $x = (x_1, x_2, \ldots, x_n) \in \mathbb{R}^n$, $n > 1$, we consider the transformation given by

$$
\begin{aligned}
x_1(r, \theta_1, ..., \theta_{n-1}) &= r\cos\theta_1 \\
x_2(r, \theta_1, ..., \theta_{n-1}) &= r\sin\theta_1 \cos\theta_2 \\
x_3(r, \theta_1, ..., \theta_{n-1}) &= r\sin\theta_1 \sin\theta_2 \cos\theta_3 \\
&\ \ \vdots \\
x_k(r, \theta_1, ..., \theta_{n-1}) &= r\sin\theta_1 \sin\theta_2 ... \sin\theta_{k-1}\cos\theta_k, 2 \leq k \leq n-2 \\
x_n(r, \theta_1, ..., \theta_{n-1}) &= r\sin\theta_1 \sin\theta_2 ... \sin\theta_{n-1},
\end{aligned}
$$

where $0 \leq r < \infty$, $0 \leq \theta_k \leq \pi$, $k = 1,...,n-2$, $0 \leq \theta_{n-1} \leq 2\pi$ and $x' = (\theta_1,...,\theta_{n-1}) \in S^{n-1}$.

The Jacobian associated with the above transformation is equal

$$J(r, \theta_1, ..., \theta_{n-1}) = r^{n-1}\prod_{j=1}^{n-1}(\sin\theta_j)^{n-1-j},$$

thus $f$ is integrable in $\mathbb{R}^n$,

$$\int_{\mathbb{R}^n} f(|x|)\,dx = \int_0^\infty \int_0^\pi \int_0^\pi \cdots \int_0^{2\pi} f(r)\, J(r,\theta)\,dr\,d\theta_1 \ldots d\theta_{n-1}$$

$$= \int_0^\infty r^{n-1} f(r)\,dr \int_0^\pi \int_0^\pi \cdots \int_0^{2\pi} \prod_{j=1}^{n-1} (\sin\theta_j)^{n-1-j}\,d\theta_1 \ldots d\theta_{n-1}$$

$$= |S^{n-1}| \int_0^\infty f(r)\, r^{n-1}\,dr,$$

where

$$\int_0^\pi \int_0^\pi \cdots \int_0^{2\pi} \prod_{j=1}^{n-1} (\sin\theta_j)^{n-1-j}\,d\theta_1 \ldots d\theta_{n-1} = \int_{S^{n-1}} dx' = |S^{n-1}|$$

is the surface area of the unit sphere. And thus

$$\int_{\mathbb{R}^n} f(|x|)\,dx = \int_{S^{n-1}} \int_0^\infty r^{n-1} f(rx')\,dr\,dx', \tag{5.1}$$

here $dx'$ is called the surface area element on $S^{n-1}$.

It is generally written in the form

$$\int_{\mathbb{R}^n} f(x)\,dx = \int_0^\infty \int_{S^{n-1}} f(r\sin\theta_1,\ldots,r\sin\theta_1 \ldots \sin\theta_{n-1})\, r^{n-1}\,dr\,d\theta_1 \ldots d\theta_{n-1}$$

$$= \int_0^\infty \int_{S^{n-1}} f(r,\theta)\, r^{n-1}\,d\sigma\,dr.$$

Here, the volume element $dx$ is written as $dx = r^{n-1}\,dr\,d\sigma$. Also, $d\sigma$ is the surface measure determined by $dx$ on $S^{n-1}$.

As a special case, if $dm$ is the Lebesgue measure on $\mathbb{R}^n$ and $d\sigma$ is the standard surface measure on $S^{n-1}$, then we have the formula

$$\int_{B(x,R)} f(y)\,dm(y) = \int_0^R \int_{S^{n-1}} f(x+rt)\,d\sigma(t)\, r^{n-1}\,dr.$$

We define the surface-mean-value of $f$ over $S(x,r)$ by

$$\mathcal{M}_f^r(x) = \frac{1}{|S^{n-1}|} \int_{S^{n-1}} f(x+rt)\,d\sigma(t) = \frac{1}{|S^{n-1}|\, r^{n-1}} \int_{S(x,r)} f(y)\,dS(y) \tag{5.2}$$

for all $f$ integrable with respect to $dS$, the surface measure in $S(x,r)$.

We also define the **space-mean-value** of $f$ over $B(x,r)$ by

$$\mathcal{A}_f^r(x) = \frac{1}{|S^n|} \int\limits_{B(0,1)} f(x+ry)\,dm(y) = \frac{1}{|S^n|\,r^n} \int\limits_{B(x,r)} f(y)\,dm(y)$$

for all $f$ integrable with respect to $dm$ in $B(x,r)$.

By the formula in (5.2), we get

$$\mathcal{A}_f^R(x) = \frac{n}{R^n} \int\limits_0^R \mathcal{M}_f^r(x)\,r^{n-1}\,dr.$$

Define $\bar{\mathbb{R}}^n$, the one-point compactification of $\mathbb{R}^n$, by adjoining the point at $\infty$ to $\mathbb{R}^n$ : $\bar{\mathbb{R}}^n = \mathbb{R}^n \cup \{\infty\}$. The $\varepsilon$-neighborhoods of points $x \in \mathbb{R}^n$ are the usual balls $B(x,\varepsilon)$, while the $\varepsilon$-neighborhood of $\infty$ is defined to be the set $\left\{x \in \mathbb{R}^n : |x| > \frac{1}{\varepsilon}\right\} \cup \{\infty\}$. We define open sets in $\bar{\mathbb{R}}^n$ through these neighborhoods, in the usual way, and we, also, define closed sets (complements of open sets) and the notion of convergent sequence: a sequence in $\bar{\mathbb{R}}^n$ converges to some point of $\bar{\mathbb{R}}^n$, if the sequence is, eventually, contained in every $\varepsilon$-neighborhood of this point. Hence, if the limit point is in $\mathbb{R}^n$, then the new notion of convergence coincides with the usual one, while $x_m \to \infty$ is equivalent to $|x_m|$.

Recall that the closure of a set $A \subset \mathbb{R}^n$ is the smallest closed set in $\mathbb{R}^n$ that contains $A$. We denote the closure by $\bar{A}$. Thus, $\bar{\bar{A}} = \bar{A}$. The support of a continuous function $f : \mathbb{R}^n \to \mathbb{C}$, denoted by supp $\{f\} = \overline{\{x \in \mathbb{R}^n : f(x) \neq 0\}}$ , is the closure of the set of points $x \in \mathbb{R}^n$ where $f(x)$ is nonzero, i.e.,

$$\text{supp}\,\{f\} = \overline{\{x \in \mathbb{R}^n : f(x) \neq 0\}}.$$

The support of $f = $supp $f$ is the closure of $\{x : f(x) \neq 0\}$ and denoted by supp $f = \overline{\{x \in \mathbb{R}^n : f(x) \neq 0\}}$. If *supp* $f$ is bounded, then we say $f \in \mathcal{C}_0$; $\mathcal{C}^n$ denotes the space of n-times continuously differentiable functions; $\mathcal{C}_0^n = \mathcal{C}_0 \cap \mathcal{C}^n$; $\mathcal{C}^\infty$ and $\mathcal{C}_0^\infty$ are defined analogously.

Saying that the support of a measurable function is bounded is equivalent to saying that it is a compact subset of $\mathbb{R}^n$ and equivalent to saying that the function vanishes almost everywhere outside a compact subset of $\mathbb{R}^n$. In this case, we say that the function **is compactly supported** or that it **has compact support**.

In case $f$ is continuous in $\mathbb{R}^n$, then it is easy to see that its support is the smallest closed set in $\mathbb{R}^n$ outside of which $f$ is everywhere 0.

If $d\mu$ is a Borel measure (of any kind), we say that the point $x \in \mathbb{R}^n$ is a **support-point** of $d\mu$, if $d\mu$ is the zero measure in no neighborhood of $x$. The **support** or, sometimes called, **closed-support** of $d\mu$ is defined by

$$\text{supp}\,\{d\mu\} = \{x \in \mathbb{R}^n : x \text{ is a support-point of } d\mu\}.$$

It is easy to see, using the regularity of $d\mu$, that supp$\{d\mu\}$ is the smallest closed set in $\mathbb{R}^n$ outside of which $d\mu$ is the zero measure.

**Example 5** *The finite nonnegative Borel measure* $d\mu = \sum\limits_{k=1}^{\infty} 2^{-k} d\delta_{\frac{1}{k}}$ *is supported in the set* $\{\frac{1}{k} : k \in \mathbb{N}\}$, *but* supp$\{d\mu\} = \{\frac{1}{k} : k \in \mathbb{N}\} \cup \{0\}$.

Whenever we say that $d\mu$ is compactly supported or that it has compact support, we mean that supp$\{d\mu\}$ is a compact subset of $\mathbb{R}^n$. In case $d\mu = f\,dm$ is an absolutely continuous complex Borel measure with density function $f \in L(\mathbb{R}^n)$, then, clearly, supp$\{f\,dm\}$ =supp$\{f\}$.

Here is a classic example of a compactly supported, infinitely differentiable function on $\mathbb{R}^n$; its support is the unit ball $\{x \in \mathbb{R}^n : |x| = 1\}$ :

$$h(x) = \begin{cases} \exp\left[-\frac{1}{1-|x|^2}\right] & \text{if } |x| < 1 \ (x \in B(0,1)), \\ 0 & \text{if } |x| \geq 1 \ (x \notin B(0,1)). \end{cases}$$

This function has the properties:

$1-h$ is in $C_0^{\infty}(\mathbb{R}^n)$,

$2-h$ is nonnegative and supp $\{h\} = \overline{B(0,1)}$,

$3-h$ is radial. I.e. $h(x) = h(y)$ whenever $|x| = |y|$,

$4-\int_{\mathbb{R}^n} h(x)\,dm(x) = 1$, if we choose the constant $C$ appropriately.

The verification that $h$ is actually in $C_0^{\infty}$ is left as an exercise.

Now, the functions $(\delta > 0)$

$$h_\delta(x) = \frac{1}{\delta^n} h\left(\frac{x}{\delta}\right), \quad x \in \mathbb{R}^n,$$

have the same properties 1, 3 and 4 and property 2 replaced by supp $(h_\delta) = B(0,\delta)$.

For $n = 1$, a useful technical tool is the function

$$h_0(t) = \begin{cases} \exp\left[-\frac{1}{t}\right] & \text{if } t > 0, \\ 0 & \text{if } t \leq 0. \end{cases}$$

It is easy to prove that $h_0$ is in $C_0^{\infty}(\mathbb{R})$ and supp$\{h_0\} = \mathbb{R}_0^+$.

**Definition 5.2**   (**Fubini's theorem**) Let $X \times Y$ be an interval in $R^{m+n}$, which is the direct product of intervals $X \subset R^m$ and $Y \subset R^n$. If the function $f : X \times Y \to \mathbb{R}$ is integrable over $X \times Y$, then all three of the integrals

$$\iint_{X \times Y} f(x,y)\,dxdy, \quad \int_X dx \int_Y f(x,y)\,dy, \quad \int_Y dy \int_X f(x,y)\,dx \tag{5.3}$$

exist and are equal. As a consequence, Fubini's theorem allows the order of integration to be changed in certain iterated integrals. Fubini's theorem implies that two iterated integrals are equal to the corresponding double integral across its integrands.

**Example 6** *Calculate* $\iint\limits_R f(x,y)\,dA$ *for* $f(x,y) = 1 - 6x^2y$ *and* $R : 0 \leq x \leq 2,$ $-1 \leq y \leq 1.$

**Solution** $\quad \iint\limits_R f(x,y)\,dA = \int\limits_0^2 \int\limits_{-1}^1 \left(1 - 6x^2y\right)\,dy\,dx$

$$= \int_0^2 \left[y - 3x^2y^2\right]_{y=-1}^{y=1}\,dx = \int_0^2 \left[(1 - 3x^2) - (-1 - 3x^2)\right]\,dx = \int_0^2 2\,dx = \mathbf{4}.$$

$$\iint\limits_R f(x,y)\,dA = \int\limits_{-1}^1 \int\limits_0^2 \left(1 - 6x^2y\right)\,dx\,dy$$

$$= \int_{-1}^1 \left[x - 2x^3y\right]_{x=-0}^{x=2}\,dy = \int_{-1}^1 (2 - 16y)\,dy = \left[2y - 8y^2\right]_{-1}^1 = \mathbf{4}. \quad \blacksquare$$

**Example 7** *Calculate* $\iint\limits_R f(x,y)\,dA$ *for* $f(x,y) = \frac{y^2 - x^2}{(x^2 + y^2)^2}$ *and* $R : 0 \leq x \leq 1,$ $-1 \leq y \leq 1.$

**Solution** The function is discontinuous at the point $(0,0)$, so the formula (5.3) cannot be used. Indeed, checking it we get:

$$\iint\limits_R f(x,y)\,dA = \int\limits_0^1 \int\limits_0^1 \frac{y^2 - x^2}{(x^2 + y^2)^2}\,dx\,dy$$

$$= \int_0^1 \left[\frac{x}{x^2 + y^2}\right]_{x=0}^{x=1}\,dy$$

$$= \int_0^1 \frac{1}{1 + y^2}\,dy$$

$$= \arctan y\,|_0^1 = \frac{\pi}{4}.$$

$$\iint_R f(x,y)\, dA = \int_0^1 \int_0^1 \frac{y^2 - x^2}{(x^2 + y^2)^2}\, dy\, dx$$

$$= \int_0^1 \left[ \frac{y}{x^2 + y^2} \right]_{y=0}^{y=1} dx$$

$$= -\int_0^1 \frac{1}{1 + x^2}\, dx$$

$$= -\arctan x \,|_0^1 = -\frac{\pi}{4}.$$

■

**Example 8** *Calculate* $\displaystyle\iint_R f(x,y)\, dA$ *for* $f(x,y) = \frac{xy}{(x^2+y^2)^2}$ *and* $R: -1 \le x \le 1,$ $-1 \le y \le 1.$

**Solution**

$$\iint_R f(x,y)\, dA = \int_{-1}^1 \left( \int_{-1}^1 \frac{xy}{(x^2 + y^2)^2}\, dx \right) dy, \; y \ne 0$$

$$= \int_{-1}^1 0\, dy = 0$$

and

$$\iint_R f(x,y)\, dA = \int_{-1}^1 \left( \int_{-1}^1 \frac{xy}{(x^2 + y^2)^2}\, dy \right) dx, \; x \ne 0$$

$$= \int_{-1}^1 0\, dx = 0,$$

Thus

$$\int_{-1}^1 \left( \int_{-1}^1 \frac{xy}{(x^2 + y^2)^2}\, dx \right) dy = \int_{-1}^1 \left( \int_{-1}^1 \frac{xy}{(x^2 + y^2)^2}\, dy \right) dx = 0;$$

but the Lebesgue double integral over the square does not exist, since

$$\int_{-1}^1 \int_{-1}^1 \left| \frac{xy}{(x^2 + y^2)^2} \right| dx\, dy \ge \int_0^1 dr \int_0^{2\pi} (\sin\phi \cos\phi / r)\, d\phi = 2 \int_0^1 dr/r = \infty.$$

■

**Example 9** *For $i = 1, 2$, let $X_i = \mathbb{N}$ (the natural numbers), $M_i = 2^{\mathbb{N}}$ (the $\sigma$-algebras of all subsets of $\mathbb{N}$) and let $\mu_i$ be the counting measure. For the function $f : X_i \times X_2 \to \mathbb{R}$ defined by*

$$f(i, j) = \begin{cases} -2^{-i} &, & j = i \\ 2^{-i} &, & j = i+1 \\ 0 &, & \text{otherwise.} \end{cases}$$

*Compute the iterated integrals*

$$\int_{X_i} \left( \int_{X_2} f \, d\mu_2 \right) d\mu_1$$

*and*

$$\int_{X_2} \left( \int_{X_1} f \, d\mu_1 \right) d\mu_2.$$

*How do you reconcile your answers with Fubini's theorem?*

**Solution** For any $i \in X_i$, $j \to f(i, j)$ is integrable and

$$\int_{X_2} f(i, j) \, d\mu_2(j) = -2^{-i} + 2^{-i} = 0.$$

Thus,

$$\int_{X_i} \left( \int_{X_2} f \, d\mu_2 \right) d\mu_1 = 0.$$

For any $j \in X_2$, $i \to f(i, j)$ is integrable and

$$\int_{X_1} f(i, 1) \, d\mu_1(i) = -\frac{1}{2}.$$

■

**Lemma 5.1**

*For $n > 1$, $\left| S^{n-1} \right| = \dfrac{2\pi^{\frac{n}{2}}}{\Gamma\left(\frac{n}{2}\right)}$ and $\left| S^n \right| = \dfrac{2\pi^{\frac{n}{2}}}{n\Gamma\left(\frac{n}{2}\right)}$. Hence, $\left| S^{2m} \right| = \dfrac{\pi^m}{m!}$ and $\left| S^{2m+1} \right| = \dfrac{2(2\pi)^m}{1.3.5...(2m+1)}$. Also, $\left| S^{n-1} \right| = n \left| S^n \right|$ is the standard surface area of $S^{n-1} = S(0, 1)$.*

**Proof 5.1**  By Fubuni's theorem, since

$$\int_{-\infty}^{\infty}\int_{-\infty}^{\infty} \exp\left(-\left(x_1^2+x_2^2\right)\right) dx_1 dx_2 = \left(\int_{-\infty}^{\infty}\exp\left(-x_1^2\right) dx_1\right)\cdot\left(\int_{-\infty}^{\infty}\exp\left(-x_2^2\right) dx_2\right)$$

$$= \left(\int_{-\infty}^{\infty}\exp\left(-t^2\right) dt\right)^2 = \iint_{\mathbb{R}^2}\exp\left(-|x|^2\right) dx$$

$$= \int_0^{\infty}\int_0^{2\pi} re^{-r^2}\,d\theta dr = \int_0^{2\pi}\int_0^{\infty} re^{-r^2}\,dr d\theta = \int_0^{2\pi}\frac{d\theta}{2} = \pi,$$

then we have

$$F = \int_{\mathbb{R}^n}\exp\left(-|x|^2\right) dx = \prod_{k=1}^{n}\int_{-\infty}^{\infty}\exp\left(-x_k^2\right) dx_k = \left(\int_{-\infty}^{\infty}\exp\left(-t^2\right) dt\right)^n = \pi^{n/2}.$$

On the other hand, by (5.1),

$$F = \int_{S^{n-1}}\int_0^{\infty}\exp\left(-r^2\right) r^{n-1} dr dx'$$

$$= \int_{S^{n-1}} dx'\int_0^{\infty}\exp\left(-r^2\right) r^{n-1} dr$$

$$= \left|S^{n-1}\right|\int_0^{\infty}\exp\left(-\omega\right)\omega^{\frac{n}{2}-1} d\omega$$

$$= \frac{1}{2}\left|S^{n-1}\right|\Gamma(n/2),$$

where $\Gamma(n)$ is is the gamma function and is defined by

$$\Gamma(n) = \int_0^{\infty} x^{n-1}e^{-x} dx.$$

Moreover, since $\Gamma(1+\omega) = \omega\Gamma(\omega)$, then we obtain

$$|S^n| = \int_{|x|\leq 1} dx = \int_{S^{n-1}}\int_0^{1} r^{n-1} dr dx' = \frac{1}{n}\left|S^{n-1}\right|$$

$$= \frac{1}{n}\frac{2\pi^{\frac{n}{2}}}{\Gamma\left(\frac{n}{2}\right)} = \frac{\pi^{\frac{n}{2}}}{\frac{n}{2}\Gamma\left(\frac{n}{2}\right)} = \frac{\pi^{\frac{n}{2}}}{\Gamma\left(1+\frac{n}{2}\right)}.$$

Observe that for any $x \in \mathbb{R}^n$ and $\rho > 0$ the area of the sphere $S_\rho(x)$ of center $x$ and radius $\rho$ is equal to $\rho^{n-1} \left| S^{n-1} \right|$ and its volume is equal to $\rho^n |S^n| = \rho^n \frac{1}{n} \left| S^{n-1} \right|$. As we know, for the univariate integral, we have

$$\int_0^1 \frac{dx}{|x|^\alpha} = \begin{cases} \text{convergent, } \alpha < 1 \\ \text{divergent, } \alpha \geq 1 \end{cases}$$

and

$$\int_1^\infty \frac{dx}{|x|^\alpha} = \begin{cases} \text{convergent, } \alpha > 1 \\ \text{divergent, } \alpha \leq 1 \end{cases}.$$

Now let's look at the situation in $\mathbb{R}^n$. Letting

$$|x| = \left( \sum_{i=1}^n x_i^2 \right)^{1/2}$$

denote the norm of $x$, consider the integrals

$$\int_{\mathbb{R}^n} \frac{dx}{|x|^\alpha}, \ \alpha > 0.$$

The integrands may have trouble at 0 and at infinity, that is this integral has singularity at $x = 0$ and $x = \infty$, so we split this integral into two parts:

$$\int_{\mathbb{R}^n} \frac{dx}{|x|^\alpha} = \int_{|x| \leq 1} \frac{dx}{|x|^\alpha} + \int_{|x| \geq 1} \frac{dx}{|x|^\alpha}$$
$$= F_1 + F_2,$$

where $r = |x|$, the element of volume $dx$ can be written in "polar coordinates" as $dx = r^{n-1} dr d\sigma$ and $d\sigma$ is the surface measure determined by $dx$ on the unit sphere $S^{n-1} = \{x \in \mathbb{R}^n : |x| = 1\}$.

Let's guess $F_1$ first.

$$F_1 = \int_{|x|\leq 1} \frac{dx}{|x|^\alpha} = \int_{S^{n-1}} \int_0^1 \frac{r^{n-1}}{r^\alpha} dr d\sigma$$

$$= \int_{S^{n-1}} d\sigma \int_0^1 \frac{r^{n-1}}{r^\alpha} dr$$

$$= |S^{n-1}| \int_0^1 \frac{1}{r^{\alpha-n+1}} dr$$

$$= |S^{n-1}| \begin{cases} \alpha - n + 1 < 1, \text{ for } \alpha < n \text{ convergent} \\ \alpha - n + 1 \geq 1, \text{ for } \alpha \geq n \text{ divergent} \end{cases}$$

$$= |S^{n-1}| \int_0^1 r^{n-1-\alpha} dr$$

$$= \frac{|S^{n-1}|}{n-\alpha}, \text{ only if } n - \alpha > 0.$$

Now let's look at the case of $F_2$.

$$F_2 = \int_{|x|\geq 1} \frac{dx}{|x|^\alpha} = \int_{S^{n-1}} \int_1^\infty \frac{r^{n-1}}{r^\alpha} dr d\sigma$$

$$= \int_{S^{n-1}} d\sigma \int_1^\infty \frac{r^{n-1}}{r^\alpha} dr$$

$$= |S^{n-1}| \int_1^\infty \frac{1}{r^{\alpha-n+1}} dr$$

$$= |S^{n-1}| \begin{cases} \alpha - n + 1 > 1, \text{ for } \alpha > n \text{ convergent} \\ \alpha - n + 1 \leq 1, \text{ for } \alpha \leq n \text{ divergent} \end{cases}$$

$$= |S^{n-1}| \int_1^\infty r^{n-1-\alpha} dr$$

$$= \frac{|S^{n-1}|}{\alpha-n}, \text{ only if } \alpha - n > 0.$$

Moreover, if $\alpha = n$, then $F_1 = F_2 = \infty$.

Finally, let's find the volume of the sphere for $n = 3$. Indeed, we have

$$|S^3| = \frac{\pi^{\frac{3}{2}}}{\frac{3}{2}\Gamma\left(\frac{3}{2}\right)}$$

$$= \frac{\pi^{\frac{3}{2}}}{\frac{3}{2} \cdot \frac{1}{2}\Gamma\left(\frac{1}{2}\right)}$$

$$= \frac{\pi^{\frac{3}{2}}}{\frac{3}{4}\sqrt{\pi}}$$

$$= \frac{4}{3}\pi$$

and similarly, the surface area of the unit sphere is found as follows:

$$|S^2| = \frac{2\pi^{\frac{3}{2}}}{\Gamma\left(\frac{3}{2}\right)} = \frac{2\pi^{\frac{3}{2}}}{\frac{1}{2}\Gamma\left(\frac{1}{2}\right)} = 4\frac{\pi^{\frac{3}{2}}}{\sqrt{\pi}} = 4\pi$$

where

$$\frac{1}{2}\Gamma\left(\frac{1}{2}\right) = \Gamma\left(1 + \frac{1}{2}\right) = \Gamma\left(\frac{3}{2}\right)$$

and $\Gamma\left(\frac{1}{2}\right) = \sqrt{\pi}$.

$B(x,r) = \{y \in \mathbb{R}^n : |x - y| < r\}$ denotes $x$-centred Euclidean ball with radius $r$, $B^C(x,r)$ denotes its complement and $|B(x,r)|$ is the Lebesgue measure of the ball $B(x,r)$, $|B(x,r)| = v_n r^n$, where $v_n = |B(0,1)| = |S^n| = \frac{2\pi^{\frac{n}{2}}}{n\Gamma\left(\frac{n}{2}\right)}$ is the volume of the unit sphere and $\tilde{B}(x,r) = B(x,r) \cap E$, where $E \subset \mathbb{R}^n$ is an open set.

For example,

$$|B(0,2)| = \int\limits_{|x| \leq 2} dx = \int\limits_{S^{n-1}} \int\limits_0^2 r^{n-1} dr dx' = \int\limits_{S^{n-1}} dx' \int\limits_0^2 r^{n-1} dr = |S^{n-1}|\frac{2^n}{n},$$

where $|S^{n-1}| = \frac{2\pi^{\frac{n}{2}}}{\Gamma\left(\frac{n}{2}\right)}$ is the surface area of the unit sphere. Moreover,

$$|B(x,r)| = |S^n| r^n = \frac{\pi^{\frac{n}{2}}}{\Gamma\left(1 + \frac{n}{2}\right)} r^n = \frac{1}{n}|S^{n-1}| r^n,$$

where $\Gamma(n)$ is is the gamma function and is defined by

$$\Gamma(n) = \int\limits_0^\infty x^{n-1} e^{-x} dx.$$

Finally, if $f$ is locally integrable on $\mathbb{R}^n$ and $E$ is a measurable set of finite Lebesgue measure, then we use the notation

$$f_E = \frac{1}{|E|} \int\limits_E f(x)\, dx.$$

The exponents $p'(\cdot)$ and $s'(\cdot)$ always denote the conjugate index of any exponent $1 < p(x) < \infty$ and $1 < s(x) < \infty$, that is, $\frac{1}{p'(x)} := 1 - \frac{1}{p(x)}$ and $\frac{1}{s'(x)} := 1 - \frac{1}{s(x)}$.

Given an open set $E \subset \mathbb{R}^n$ and we define $\mathcal{P}(E)$ to be the set of measurable function $p(\cdot) : E \to [1, \infty)$ such that

$$p_-(E) = \operatorname*{essinf}_{x \in E} p(x), \ p_+(E) = \operatorname*{esssup}_{x \in E} p(x)$$

and

$$1 \le p_-(E) \le p_+(E) < \infty.$$

Define $\mathcal{P}(\mathbb{R}^n)$ to be the set of $p(\cdot) : \mathbb{R}^n \to [1, \infty)$ such that $1 \le p_- := \operatorname*{essinf}_{x \in \mathbb{R}^n} p(x)$ and $p_+ := \operatorname*{esssup}_{x \in \mathbb{R}^n} p(x) < \infty$. Let $\mathcal{B}(\mathbb{R}^n)$ denote the set of $p(\cdot) \in \mathcal{P}(\mathbb{R}^n)$ which satisfies the following conditions

$$|p(x) - p(y)| \le \frac{-C}{\log(|x - y|)} \qquad |x - y| \le \frac{1}{2}, \qquad \forall x, y \in \mathbb{R}^n, \qquad (5.4)$$

and

$$|p(x) - p(y)| \le \frac{C}{\log(e + |x|)} \qquad |y| \ge |x|, \qquad \forall x, y \in \mathbb{R}^n, \qquad (5.5)$$

where $C = C(p) > 0$ does not depend on $x$, $y$, then we call that $p(\cdot)$ is log-Hölder continuous and denoted by $p(\cdot) \in \mathcal{P}^{\log}(\mathbb{R}^n)$. It is immediate that if $p(x) \in \mathcal{P}^{\log}(\mathbb{R}^n)$, then $p'(\cdot) \in \mathcal{P}^{\log}(\mathbb{R}^n)$.

For unbounded set, we say $E = \mathbb{R}^n$. If $E$ is an unbounded set, we shall use the assumption: There exists $p_\infty =: \lim_{|x| \to \infty} p(x)$. Denote by $\mathcal{P}_\infty^{\log}(\mathbb{R}^n)$ the set of all log-Hölder continuous functions at infinity. Also, we denote the subset of $\mathcal{P}^{\log}(\mathbb{R}^n)$ by $\mathcal{P}_\infty^{\log}(\mathbb{R}^n)$ with the exponents satisfying the following decay condition that there exist a number $p_\infty \in [1, \infty)$ and a constant $C_\infty$ such that

$$|p(x) - p_\infty| \le \frac{C_\infty}{\log(e + |x|)} \qquad \forall x \in \mathbb{R}^n. \qquad (5.6)$$

Note that if $E$ is an unbounded set and $p_\infty$ exists, then (5.6) is equivalent to condition (5.5). We would also like to remark that $p(x) \in \mathcal{P}_\infty^{\log}(\mathbb{R}^n)$ if and if only $p'(\cdot) \in \mathcal{P}_\infty^{\log}(\mathbb{R}^n)$ and $(p_\infty)' = p'_\infty$ (see [10] for details). We will also make use of the estimate provided by the following fact (see [10], Corollary 4.5.9).

$$\left\| \chi_{\bar{B}(x,r)} \right\|_{L^{p(\cdot)}(\mathbb{R}^n)} \lesssim r^{\psi_p(x,r)}, \qquad x \in \mathbb{R}^n, \ p(x) \in \mathcal{P}_\infty^{\log}(\mathbb{R}^n), \qquad (5.7)$$

where

$$\psi_p(x,r) = \begin{cases} \frac{n}{p(x)}, & r \le 1 \\ \frac{n}{p(\infty)}, & r > 1 \end{cases}.$$

$C$ stands for a positive constant that can change its value in each statement without explicit mention.

The notation $F \lesssim G$ means $F \le CG$, where $C$ is independent of $G$.

Let $\Omega \in L^s(S^{n-1})$ with $1 < s \le \infty$ be homogeneous function of degree 0 on $\mathbb{R}^n$ and satisfy the integral zero property over the unit sphere $S^{n-1}$. Moreover,

note that $\|\Omega\|_{L^s(S^{n-1})} := \left( \int_{S^{n-1}} |\Omega(z')|^s d\sigma(z') \right)^{\frac{1}{s}}$ and

$$\|\Omega(z-y)\|_{L^s(B(x,r))} = \left( \int_{B(x,r)} \Omega((z-y))^s dz \right)^{\frac{1}{s}}$$

$$\lesssim \left( \int_{B(x,r)} \Omega(\sigma)^s \int_0^r \rho^{n-1} d\rho d\sigma \right)^{\frac{1}{s}}$$

$$\lesssim \|\Omega\|_{L^s(S^{n-1})} r^{\frac{n}{s}}, \tag{5.8}$$

for $z \in B(x,r)$.

**(Rough $(p,q)$-admissible operator)** A rough sublinear potential type operator $T_{\Omega,\alpha}$, i.e., $|T_{\Omega,\alpha}(f+g)| \le |T_{\Omega,\alpha}(f)| + |T_{\Omega,\alpha}(g)|$ and for $\forall \lambda \in \mathbb{C}$ $|T_{\Omega,\alpha}(\lambda f)| = |\lambda||T_{\Omega,\alpha}(f)|$, will be called rough $(p,q)$-admissible operator if $T_{\Omega,\alpha}$ fullfills the following size condition:

$$\chi_{B(z,r)}(x)|T_{\Omega,\alpha}(f\chi_{\mathbb{R}^n\setminus B(z,2r)})(x)| \le C\chi_{B(z,r)}(x) \int_{\mathbb{R}^n\setminus B(z,2r)} \frac{|\Omega(x-y)|}{|x-y|^{n-\alpha}}|f(y)|dy \tag{5.9}$$

for $z \in \mathbb{R}^n$ and $r > 0$;

$T_{\Omega,\alpha}$ is $\left(L^{p(\cdot)}(\mathbb{R}^n) \to L^{q(\cdot)}(\mathbb{R}^n)\right)$-bounded. Note that rough $(p,q)$-admissible operators were introduced to study their boundedness on Morrey spaces with variable exponents in [25]. On the other hand, suppose that $0 < \alpha < n$. Then, the rough Riesz type potential operator $I_{\Omega,\alpha}$ and the corresponding rough fractional maximal operator $M_{\Omega,\alpha}$ are defined, respectively, by

$$I_{\Omega,\alpha}f(x) = \int_{\mathbb{R}^n} \frac{\Omega(x-y)}{|x-y|^{n-\alpha}} f(y)dy$$

and

$$M_{\Omega,\alpha}f(x) = \sup_{r>0} |B(x,r)|^{\frac{\alpha}{n}-1} \int_{B(x,r)} |\Omega(x-y)||f(y)|dy.$$

The operators $M_{\Omega,\alpha}$ and $I_{\Omega,\alpha}$ are also rough $(p,q)$-admissible operators. Moreover, these operators satisfy (5.9). In fact, we can easily see that when $\Omega \equiv 1$; $M_{1,\alpha} \equiv M_\alpha$ and $I_{1,\alpha} \equiv I_\alpha$ are the fractional maximal operator and the Riesz potential operator.

**(Rough $p(x)$-admissible operator)** Let $T_\Omega$ be a sublinear operator, that is $|T_\Omega (f+g)| \leq |T_\Omega (f)| + |T_\Omega (g)|$. ($p(x)$-admissible singular operators). Let a rough sublinear singular type operator $T_\Omega$ will be called rough $p(x)$-admissible singular operators, if :$T_\Omega$ satisfies the size condition of the form

$$\chi_{B(z,r)}(x)\left|T_\Omega\left(f\chi_{\mathbb{R}^n\setminus B(z,2r)}\right)(x)\right| \leq C\chi_{B(z,r)}(x) \int\limits_{\mathbb{R}^n\setminus B(z,2r)} \frac{|\Omega(x-y)|}{|x-y|^n}|f(y)|\,dy,$$

$$(5.10)$$

for $z \in \mathbb{R}^n$ and $r > 0$;

$T_\Omega$ is bounded in $L^{p(\cdot)}(\mathbb{R}^n)$. On the other hand, the rough Calderón-Zygmund type singular integral operator $T_\Omega$ in the sense of principal value Cauchy integral is defined by

$$T_\Omega f(x) = p.v. \int\limits_{\mathbb{R}^n} \frac{\Omega(x-y)}{|x-y|^n} f(y)\,dy,$$

and the rough Hardy-Littlewood maximal operator $M_\Omega$ is also defined by

$$M_\Omega f(x) = \sup_{r>0} \frac{1}{|B(x,r)|} \int\limits_{B(x,r)} |\Omega(y)|\,|f(x-y)|\,dy.$$

The operators $M_\Omega$ and $T_\Omega$ are also rough $p(x)$-admissible singular operatorss. Moreover, these operators satisfy (5.10). In fact, we can easily see that when $\Omega \equiv 1$; $M$ and $T$ are the Hardy-Littlewood maximal operator and the standard Calderón-Zygmund type singular integral operator, respectively.

Among all the function spaces, the $L^p(\mathbb{R}^n)$ spaces are the most fundamental and are always of great interest. Let $f : \mathbb{R}^n \to \mathbb{R}$ be a measurable function. The set

$$L^p(\mathbb{R}^n) = \left\{ f : \int\limits_{\mathbb{R}^n} |f(x)|^p\,dx < \infty,\ 1 \leq p \leq \infty \right\}$$

is called the class of functions integrable to the $p$-th power, and the norm of a function $f$ on $L^p(\mathbb{R}^n)$ space is defined by

$$\|f\|_{L^p(\mathbb{R}^n)} = \begin{cases} \left(\int\limits_{\mathbb{R}^n} |f(x)|^p\,dx\right)^{\frac{1}{p}}, & 1 \leq p < \infty \\ \operatorname*{esssup}\limits_{x\in\mathbb{R}^n} |f(x)|, & p = \infty \end{cases},$$

where

$$\|f\|_{L^\infty(\mathbb{R}^n)} = \operatorname*{esssup}_{x\in\mathbb{R}^n}|f(x)| = \inf\{\lambda > 0 : |\{x\in\mathbb{R}^n : |f(x)| > \lambda\}| = 0\}$$
$$= \inf\{\lambda > 0 : |f(x)| \le \lambda,\ \text{almost everywhere}\}.$$

For example, let

$$f(x) = \begin{cases} 2x, & x\in\mathbb{Q} \\ 2, & x\notin\mathbb{Q} \end{cases},$$

where $\mathbb{Q}$ is rational numbers. Then, since

$$|f(x)| \le 2$$

almost everywhere, $\operatorname{esssup}|f(x)| = 2$. Thus, $f\in L^\infty(\mathbb{R}^n)$.

$L^p(\mathbb{R}^n)$ is a linear space. Indeed, let $f,g\in L^p(\mathbb{R}^n)$ and $\theta\in\mathbb{R}$. Since

$$\int_{\mathbb{R}^n} |\theta f(x)|^p\,dx = |\theta|^p \int_{\mathbb{R}^n} |f(x)|^p\,dx < \infty$$

and

$$\int_{\mathbb{R}^n} |f(x)+g(x)|^p\,dx$$
$$\le \int_{\mathbb{R}^n} (|f(x)|+|g(x)|)^p\,dx$$
$$\le \int_{\mathbb{R}^n} [2\max(|f(x)|,|g(x)|)]^p\,dx$$
$$\le 2^p \int_{\mathbb{R}^n} (|f(x)|^p + |g(x)|^p)\,dx$$
$$= 2^p\left(\int_{\mathbb{R}^n} |f(x)|^p\,dx + \int_{\mathbb{R}^n} |g(x)|^p\,dx\right) < \infty,$$

then $L^p(\mathbb{R}^n)$ is a linear space.

**Example 10** *For every* $1 \le p < \infty$, $f(x) = |x| \notin L^p(\mathbb{R}^n)$.

**Solution**  For every $p \geq 1$,

$$\left( \int_{\mathbb{R}^n} |x|^p \, dx \right)^{1/p}$$

$$= \left( \int_{|x| \leq 1} |x|^p \, dx + \int_{|x| > 1} |x|^p \, dx \right)^{1/p}$$

$$= \left| S^{n-1} \right|^{1/p} \left( \int_0^1 r^{n-1} r^p \, dr + \int_1^\infty r^{n-1} r^p \, dr \right)^{1/p}$$

$$= \left| S^{n-1} \right|^{1/p} \left( \int_0^1 r^{n+p-1} \, dr + \int_1^\infty r^{n+p-1} \, dr \right)^{1/p}$$

$$= \left| S^{n-1} \right|^{1/p} \left[ \frac{1}{n+p} + \lim_{t \to \infty} \left( \frac{r^{n+p}}{n+p} \Big|_1^t \right) \right]^{1/p}$$

$$= \left| S^{n-1} \right|^{1/p} \left[ \frac{1}{n+p} + \lim_{t \to \infty} \left( \frac{t^{n+p}}{n+p} - \frac{1}{n+p} \right) \right]^{1/p}$$

$$= \infty, \text{ since } n + p > 0.$$

Thus, $f(x) = |x| \notin L^p(\mathbb{R}^n)$.  ∎

**Example 11**  *For every $pq + n > 0$, $f(x) = |x|^q \, \chi_{B(0,1)}(x) \in L^p(\mathbb{R}^n)$.*

**Solution**  For every $pq + n > 0$,

$$\left( \int_{\mathbb{R}^n} |x|^{pq} \, \chi_{B(0,1)}(x) \, dx \right)^{1/p} = \left( \int_{|x| \leq 1} |x|^{pq} \, dx \right)^{1/p}$$

$$= \left| S^{n-1} \right|^{1/p} \left( \int_0^1 r^{n-1} r^{pq} \, dr \right)^{1/p}$$

$$= \left| S^{n-1} \right|^{1/p} \left( \int_0^1 r^{n+pq-1} \, dr \right)^{1/p}$$

$$= \left| S^{n-1} \right|^{1/p} \left[ \frac{1}{n+pq} \right]^{1/p} < \infty.$$

Thus, $f(x) = |x|^q \, \chi_{B(0,1)}(x) \in L^p(\mathbb{R}^n)$.  ∎

**Example 12** *For every $pq + n < 0$, $f(x) = |x|^q \chi_{B^C(0,1)}(x) \in L^p(\mathbb{R}^n)$.*

**Solution**  For every $pq + n < 0$,

$$
\left( \int_{\mathbb{R}^n} |x|^{pq} \chi_{B(0,1)}(x)\, dx \right)^{1/p} = \left( \int_{|x|>1} |x|^{pq}\, dx \right)^{1/p}
$$

$$
= \left| S^{n-1} \right|^{1/p} \left( \int_1^\infty r^{n-1} r^{pq}\, dr \right)^{1/p}
$$

$$
= \left| S^{n-1} \right|^{1/p} \left( \int_1^\infty r^{n+pq-1}\, dr \right)^{1/p}
$$

$$
= \left| S^{n-1} \right|^{1/p} \left[ \lim_{t \to \infty} \left( \frac{r^{n+pq}}{n+pq} \Big|_1^t \right) \right]^{1/p}
$$

$$
= \left| S^{n-1} \right|^{1/p} \left[ \lim_{t \to \infty} \left( \frac{t^{n+pq}}{n+pq} - \frac{1}{n+pq} \right) \right]^{1/p}
$$

$$
= \left| S^{n-1} \right|^{1/p} \left( \frac{1}{n+pq} \right)^{1/p} < \infty.
$$

Thus, $f(x) = |x|^q \chi_{B^C(0,1)}(x) \in L^p(\mathbb{R}^n)$.  ∎

**Definition 5.3**  **(Weak $L^p(\mathbb{R}^n)$ Space ($L^{p,\infty}(\mathbb{R}^n)$))** Let $f : \mathbb{R}^n \to \mathbb{R}$ be measurable function and $1 \leq p < \infty$. The function $f$ is said to belong to the weak Lebesgue space on $\mathbb{R}^n$, if there is a constant $C > 0$ such that

$$
\sup_{\lambda > 0} \lambda \left| \{ x \in \mathbb{R}^n : |f(x)| > \lambda \} \right|^{1/p} \leq C < \infty.
$$

Equivalently, the weak $L^p(\mathbb{R}^n)$ is defined by

$$
L^{p,\infty}(\mathbb{R}^n) = \{ f : \|f\|_{L^{p,\infty}} < \infty \},
$$

where

$$
\|f\|_{L^{p,\infty}} = \sup_{\lambda > 0} \lambda \left| \{ x \in \mathbb{R}^n : |f(x)| > \lambda \} \right|^{1/p}
$$

denotes the seminorm of $f$ in the weak $L^p(\mathbb{R}^n)$.

It is easy to verify that for $1 \leq p < \infty$, $L^p(\mathbb{R}^n) \subset L^{p,\infty}(\mathbb{R}^n)$. That is, for every $p \geq 1$ $\|f\|_{L^{p,\infty}} \leq \|f\|_{L^p}$. Indeed,

$$\|f\|_{L^{p,\infty}} = \sup_{\lambda > 0} \lambda \left| \{x \in \mathbb{R}^n : |f(x)| > \lambda\} \right|^{1/p}$$

$$= \sup_{\lambda > 0} \lambda \left( \int_{\{x \in \mathbb{R}^n : |f(x)| > \lambda\}} dx \right)^{1/p}$$

$$= \sup_{\lambda > 0} \left( \int_{\{x \in \mathbb{R}^n : |f(x)| > \lambda\}} \lambda^p dx \right)^{1/p}$$

$$\leq \sup_{\lambda > 0} \left( \int_{\{x \in \mathbb{R}^n : |f(x)| > \lambda\}} |f(x)|^p dx \right)^{1/p}$$

$$\leq \left( \int_{\mathbb{R}^n} |f(x)|^p dx \right)^{1/p} = \|f\|_{L^p}.$$

Thus, for $1 \leq p < \infty$, $L^p(\mathbb{R}^n) \subset L^{p,\infty}(\mathbb{R}^n)$ is obtained.

**Example 13** *Let the function $f(x) = |x|^{-n/p}$ be given on $\mathbb{R}^n$. $f \notin L^p(\mathbb{R}^n)$, but $f \in L^{p,\infty}(\mathbb{R}^n)$.*

**Solution**  For $\forall p \geq 1$, since

$$\|f\|_{L^p}^p = \int_{\mathbb{R}^n} |x|^{-n} dx = \int_{|x| \leq 1} |x|^{-n} dx + \int_{|x| > 1} |x|^{-n} dx$$

$$= |S^{n-1}| \left( \int_0^1 r^{-n} r^{n-1} dr + \int_1^\infty r^{-n} r^{n-1} dr \right)$$

$$= |S^{n-1}| \left( \lim_{\varepsilon \to 0^+} \int_\varepsilon^1 \frac{dr}{r} + \lim_{t \to \infty} \int_1^t \frac{dr}{r} \right)$$

$$= |S^{n-1}| \left[ \lim_{\varepsilon \to 0^+} \left( \ln r \big|_\varepsilon^1 \right) + \lim_{t \to \infty} \left( \ln r \big|_1^t \right) \right]$$

$$= \infty,$$

then $f \notin L^p(\mathbb{R}^n)$. Let us now show that $f \in L^{p,\infty}(\mathbb{R}^n)$ for every $p \geq 1$. Since

$$
\begin{aligned}
\|f\|_{L^{p,\infty}}^p &= \sup_{\lambda>0} \lambda^p \left| \left\{ x \in \mathbb{R}^n : |x|^{-n/p} > \lambda \right\} \right| \\
&= \sup_{\lambda>0} \lambda^p \int_{\{x \in \mathbb{R}^n : |x|^{-n/p} > \lambda\}} dx \\
&= \sup_{\lambda>0} \lambda^p \int_{|x| < \lambda^{-p/n}} dx \\
&= \sup_{\lambda>0} \lambda^p \left| S^{n-1} \right| \left( \int_0^{\lambda^{-p/n}} r^{n-1} dr \right) \\
&= \sup_{\lambda>0} \lambda^p \left| S^{n-1} \right| \left( \frac{r^n}{n} \Big|_0^{\lambda^{-p/n}} \right) \\
&= \sup_{\lambda>0} \lambda^p \left| S^{n-1} \right| \frac{\lambda^{-p}}{n} \\
&= \sup_{\lambda>0} \frac{\left| S^{n-1} \right|}{n} = |S^n| = |B(0,1)| < \infty,
\end{aligned}
$$

then $f \in L^{p,\infty}(\mathbb{R}^n)$. ∎

**Example 14** *Let the function* $f(x) = \frac{1}{|x|}$ *be given on* $\mathbb{R}^n$. $f \in L^{p,\infty}(\mathbb{R}^n)$, *but* $f \notin L^p(\mathbb{R}^n)$.

**Solution**

$$
\begin{aligned}
\|f\|_{L^{p,\infty}}^p &= \sup_{\lambda>0} \lambda^p \left| \left\{ x \in \mathbb{R}^n : \frac{1}{|x|} > \lambda \right\} \right| \\
&= \sup_{\lambda>0} \lambda^p \int_{\{x \in \mathbb{R}^n : \frac{1}{|x|} > \lambda\}} dx \\
&= \sup_{\lambda>0} \lambda^p \int_{\{x \in \mathbb{R}^n : |x| < \frac{1}{\lambda}\}} dx \\
&= \sup_{\lambda>0} \lambda^p \left| B\left(0, \frac{1}{\lambda}\right) \right| \\
&= \sup_{\lambda>0} \lambda^p \int_{S^{n-1}} \int_0^{1/t} r^{n-1} dr d\sigma \\
&= \sup_{\lambda>0} \lambda^p \left| S^{n-1} \right| \int_0^{1/t} r^{n-1} dr.
\end{aligned}
$$

∎

**Example 15** *Let the function $f(x) = \frac{1}{|x|}$ be given on $\mathbb{R}$. $f \in L^{1,\infty}(\mathbb{R})$, but $f \notin L(\mathbb{R})$.*

**Solution**  Let $p = 1$. Since

$$\|f\|_{L^{1,\infty}} = \sup_{\lambda > 0} \lambda \left| \left\{ x \in \mathbb{R} : \frac{1}{|x|} > \lambda \right\} \right|$$

$$= \sup_{\lambda > 0} \lambda \int\limits_{\left\{ x \in \mathbb{R} : \frac{1}{|x|} > \lambda \right\}} dx$$

$$= \sup_{\lambda > 0} \lambda \int\limits_{\left\{ x \in \mathbb{R} : |x| < \frac{1}{\lambda} \right\}} dx$$

$$= \sup_{\lambda > 0} \lambda \int\limits_{-1/\lambda}^{1/\lambda} dx = 2 < \infty,$$

then $f \in L^{1,\infty}(\mathbb{R})$. But, $\frac{1}{|x|} \notin L(\mathbb{R})$. Indeed,

$$\int\limits_{\mathbb{R}} \frac{dx}{|x|} = \int\limits_{-\infty}^{0} -\frac{1}{x} dx + \int\limits_{0}^{\infty} \frac{1}{x} dx = \infty.$$

◼

**Example 16** *Let $f : \mathbb{R} \to \mathbb{R}$ function be given as follows:*

$$f(x) = \begin{cases} \frac{1}{\sqrt{|x|}} & , & x \in (-1,1) \setminus \{0\} \\ 0 & , & x = 0 \\ \frac{1}{x^2} & , & x \in \mathbb{R} \setminus (-1,1) \end{cases}.$$

*Then, $f \in L(\mathbb{R})$.*

**Solution**  Since

$$\int\limits_{\mathbb{R}} |f(x)| dx = \int\limits_{-\infty}^{-1} \frac{1}{x^2} dx + \int\limits_{-1}^{0} \frac{1}{\sqrt{-x}} dx + \int\limits_{0}^{1} \frac{1}{\sqrt{x}} dx + \int\limits_{1}^{\infty} \frac{1}{x^2} dx,$$

$$\int\limits_{-\infty}^{-1} \frac{1}{x^2} dx = \lim_{a \to -\infty^+} \int\limits_{a}^{-1} \frac{1}{x^2} dx = \lim_{a \to -\infty^+} \left( 1 - \frac{1}{a} \right) = 1,$$

$$\int_{-1}^{0} \frac{1}{\sqrt{-x}}\, dx = \lim_{b \to 0^-} \int_{-1}^{b} \frac{1}{\sqrt{-x}}\, dx = \lim_{b \to 0^-} \left(2 - 2\sqrt{a}\right) = 2,$$

$$\int_{0}^{1} \frac{1}{\sqrt{x}}\, dx = \lim_{c \to 0^+} \int_{c}^{1} \frac{1}{\sqrt{x}}\, dx = \lim_{c \to 0^+} \left(2\sqrt{a} - 2\right) = 2,$$

$$\int_{1}^{\infty} \frac{1}{x^2}\, dx = \lim_{d \to \infty^-} \int_{1}^{d} \frac{1}{x^2}\, dx = \lim_{d \to \infty^-} \left(1 - \frac{1}{d}\right) = 1$$

and

$$\int_{\mathbb{R}} |f(x)|\, dx = 1 + 2 + 2 + 1 = 6 < \infty,$$

then $f \in L(\mathbb{R})$.  ■

**Example 17** *Let $f : \mathbb{R} \to \mathbb{R}_+$ be measurable, and let $\varepsilon > 0$. Show that there exists $g : \mathbb{R} \to \mathbb{R}_+$ measurable such that*

**Solution**   (*i*) $\|f - g\|_{L^\infty(\mathbb{R})} \le \varepsilon$
   (*ii*) For every $\rho \in \mathbb{R}$, $|\{x \in \mathbb{R}^n : g(x) = r\}| = 0$.  ■

**Solution**   (*i*) Take $\{\rho_n\}$ such that $0 < \rho_1 < \rho_2 < \cdots < \rho_n < \cdots$, $\lim\limits_{n \to \infty} \rho_n = \infty$, $\rho_{n+1} - \rho_n < \varepsilon$, for all $n$. Set $f_1(x) = f(x) + arcctg x$, denote

$$A_n = \{x : \rho_{n-1} < f_1(x) \le \rho_n\}.$$

Set $g_1 = \sum \rho_n$.  ■

**Definition 5.4**   Let $f$ and $g$ be measurable funcitons on $\mathbb{R}^n$. The convolution $h = f * g$ is defined by

$$h(x) = (f * g)(x) = \int_{\mathbb{R}^n} f(y) g(x - y)\, dy.$$

**Remark 5.1**   Convolution is a kind of multiplication. For $f(x) = g(x) = \frac{1}{\sqrt{x}}$, since since $\int_{0}^{1} \frac{1}{\sqrt{x}}\, dx < \infty$, then $f, g \in L(0,1)$. But, since $\int_{0}^{1} \frac{1}{\sqrt{x}} \cdot \frac{1}{\sqrt{x}}\, dx = \infty$, then $f(x) \cdot g(x) \notin L(0,1)$. This means that the $L^p$ space is not closed according to the multiplication operation. Convolution is defined to eliminate this problem.  ■

**Example 18** *For $f(x) = g(x) = \frac{1}{\sqrt{x}}$, $(f * g)(x) \in L(0,1)$.*

**Solution**  Let's first find the function $(f * g)(x)$. The integral

$$(f * g)(x) = \int_0^1 f(y) g(x - y)\, dy$$

$$= \int_0^1 \frac{1}{\sqrt{y}} \cdot \frac{1}{\sqrt{x - y}}\, dy$$

$$= \int_0^1 y^{-1/2} (x - y)^{-1/2}\, dy$$

is the binomial integral.

If the above integral is solved, then

$$(f * g)(x) = \int_{\sqrt{x-1}}^{\infty} \frac{2}{t^2 + 1}\, dt$$

$$= 2. \lim_{u \to \infty} \left( \frac{\pi}{2} - \arctan \sqrt{x - 1} \right) < \infty$$

is obtained. Thus, $(f * g)(x) \in L(0,1)$.  ∎

### *Lemma 5.2*

*(Real Number Conjugate Indices Inequality) Let $1 < p < \infty$ and $p'$ be the corresponding conjugate index. Then if $A$ and $B$ are positive numbers,*

$$AB \leq \frac{A^p}{p} + \frac{B^{p'}}{p'}.$$

**Proof 5.2**  This proof is standard in any linear analysis book and so we will not repeat it here.

### *Theorem 5.1*

*(Hölder's inequality for integrals) If $p \geq 1$, and if $p$ and $p'$ are nonnegative extended real numbers such that*

$$\frac{1}{p} + \frac{1}{p'} = 1,$$

*and if $f \in L^p(\mathbb{R}^n)$, $g \in L^{p'}(\mathbb{R}^n)$, then $f \cdot g \in L(\mathbb{R}^n)$ and*

$$\int |fg|\, dx = \|fg\|_{L(\mathbb{R}^n)} \leq \|f\|_{L^p(\mathbb{R}^n)} \|g\|_{L^{p'}(\mathbb{R}^n)};$$

*if $0 < p < 1$, $f \in L^p(\mathbb{R}^n)$ and $0 < \int |g|^{p'} dx < \infty$, then*

$$\int |fg| \, dx = \|fg\|_{L(\mathbb{R}^n)} \geq \|f\|_{L^p(\mathbb{R}^n)} \|g\|_{L^{p'}(\mathbb{R}^n)}.$$

**Proof 5.3**  The result is clearly true if $f = g = 0$ *a.e.* Also, if $\int_{\mathbb{R}^n} |f(x)|^p \, dx = 0$,

then $|f(x)|^p = 0$ *a.e.* which tells us $f = 0$ *a.e.* and the result follows again. We

handle the case where $\int_{\mathbb{R}^n} |g(x)|^{p'} \, dx = 0$ in a similar fashion. Thus, we will assume

both $A^p = \int_{\mathbb{R}^n} |f(x)|^p \, dx > 0$ and $B^{p'} = \int_{\mathbb{R}^n} |g(x)|^{p'} \, dx > 0$. $E_f$ and $E_g$ be the sets where

$f$ and $g$ are not finite. By our assumption, we know the measure of these sets is 0.
Hence, for all $x$ in $E_f^C \cap E_g^C$, the values $f(x)$ and $g(x)$ are finite. We apply Lemma 5.2
to conclude

$$\frac{|f(x)|}{A} \frac{|g(x)|}{B} \leq \frac{1}{p} \frac{|f(x)|^p}{A^p} + \frac{1}{p'} \frac{|g(x)|^{p'}}{A^{p'}}$$

holds on $E_f^C \cap E_g^C$. Off of this set, we have that the left-hand side is $\infty$ and so is
the right-hand side. Hence, even on $E_f^C \cup E_g^C$, the inequality is satisfied. On the other
hand, we know that $h$ is measurable and $k \in L(\mathbb{R}^n)$ with $|h| \leq |k|$ implies $h$ is also
summable and $\int |h| \, dx \leq \int |k| \, dx$.

Thus, since the function on the right-hand side is summable, we must have the
left-hand side as a summable function too from the above information. Hence, $fg \in
L(\mathbb{R}^n)$. We then have

$$\int \frac{|f(x)|}{A} \frac{|g(x)|}{B} dx \leq \int \frac{1}{p} \frac{|f(x)|^p}{A^p} dx + \int \frac{1}{p'} \frac{|g(x)|^{p'}}{A^{p'}} dx$$

$$= \frac{1}{pA^p} \int |f(x)|^p \, dx + \frac{1}{p'A^{p'}} \int |g(x)|^{p'} \, dx$$

$$= \frac{1}{p} + \frac{1}{p'} = 1.$$

Thus, we have

$$\int |fg| \, dx \leq AB = \left( \int |f(x)|^p \, dx \right)^{1/p} + \left( |g(x)|^{p'} \, dx \right)^{1/p'},$$

which gives the required inequality. If $p = \infty$ and $f \in L^\infty(\mathbb{R}^n)$, then since $|f| \leq
\|f\|_{L^\infty(\mathbb{R}^n)}$ for almost all $x$, we have

$$|f(x) g(x)| \leq \|f\|_{L^\infty(\mathbb{R}^n)} |g(x)|. \tag{5.11}$$

Then, integrating (5.11), we obtain

$$\int |fg|\,dx = \int |f(x)g(x)|\,dx \le \|f\|_{L^\infty(\mathbb{R}^n)} \int |g(x)|\,dx = \|f\|_{L^\infty(\mathbb{R}^n)} \|g\|_{L(\mathbb{R}^n)}.$$

If $p = 1$, then $p' = \infty$, and the same argument applies.

The case $0 < p < 1$ can be shown by applying the above Hölder's inequality with $\varphi = |fg|^p$, $\psi = |g|^{-p}$ and $q = \frac{1}{p}$.

**Example 19**  *Let $f : \mathbb{R} \to \mathbb{R}_+$ be measurable, and let $0 < \rho < \infty$. Show that*

$$\frac{1}{\frac{1}{|I|}\int_I f\,dx} \le \left( \frac{1}{|I|}\int_I \frac{1}{f^\rho}\,dx \right)^{\frac{1}{\rho}}.$$

**Solution**   Set $p = 1 + \frac{1}{\rho}$. Then $p > 1$ and $\frac{1}{p} + \frac{1}{\rho p} = 1$. Thus, by Hölder's inequality

$$|I| = \int_I f^{\frac{1}{p}} f^{-\frac{1}{p}}\,dx$$

$$\le \left( \int_I f^{\frac{1}{p}\cdot p}\,dx \right)^{\frac{1}{p}} \cdot \left( \int_I f^{-\frac{1}{p}\cdot \rho p}\,dx \right)^{\frac{1}{\rho p}}$$

$$= \left( \int_I f\,dx \right)^{\frac{1}{p}} \cdot \left( \int_I f^{-\rho}\,dx \right)^{\frac{1}{\rho p}},$$

we have

$$|I|^{1+\frac{1}{\rho}} = \left( \int_I f\,dx \right) \cdot \left( \int_I f^{-\rho}\,dx \right)^{\frac{1}{\rho}},$$

which implies

$$\frac{1}{\frac{1}{|I|}\int_I f\,dx} \le \left( \frac{1}{|I|}\int_I f^{-\rho}\,dx \right)^{\frac{1}{\rho}}.$$

■

**Definition 5.5**   If T is a linear operator defined on $L^p(\mathbb{R}^n)$ $(1 \le p < \infty)$, we say T is bounded in $L^p(\mathbb{R}^n)$ or T is of strong-type $(p, p)$, if there exists a constant $C$, independent of $f$, such that $\|Tf\|_{L^p(\mathbb{R}^n)} \le C\|f\|_{L^p(\mathbb{R}^n)}$ for all $f \in L^p(\mathbb{R}^n)$. The smallest value of $C$ for which the preceding inequality holds is called the norm of T. If there exists a constant $C$ such that $|\{x \in \mathbb{R}^n : |Tf(x)| > \lambda\}| \leq C \left( \frac{\|f\|_{L^p(\mathbb{R}^n)}}{\lambda} \right)^p$, then T is said to be weak-type $(p,p)$. Suppose that T is a sublinear operator and $1 \leq p,q \leq \infty$. T is said to be of weak type $(p,q)$ if T is a bounded operator from $L^p(\mathbb{R}^n)$ to $L^{q,\infty}(\mathbb{R}^n)$. That is, there exists a constant $C > 0$ such that $|\{x \in \mathbb{R}^n : |Tf(x)| > \lambda\}| \leq C \left( \frac{\|f\|_{L^p(\mathbb{R}^n)}}{\lambda} \right)^q$ for any $\lambda > 0$ and $f \in L^p(\mathbb{R}^n)$; T is said to be of type $(p,q)$ if T is a bounded operator from $L^p(\mathbb{R}^n)$ to $L^q(\mathbb{R}^n)$. That is, there exists a constant $C > 0$ such that $\|Tf\|_{L^p(\mathbb{R}^n)} \leq C\|f\|_{L^q(\mathbb{R}^n)}$ for all $f \in L^p(\mathbb{R}^n)$.

**Definition 5.6** **(The class of locally integrable functions)** Let $f : \mathbb{R}^n \to \mathbb{R}$ be a measurable function. For every compact set $K \subset \mathbb{R}^n$, if $\int_K |f(x)|\,dx < \infty$, then a function $f$ is said to be locally integrable and denoted by $f \in L_{loc}(\mathbb{R}^n)$. It is shown as follows:

$$L_{loc}(\mathbb{R}^n) = \left\{ f : \int_K |f(x)|\,dx < \infty;\ K \subset \mathbb{R}^n,\ K \text{ compact} \right\}.$$

Similarly, for every $p \geq 1$, the space $L_{loc}^p(\mathbb{R}^n)$ is represented as

$$L_{loc}^p(\mathbb{R}^n) = \left\{ f : \left( \int_K |f(x)|^p\,dx \right)^{1/p} < \infty;\ K \subset \mathbb{R}^n,\ K \text{ compact} \right\}.$$

Morrey spaces can complement the boundedness properties of operators that Lebesgue spaces can not handle. Morrey spaces which we have been handling are called classical Morrey spaces (see [31]). Especially, because of the need for the study of the local behavior of solutions of second order elliptic partial differential equations (PDEs) and together with the now well-studied Sobolev spaces, constitude a formidable three parameter family of spaces useful for proving regularity results for solutions to various PDEs, especially for non-linear elliptic systems, in 1938, Morrey [31] introduced the classical Morrey spaces which are natural generalizations of the classical Lebesgue spaces by using a lemma. Next, his lemma was refined by Peetre [33]. Also, Morrey's lemma gave rise to the theory of function spaces; see [33]. Thius, the function spaces dealt with are called Morrey spaces. We also refer to [1] for the latest research on the theory of Morrey spaces associated with Harmonic Analysis. After Morrey [31] introduced Morrey spaces, it is realized that Morrey spaces are used for various purposes. One of the reasons is that the Morrey spaces describe local regularity more precisely than the Lebesgue spaces. As a result, one can use Morrey spaces widely not only in Harmonic Analysis, but also in PDEs. In recent years, more and more research focuses on function spaces based on Morrey spaces to fill in some gaps in the theory of Morrey type spaces. In this sense, the classical Morrey spaces (see [31]) ever were applied to study the local regularity behavior of solutions to second order elliptic partial differential equations (see [14] and [37]). For the boundedness of various classical operators in Morrey or Morrey type spaces, refer to for maximal, potential, singular integral and others, [1, 2, 7, 28, 30, 36, 41] and references therein. In [38] the vanishing Morrey space was introduced by Vitanza to characterize the regularity results for elliptic partial differential equations. Moreover, Gürbüz ([17, 18, 19, 20, 21, 22]) ever systematically obtained the boundedness of various classical operators in such these spaces. Moreover, various Morrey spaces are defined in the process of study. Also, these spaces are useful in harmonic analysis and PDEs. But, this topic exceeds the scope of this paper. Thus, we omit the details here.

In 1938, Morrey [31] observed the following fact on real analysis and applied it to partial differential equations. That is, let $n < p$. For any ball $B = B(x, r)$, if

$$\sup_{x \in \mathbb{R}^n, r > 0} |B(x, r)|^{\frac{1}{p} - 1} \left( \|f\|_{L(B)} + \|\nabla f\|_{L(B)} \right) < \infty,$$

then $f \in Lip^{1 - \frac{n}{p}}(\mathbb{R}^n)$. His main claim is that: we do not have to assume

$$\|f\|_{L^p} + \|\nabla f\|_{L^p} < \infty.$$

Based on this observation, modulo the change of notation, the Morrey space $M_q^p(\mathbb{R}^n)$ is defined by

**Definition 5.7**    Let $0 < q \le p < \infty$. For an $L_{loc}^q(\mathbb{R}^n)$-function $f$ and any ball $B = B(x, r)$, the Morrey space $M_q^p(\mathbb{R}^n)$ is the collection of all measurable functions $f$ whose Morrey space norm is

$$\|f\|_{M_q^p(\mathbb{R}^n)} = \sup_{(x, r) \in \mathbb{R}^n \times (0, \infty)} |B|^{\frac{1}{p} - \frac{1}{q}} \|f\|_{L^q(B)} < \infty.$$

If $p = q$, then $M_q^p(\mathbb{R}^n) = L^p(\mathbb{R}^n)$ is a Lebesgue space. Indeed, we write down the definition of the norm $\|f\|_{M_p^p(\mathbb{R}^n)}$

$$\|f\|_{M_p^p(\mathbb{R}^n)} = \sup_{(x, r) \in \mathbb{R}^n \times (0, \infty)} |B|^{\frac{1}{p} - \frac{1}{p}} \|f\|_{L^p(B)} = \sup_{(x, r) \in \mathbb{R}^n \times (0, \infty)} \|f\|_{L^p(B)}. \tag{5.12}$$

Thus, since $B = B(x, r) \subset \mathbb{R}^n$, we get

$$\|f\|_{M_p^p(\mathbb{R}^n)} \le \left( \int_{\mathbb{R}^n} |f(y)|^p \, dy \right)^{\frac{1}{p}} = \|f\|_{L^p(\mathbb{R}^n)}.$$

On the other hand, by the monotone convergence theorem, we obtain

$$\|f\|_{L^p(\mathbb{R}^n)} = \left( \int_{\mathbb{R}^n} |f(y)|^p \, dy \right)^{\frac{1}{p}} = \lim_{\mu \to \infty} \left( \int_{B(x,\mu)} |f(y)|^p \, dy \right)^{\frac{1}{p}}$$

$$\leq \sup_{r \in (0,\infty)} \left( \int_{B(x,r)} |f(y)|^p \, dy \right)^{\frac{1}{p}} = \|f\|_{M_p^p(\mathbb{R}^n)}. \tag{5.13}$$

By (5.12) and (5.13), we get the desired result.

Write out the definition of the norm $\|\cdot\|_{M_p^p(\mathbb{R}^n)}$. We thus notice $L^p(\mathbb{R}^n) = M_p^p(\mathbb{R}^n)$. Moreover, for all measurable functions $f$, applying Hölder's inequality, it is easy to see that

$$\|f\|_{M_{q_1}^p(\mathbb{R}^n)} \leq \|f\|_{M_{q_0}^p(\mathbb{R}^n)}, \tag{5.14}$$

where $1 \leq q_1 \leq q_0 < p$. Indeed, we write out the norms in full as follows:

$$\|f\|_{M_{q_1}^p(\mathbb{R}^n)} = \sup_{(x,r) \in \mathbb{R}^n \times (0,\infty)} |B(x,r)|^{\frac{1}{p} - \frac{1}{q_1}} \left( \int_{B(x,r)} |f(y)|^{q_1} \, dy \right)^{\frac{1}{q_1}} \tag{5.15}$$

and

$$\|f\|_{M_{q_0}^p(\mathbb{R}^n)} = \sup_{(x,r) \in \mathbb{R}^n \times (0,\infty)} |B(x,r)|^{\frac{1}{p} - \frac{1}{q_0}} \left( \int_{B(x,r)} |f(y)|^{q_0} \, dy \right)^{\frac{1}{q_0}}. \tag{5.16}$$

On the other hand, by the Hölder's inequality (for probability measure), we obtain

$$\left( \int_{B(x,r)} |f(y)|^{q_1} \, dy \right)^{\frac{1}{q_1}} \leq \left( \int_{B(x,r)} |f(y)|^{q_0} \, dy \right)^{\frac{1}{q_0}}. \tag{5.17}$$

Inserting (5.15) and (5.16) to inequality (5.17), we get (5.14).

On the other hand, the Morrey space is defined in another way. That is, based on observation above, Peetre [33] defined the space as follows.

**Definition 5.8** Let $0 \leq \lambda \leq n$ and $0 < p < \infty$. Then for $f \in L_{loc}^p(\mathbb{R}^n)$ and any ball $B = B(x,r)$, the Morrey space $\mathcal{L}^{p,\lambda}(\mathbb{R}^n)$ is defined by

$$\|f\|_{\mathcal{L}^{p,\lambda}(\mathbb{R}^n)} = \sup_{x \in \mathbb{R}^n} \sup_{r > 0} r^{-\frac{\lambda}{p}} \|f\|_{L^p(B)} \equiv \sup_B r^{-\frac{\lambda}{p}} \|f\|_{L^p(B)} < \infty.$$

[23] Recall that $0 < q \leq p < \infty$ and $0 \leq \lambda \leq n$. By checking the definitions of $M_q^p\left(\mathbb{R}^n\right)$ and $\mathcal{L}^{q,\lambda}\left(\mathbb{R}^n\right)$, it is easy to see that if we take $\lambda = \left(1 - \frac{q}{p}\right)n \in [0,n]$, then $\mathcal{L}^{q,\left(1-\frac{q}{p}\right)n}\left(\mathbb{R}^n\right) = M_q^p\left(\mathbb{R}^n\right)$. Moreover, if we choose $p = \frac{qn}{n-\lambda} \leq q$, $M_q^{\frac{qn}{n-\lambda}}\left(\mathbb{R}^n\right) = \mathcal{L}^{q,\lambda}\left(\mathbb{R}^n\right)$. Thus, we conclude that $M_q^p\left(\mathbb{R}^n\right)$ is equivalent to $\mathcal{L}^{q,\lambda}\left(\mathbb{R}^n\right)$.

**Remark 5.2**    Obviously, the Morrey space is the generalization of the Lebesgue space that can be seen from the special case $M_q^q\left(\mathbb{R}^n\right) = L^q\left(\mathbb{R}^n\right)$ with $1 \leq q < \infty$.

∎

Recently, while we try out to resolve somewhat modern problems emerging inherently such that nonlinear elasticity theory, fluid mechanics etc., it has become that classical function spaces are no longer suitable spaces. It thus became essential to introduce and analyse the diverse function spaces from diverse viewpoints. One of such spaces is the variable exponent Lebesgue space $L^{p(\cdot)}$. This space is a generalization of the classical $L^p\left(\mathbb{R}^n\right)$ space, in which the constant exponent $p$ is replaced by an exponent function $p(\cdot): \mathbb{R}^n \to (0,\infty)$, it consists of all functions $f$ such that $\int_{\mathbb{R}^n} |f(x)|^{p(x)}\,dx$. This theory got a boost in 1931 when Orlicz published his seminal paper [32]. The next major step in the investigation of variable exponent spaces was the comprehensive paper by Kováčik and Rákosník in the early 90's [27]. Since then, the theory of variable exponent spaces was applied to many fields, refer to [6, 40] for the image processing, [4] for thermorheological fluids, [35] for electrorheological fluids and [24] for the differential equations with nonstandard growth. For the nonweighted and weighted variable exponent settings, refer to [10, 11, 13].

To get a sense of the variable Lebesgue spaces, we begin with an elementary example. On the real line, consider the function $f(x) = |x|^{-\frac{1}{3}}$. The function $f$ is extremely well-behaved, but it is not in $L^p(\mathbb{R})$ for any $p$, $1 \leq p \leq \infty$. Given a single value of $p$ it either grows too quickly at the origin or decays too slowly at infinity.

To more fully describe the behavior of $f$ we must bring to bear two different $L^p$ spaces, for instance, $L^2$ and $L^4$. We can split up the domain of $f$ say that $f \in L^2([-2,2])$ and $f \in L^4(\mathbb{R} \setminus [-2,2])$. The drawback of this approach is that for more complicated functions we need to introduce additional $L^p$ spaces or lose information. If we let $g(x) = |x|^{-\frac{1}{3}} + |x-1|^{-\frac{1}{4}}$, then $g \in L^2([-2,2])$, or more generally in $L^p([-2,2])$ for any $p < 3$, but we have lost information about the local behavior of the singularity at $x = 1$. On the other hand, $g$ is no longer in $L^4(\mathbb{R} \setminus [-2,2])$: we have $g \in L^4(\mathbb{R} \setminus [-2,2])$ for $p > 4$. To capture this behavior we must subdivide the domain further, for example, writing $g \in L^2\left(\left[-1,\frac{1}{2}\right]\right)$, $g \in L^3\left(\left[\frac{1}{2},2\right]\right)$ and $g \in L^{\frac{9}{2}}(\mathbb{R} \setminus [-1,2])$.

The variable Lebesgue spaces give a different approach: we leave the domain intact and instead allow the exponent to vary. Define the "exponent function" $p(x) = \frac{9|x|+2}{2|x|+1} = \frac{9}{2} - \frac{\frac{5}{2}}{2|x|+1}$. Then $p(0) = 2$, $p(1) = \frac{11}{3}$ and $p(x) \to \frac{9}{2}$ as $|x| \to \infty$, and it is easy to see that $\int_{\mathbb{R}} |f(x)|^{p(x)} dx < \infty$ and $\int_{\mathbb{R}} |g(x)|^{p(x)} dx < \infty$.

In other words, the single variable exponent $p(\cdot)$ allows us to describe more precisely the behavior of each function. Moreover, we can distinguish between them at infinity by modifying the exponent function. For instance, if we let $q(x) = \frac{8|x|+2}{2|x|+1} = 4 - \frac{4}{2|x|+1}$, then $\int_{\mathbb{R}} |f(x)|^{q(x)} dx < \infty$ and $|g(\cdot)|^{q(\cdot)}$ is locally integrable, but $\int_{\mathbb{R}} |g(x)|^{q(x)} dx = \infty$. These examples motivate the definition of the variable Lebesgue spaces (see [9] for more details). Unfortunately the variable exponent Lebesgue spaces $L^{p(\cdot)}$ and the classical cases have some undesired properties. For example, the variable $L^{p(\cdot)}$ spaces are not translation invariant. As a consequence, the variable exponent Lebesgue spaces are not rearrangement invariant Banach spaces, and so neither good-$\lambda$ techniques nor rearrangement inequalities may be applied for a generalization of some standard results in classical Lebesgue spaces to the case of $L^{p(\cdot)}$.

Now, we first define variable exponent Lebesgue space $L^{p(\cdot)}$.

**Definition 5.9**   Let $p(\cdot) \in \mathcal{P}(\mathbb{R}^n)$. Variable exponent Lebesgue space $L^{p(\cdot)}(\mathbb{R}^n)$ consists of all Lebesgue measurable functions $f$ satisfying

$$\|f\|_{L^{p(\cdot)}(\mathbb{R}^n)} = \inf\left\{ \eta > 0 : \int_{\mathbb{R}^n} \left(\frac{|f(x)|}{\eta}\right)^{p(x)} dx \le 1 \right\} < \infty.$$

Let $f \in L_{loc}(\mathbb{R}^n)$. The Hardy-Littlewood maximal operator $M$ is defined by

$$Mf(x) = \sup_{r>0} \frac{1}{|B(x,r)|} \int_{B(x,r)} |f(y)|\, dy.$$

It is easy to know that $L^{p(\cdot)}(\mathbb{R}^n)$ becomes a Banach function space when equipped with the Luxemburg-Nakano norm above. Moreover, these spaces are referred to as variable $L^p$ spaces, since they generalize the standard $L^p$ spaces: if $p(x) = p$ is constant, then $L^{p(\cdot)}(\mathbb{R}^n)$ is isometrically isomorphic to $L^p(\mathbb{R}^n)$. It is proved that the Hardy-Littlewood maximal operator $M$ is bounded on $L^{p(\cdot)}(\mathbb{R}^n)$ as $p(\cdot) \in \mathcal{B}(\mathbb{R}^n)$ in [5]. Also, for any $p(\cdot) \in \mathcal{B}(\mathbb{R}^n)$ and $\lambda > 1$, by Jensen's inequality, we have $\lambda p(\cdot) \in \mathcal{B}(\mathbb{R}^n)$. See [[8], Remark 2.13]. We say an order pair of variable exponents function $(p(\cdot), q(\cdot)) \in \mathcal{B}^{\alpha}(\mathbb{R}^n)$, if $p(\cdot) \in \mathcal{P}(\mathbb{R}^n)$, $0 < \alpha < \frac{n}{p_+}$, $\frac{1}{p(\cdot)} - \frac{1}{q(\cdot)} = \frac{\alpha}{n}$ with $\frac{q(\cdot)(n-\alpha)}{n} \in \mathcal{B}(\mathbb{R}^n)$.

We recall that the generalized Hölder's inequality on variable exponent Lebesgue spaces

$$\left| \int_{\mathbb{R}^n} f(x)\, g(x)\, dx \right| \leq \int_{\mathbb{R}^n} |f(x)\, g(x)|\, dx \leq C \|f\|_{L^{p(\cdot)}(\mathbb{R}^n)} \|g\|_{L^{p'(\cdot)}(\mathbb{R}^n)}, \qquad (5.18)$$

with

$$C = \sup_{x \in \mathbb{R}^n} \frac{1}{p(x)} + \sup_{x \in \mathbb{R}^n} \frac{1}{p'(x)}$$

is known to hold for $p(\cdot) : \mathbb{R}^n \to [1, \infty)$, $f \in L^{p(\cdot)}(\mathbb{R}^n)$ and $g \in L^{p'(\cdot)}(\mathbb{R}^n)$, see Theorem 2.1 in [27].

In general, if $p_1(\cdot), p_2(\cdot), \ldots p_n(\cdot) \in \mathcal{P}^{\log}(\mathbb{R}^n)$ such that $\sum_{k=1}^{n} \frac{1}{p_n(\cdot)} \equiv 1$, $x \in E$. Then, for $f_i \in L^{p_i(\cdot)}(\mathbb{R}^n)$, $i = 1, 2, \ldots, n$, we have

$$\int_{\mathbb{R}^n} |f_1(x) \ldots f_n(x)|\, dx \leq C \|f_1\|_{L^{p_1(\cdot)}(\mathbb{R}^n)} \cdots \|f_n\|_{L^{p_n(\cdot)}(\mathbb{R}^n)},$$

where

$$C = \sum_{k=1}^{n} \sup_{x \in \mathbb{R}^n} \frac{1}{p_k(x)}.$$

Now, we recall the definitions of basic spaces, such as variable exponent Morrey space, variable exponent generalized Morrey space, vanishing generalized Morrey spaces variable exponent, respectively.

We define variable exponent Morrey space as follows.

**Definition 5.10** Let $\lambda(\cdot) : \mathbb{R}^n \to [0, n]$ be a measurable function and $1 \leq p(\cdot) < \infty$. Then, the variable exponent Morrey space $L^{p(\cdot),\lambda(\cdot)} \equiv L^{p(\cdot),\lambda(\cdot)}(\mathbb{R}^n)$ is defined by

$$L^{p(\cdot),\lambda(\cdot)} \equiv L^{p(\cdot),\lambda(\cdot)}(\mathbb{R}^n) = \left\{ \begin{array}{c} f \in L^{p(\cdot)}_{loc}(\mathbb{R}^n) : \\ \|f\|_{L^{p(\cdot),\lambda(\cdot)}} = \sup_{x \in \mathbb{R}^n, r > 0} r^{-\frac{\lambda(x)}{p(x)}} \left\| f \chi_{\tilde{B}(x,r)} \right\|_{L^{p(\cdot)}(\mathbb{R}^n)} < \infty \end{array} \right\}.$$

Note that $L^{p(\cdot),0}(\mathbb{R}^n) = L^{p(\cdot)}(\mathbb{R}^n)$ and $L^{p(\cdot),n}(\mathbb{R}^n) = L^{\infty}(\mathbb{R}^n)$. If $\lambda_- > n$, then $L^{p(\cdot),\lambda(\cdot)}(\mathbb{R}^n) = \{0\}$.

Now, we consider generalized Morrey spaces $L^{p(\cdot),w(\cdot)}(\mathbb{R}^n)$ with variable exponent $p(x)$ and a general function $w(x,r) : \mathbb{R}^n \times (0,\infty) \to (0,\infty)$ defining the Morrey type norm; see the definition of the spaces $L^{p(\cdot),w(\cdot)}(\mathbb{R}^n)$ below. Everywhere in the sequel the functions $w(x,r)$, $w_1(x,r)$, $w_2(x,r)$ used in the body of this talk, are non-negative measurable functions on $\mathbb{R}^n \times (0,\infty)$. We recall the definition of variable exponent generalized Morrey space in the following.

**Definition 5.11**    Let $1 \le p(x) \le p_+ < \infty$, $x \in \mathbb{R}^n$, $w(x,r) : \mathbb{R}^n \times (0,\infty) \to (0,\infty)$, where

$$\inf_{x \in \mathbb{R}^n} w(x,r) > 0 \qquad r > 0. \tag{5.19}$$

Then, the variable exponent generalized Morrey space $L^{p(\cdot),w} \equiv L^{p(\cdot),w}(\mathbb{R}^n)$ is defined by

$$L^{p(\cdot),w} \equiv L^{p(\cdot),w}(\mathbb{R}^n) = \left\{ \begin{array}{c} f \in L^{p(\cdot)}_{loc}(\mathbb{R}^n) : \\ \|f\|_{L^{p(\cdot),w}} = \sup_{x \in \mathbb{R}^n, r > 0} r^{-\psi_p(x,r)} w(x,r)^{-\frac{1}{p(x)}} \|f\|_{L^{p(\cdot)}(\tilde{B}(x,r))} < \infty \end{array} \right\}. \tag{5.20}$$

**Remark 5.3**    According to Definition 5.11, if $w(x,r) = r^{-\psi_p(x,r)+\frac{\lambda(x)}{p(x)}}$, then the variable exponent generalized Morrey space $L^{p(\cdot),w}(\mathbb{R}^n)$ is exactly the variable exponent Morrey space $L^{p(\cdot),\lambda(\cdot)}(\mathbb{R}^n)$.    ■

Then, recall that the concept of the variable exponent vanishing generalized Morrey space $VL^{p(\cdot),w}(\mathbb{R}^n)$ has been introduced in [29] in the following form:

**Definition 5.12**    Let $1 \le p(x) \le p_+ < \infty$, $x \in \mathbb{R}^n$, $w(x,r) : \mathbb{R}^n \times (0,\infty) \to (0,\infty)$ and $w(x,r)$ satisfies (5.19). Then, the variable exponent vanishing generalized Morrey space $VL^{p(\cdot),w} \equiv VL^{p(\cdot),w}(\mathbb{R}^n)$ is defined as the space of functions $f \in L^{p(\cdot),w}(\mathbb{R}^n)$ such that

$$\left\{ f \in L^{p(\cdot),w}(\mathbb{R}^n) : \lim_{r \to 0} \sup_{x \in \mathbb{R}^n} \mathfrak{M}_{p(\cdot),w}(f;x,r) = 0 \right\},$$

where

$$\mathfrak{M}_{p(\cdot),w}(f;x,r) := \frac{1}{r^{\psi_p(x,r)} w(x,r)^{\frac{1}{p(x)}}} \|f\|_{L^{p(\cdot)}(\tilde{B}(x,r))}.$$

In this work, we assume that

$$\lim_{r \to 0} \sup_{x \in \mathbb{R}^n} \frac{1}{\inf_{x \in \mathbb{R}^n} r^{\psi_p(x,r)} w(x,r)^{\frac{1}{p(x)}}} = 0 \tag{5.21}$$

and

$$\sup_{0 < r < \infty} \frac{1}{\inf_{x \in \mathbb{R}^n} r^{\psi_p(x,r)} w(x,r)^{\frac{1}{p(x)}}} > 0, \tag{5.22}$$

which make the spaces $VL^{p(\cdot),w}(\mathbb{R}^n)$ non-trivial.

On the other hand, Kováčik and Rákosník [27] established many of the basic properties of Lebesgue and Sobolev spaces. Moreover, since these authors clarified fundamental properties of the variable exponent Lebesgue and Sobolev spaces, there are many spaces studied, such as variable exponent Morrey, generalized Morrey, vanishing generalized Morrey, Herz-Morrey spaces, etc. see [3, 15, 25, 26, 29, 39]. In the last decade, when the parameters that define the operator have changed from point to point, there has been a strong interest in fractional type operators and the "variable setting" function spaces. The field called variable exponent analysis has become a fairly branched area with many interesting results obtained in the last decade such as harmonic analysis, approximation theory, operator theory, pseudo-differential operators, etc. But, the results in this paper lie in these spaces known as variable exponent Morrey type spaces on the rough fractional type operators with a variable order of harmonic analysis, which has been extensively developed during the last ten years and continues to attract attention of researchers from various fields of mathematics. Many of problems about such spaces have been solved both in the classical setting and in the Euclidean setting, including fractional upper and lower dimensions. For example, in 2008 variable exponent Morrey spaces $L^{p(\cdot),\lambda(\cdot)}$ were introduced to study the boundedness of $M$ and $I_{\alpha(\cdot)}$ in the Euclidean setting by Almeida et al. [3]. In 2010, variable exponent generalized Morrey spaces $L^{p(\cdot),w(\cdot)}(E)$ were introduced to consider the boundedness of $M$, $I_{\alpha(\cdot)}$, $T$ for bounded sets $E \subset \mathbb{R}^n$ on $L^{p(\cdot),w(\cdot)}(E)$ in [15]. In 2016, variable exponent vanishing generalized Morrey spaces $VL_{\Pi}^{p(\cdot),w(\cdot)}(E)$ were introduced to characterize the boundedness of $M$, $I_{\alpha(\cdot)}$, $T$ for bounded or unbounded sets $E$ on $VL_{\Pi}^{p(\cdot),w(\cdot)}(E)$ in [29]. Gürbüz et al. [20] showed the boundedness of rough Calderón-Zygmund type singular integral operator, rough Hardy-Littlewood maximal operator as well as the corresponding commutators in variable exponent vanishing generalized Morrey spaces on bounded sets. In 2021, Gürbüz et al. dwelled on Adams-Spanne type estimates applying some properties of variable exponent for rough Riesz type potential operator of variable order and rough fractional maximal operator of variable order on generalized variable exponent Morrey spaces, respectively in [21].

Inspired by the statements above, in this work, we continue to develop the results from [20, 21, 29]. The boundedness of rough $(p,q)$-admissible and rough $p(x)$-admissible operators in variable exponent vanishing generalized Morrey spaces on unbounded sets were considered, where the smoothness condition on $\Omega$ has been removed.

## 5.2 Basic Tools

In this section, we shall give some lemmas which will be used in the proofs of our main theorems. That is, we will consider both $I_{\Omega,\alpha}$ and $M_{\Omega,\alpha}$ are bounded from $L^{p(\cdot)}(\mathbb{R}^n)$ to $L^{q(\cdot)}(\mathbb{R}^n)$. Indeed, we have the following results.

**Lemma 5.3**

[8] *Suppose that* $p(\cdot) \in \mathcal{B}(\mathbb{R}^n)$, $0 < \alpha < \frac{n}{p_+}$, $\frac{1}{p(x)} - \frac{1}{q(x)} = \frac{\alpha}{n}$. *Then for any* $f \in L^{p(\cdot)}(\mathbb{R}^n)$, *we have*

$$\|M_\alpha f\|_{L^{q(\cdot)}(\mathbb{R}^n)} \lesssim \|f\|_{L^{p(\cdot)}(\mathbb{R}^n)}.$$

**Lemma 5.4**

[27] *Given* $p(\cdot) : \mathbb{R}^n \to [1, \infty)$ *such that* $p_+ < \infty$, *then* $\|f\|_{L^{p(\cdot)}(\mathbb{R}^n)} < C_1$ *if and only if* $|f|_{L^{p(\cdot)}(\mathbb{R}^n)} = \int_{\mathbb{R}^n} |f(y)|^{p(y)}\, dy < C_2$. *In particular, if either of the constants* $C_1$, $C_2$ *equals* 1 *we can take the other equal to* 1 *as well.*

**Lemma 5.5**

[12] *Given* $0 < \alpha < n$ *and* $0 < \varepsilon < \min(\alpha, n - \alpha)$. *Then there exists a constant* $C = C(\alpha, n, \varepsilon)$ *such that for all* $f \in L_{loc}(\mathbb{R}^n)$ *and* $x \in \mathbb{R}^n$

$$|I_{\Omega,\alpha} f(x)| \le C [M_{\Omega,\alpha-\varepsilon} f(x)]^{\frac{1}{2}} [M_{\Omega,\alpha+\varepsilon} f(x)]^{\frac{1}{2}}.$$

**Lemma 5.6**

*Let* $p(\cdot), q(\cdot) \in \mathcal{P}(\mathbb{R}^n)$, $0 < \alpha < \frac{n}{p_+}$, $\frac{1}{p(x)} - \frac{1}{q(x)} = \frac{\alpha}{n}$ *and* $(p')_+ \le s$. *For any* $f \in L^{p(\cdot)}(\mathbb{R}^n)$, *if* $\frac{q(\cdot)(n-\alpha)}{n} \in \mathcal{B}(\mathbb{R}^n)$, $\Omega \in L^s(S^{n-1})$ *with* $1 < s \le \infty$, *then we have*

$$\|M_{\Omega,\alpha} f\|_{L^{q(\cdot)}(\mathbb{R}^n)} \lesssim \|f\|_{L^{p(\cdot)}(\mathbb{R}^n)}, \tag{5.23}$$

$$\|I_{\Omega,\alpha} f\|_{L^{q(\cdot)}(\mathbb{R}^n)} \lesssim \|f\|_{L^{p(\cdot)}(\mathbb{R}^n)}. \tag{5.24}$$

**Proof 5.4**    We first prove (5.23). Let $f \in L^{p(\cdot)}(\mathbb{R}^n)$. Given $0 < \alpha < n$, $1 < s' < \frac{n}{\alpha}$, $\Omega \in L^s(S^{n-1})$ is homogeneous of degree zero on $\mathbb{R}^n$. By Hölder's inequality, we obtain

$$|M_{\Omega,\alpha}f(x)| = \sup_{r>0} |B(x,r)|^{\frac{\alpha}{n}-1} \int_{B(x,r)} |\Omega(x-y)||f(y)|dy$$

$$\lesssim \sup_{r>0} |B(x,r)|^{\frac{\alpha}{n}-1} \left( \int_{B(x,r)} |\Omega(y)|^s dy \right)^{\frac{1}{s}} \left( \int_{B(x,r)} |f(x-y)|^{s'} dy \right)^{\frac{1}{s'}}$$

$$\lesssim \|\Omega\|_{L^s(S^{n-1})} \sup_{r>0} |B(x,r)|^{\frac{\alpha s'}{n}-1} \left( \int_{B(x,r)} |f(x-y)|^{s'} dy \right)^{\frac{1}{s'}}$$

$$\lesssim \|\Omega\|_{L^s(S^{n-1})} \left( M_{\alpha s'} \left( |f|^{s'} \right)(x) \right)^{\frac{1}{s'}},$$

where $\frac{1}{s} + \frac{1}{s'} = 1$.

According to above inequality, we have

$$\int_{\mathbb{R}^n} \left( \frac{M_{\Omega,\alpha}f}{\phi} \right)^{q(x)} dx \lesssim \int_{\mathbb{R}^n} \left( \frac{M_{\alpha s'}\left( |f|^{s'} \right)(x)}{\phi} \right)^{\frac{q(x)}{s'}} dx.$$

Since $\frac{1}{p(x)} - \frac{1}{q(x)} = \frac{\alpha}{n}$, then we have

$$\frac{s'}{p(x)} - \frac{s'}{q(x)} = \frac{\alpha s'}{n}.$$

Moreover, we get

$$\left\{ \phi > 0 : C\int_{\mathbb{R}^n} \left( \frac{M_{\alpha s'}\left( |f|^{s'} \right)(x)}{\phi} \right)^{\frac{q(x)}{s'}} dx \leq 1 \right\} \subseteq \left\{ \phi > 0 : \int_{\mathbb{R}^n} \left( \frac{M_{\Omega,\alpha}f}{\phi} \right)^{q(x)} dx \leq 1 \right\}.$$

Thus,

$$\inf \left\{ \phi > 0 : \int_{\mathbb{R}^n} \left( \frac{M_{\Omega,\alpha}f}{\phi} \right)^{q(x)} dx \leq 1 \right\} \leq \inf \left\{ \phi > 0 : C\int_{\mathbb{R}^n} \left( \frac{M_{\alpha s'}\left( |f|^{s'} \right)(x)}{\phi} \right)^{\frac{q(x)}{s'}} dx \leq 1 \right\}.$$

By Lemma 5.3, we have

$$
\begin{aligned}
\|M_{\Omega,\alpha}f\|_{L^{q(\cdot)}(\mathbb{R}^n)}
&\lesssim \left\|\left(M_{\alpha s'}\left(|f|^{s'}\right)\right)^{\frac{1}{s'}}\right\|_{L^{q(\cdot)}(\mathbb{R}^n)} \\
&\lesssim \left\|M_{\alpha s'}\left(|f|^{s'}\right)\right\|_{L^{\frac{q(x)}{s'}}(\mathbb{R}^n)} \\
&\lesssim \left\||f|^{s'}\right\|_{L^{\frac{p(x)}{s'}}(\mathbb{R}^n)} \\
&\lesssim \|f\|_{L^{p(\cdot)}(\mathbb{R}^n)}.
\end{aligned}
$$

Now we pay attention to the proof of (5.24). It is enough to prove that the inequality

$$
|I_{\Omega,\alpha}f|_{L^{q(\cdot)}(\mathbb{R}^n)} = \int_{\mathbb{R}^n} |I_{\Omega,\alpha}f(x)|^{q(x)}\,dx \le C.
$$

Fix an epsilon with $0 < \varepsilon < \min(\alpha, n-\alpha)$ satisfying

$$
1 + \frac{\varepsilon}{n}q_+ < 2. \tag{5.25}
$$

Define $r(x) = \dfrac{2}{1+\frac{\varepsilon q(x)}{n}}$. Then, by (5.25) we have $r_- > 1$. Thus, by elementary algebra, for all $x \in \mathbb{R}^n$,

$$
\frac{1}{p(x)} - \frac{1}{\frac{r(x)q(x)}{2}} = \frac{\alpha-\varepsilon}{n}, \quad \frac{1}{p(x)} - \frac{1}{\frac{r'(x)q(x)}{2}} = \frac{\alpha+\varepsilon}{n}. \tag{5.26}
$$

By Lemma 5.5 and (5.18), then

$$
\begin{aligned}
\int_{\mathbb{R}^n} |I_{\Omega,\alpha}f(x)|^{q(x)}\,dx
&\lesssim \int_{\mathbb{R}^n} [M_{\Omega,\alpha-\varepsilon}f(x)]^{\frac{1}{2}}\,[M_{\Omega,\alpha+\varepsilon}f(x)]^{\frac{1}{2}}\,dx \\
&\lesssim \left\|[M_{\Omega,\alpha-\varepsilon}f(x)]^{\frac{q(x)}{2}}\right\|_{L^{r(\cdot)}(\mathbb{R}^n)} \left\|[M_{\Omega,\alpha+\varepsilon}f(x)]^{\frac{q(x)}{2}}\right\|_{L^{r'(\cdot)}(\mathbb{R}^n)}.
\end{aligned}
$$

Without loss of generality, we may assume that the infimum is taken over values of $\phi$ greater than 1. Since $\phi > 1$ and $x \in \mathbb{R}^n$, $\phi^{\frac{2}{q(x)}} \ge \phi^{\frac{2}{q_+}}$, we have

$$
\begin{aligned}
&\int_{\mathbb{R}^n} \left(\frac{[M_{\Omega,\alpha-\varepsilon}f(x)]^{\frac{q(x)}{2}}}{\phi}\right)^{r(x)} dx \\
&= \int_{\mathbb{R}^n} \left(\frac{M_{\Omega,\alpha-\varepsilon}f(x)}{\phi^{\frac{2}{q(x)}}}\right)^{\frac{r(x)q(x)}{2}} dx \\
&\lesssim \int_{\mathbb{R}^n} \left(\frac{M_{\Omega,\alpha-\varepsilon}f(x)}{\phi^{\frac{2}{q_+}}}\right)^{\frac{r(x)q(x)}{2}} dx.
\end{aligned}
$$

Therefore, by (5.23) and (5.26), we can obtain

$$\left\|\, [M_{\Omega,\alpha-\varepsilon} f(x)]^{\frac{q(x)}{2}}\,\right\|_{L^{r(\cdot)}(\mathbb{R}^n)} \leq \left\| [M_{\Omega,\alpha-\varepsilon} f(x)]\right\|_{L^{\frac{r(x)q(x)}{2}}(\mathbb{R}^n)}^{\frac{q+}{2}} \lesssim \|f\|_{L^{p(\cdot)}(\mathbb{R}^n)}^{\frac{q+}{2}} \leq C.$$

In the same way, we have

$$\int_{\mathbb{R}^n} \left( \frac{[M_{\Omega,\alpha+\varepsilon} f(x)]^{\frac{q(x)}{2}}}{\phi} \right)^{r'(x)} dx$$

$$= \int_{\mathbb{R}^n} \left( \frac{M_{\Omega,\alpha+\varepsilon} f(x)}{\phi^{\frac{2}{q(x)}}} \right)^{\frac{r'(x)q(x)}{2}} dx$$

$$\lesssim \int_{\mathbb{R}^n} \left( \frac{M_{\Omega,\alpha+\varepsilon} f(x)}{\phi^{\frac{2}{q+}}} \right)^{\frac{r'(x)q(x)}{2}} dx.$$

Thus, by (5.23) and (5.26), we have

$$\left\|\, [M_{\Omega,\alpha+\varepsilon} f(x)]^{\frac{q(x)}{2}}\,\right\|_{L^{r'(\cdot)}(\mathbb{R}^n)} \leq \left\| [M_{\Omega,\alpha+\varepsilon} f(x)]\right\|_{L^{\frac{r'(x)q(x)}{2}}(\mathbb{R}^n)}^{\frac{q+}{2}} \lesssim \|f\|_{L^{p(\cdot)}(\mathbb{R}^n)}^{\frac{q+}{2}} \leq C.$$

As a result,

$$|I_{\Omega,\alpha} f|_{L^{q(\cdot)}(\mathbb{R}^n)} = \int_{\mathbb{R}^n} |I_{\Omega,\alpha} f(x)|^{q(x)} dx \leq C,$$

which ends the proof.


### *Corollary 5.1*

*Let $\Omega \in L^s(S^{n-1})$ with $1 < s \leq \infty$ be homogeneous function of degree $0$ on $\mathbb{R}^n$, $\frac{p}{s'} \in \mathcal{B}(\mathbb{R}^n)$ and $(p')_+ \leq s$. Under the conditions of Lemma 5.6 (taking $\alpha = 0$ there), the operators $T_\Omega$ and $M_\Omega$ are $\left( L^{p(\cdot)}(\mathbb{R}^n) \to L^{p(\cdot)}(\mathbb{R}^n) \right)$-bounded, that is,*

$$\|T_\Omega f\|_{L^{p(\cdot)}(\mathbb{R}^n)} \lesssim \|f\|_{L^{p(\cdot)}(\mathbb{R}^n)}, \tag{5.27}$$

$$\|M_\Omega f\|_{L^{p(\cdot)}(\mathbb{R}^n)} \lesssim \|f\|_{L^{p(\cdot)}(\mathbb{R}^n)}$$

*are valid.*

Let's end this section with the following lemma.

**Lemma 5.7**
*Let $p(\cdot) \in \mathcal{P}_\infty^{\log}(\mathbb{R}^n)$ satisfies $1 \leq p_- \leq p(\cdot) \leq p_+ < \infty$, $0 < \alpha < n$ and additionally $\sup\limits_{x \in \mathbb{R}^n}(n + \alpha p(\infty)) < \infty$. Then, for $x \in \mathbb{R}^n$ and $r > 0$ we have*

$$\left\| |x - \cdot|^\alpha \chi_{\mathbb{R}^n \setminus \tilde{B}(x,r)} \right\|_{L^{p(\cdot)}(\mathbb{R}^n)} \lesssim r^{\alpha + \psi_p(x,r)}. \tag{5.28}$$

**Proof 5.5**  To prove Lemma 5.2, we shall follow the method in [16]. First, we introduce $\tilde{B}_k(x,r) := \tilde{B}\left(x, 2^k\right) \setminus \tilde{B}\left(x, 2^{k-1}\right)$. Then, by (5.7),

$$\left\| \chi_{\tilde{B}_k(x,2^k r)} \right\|_{L^{p(\cdot)}(\mathbb{R}^n)} \lesssim \left(2^k r\right)^{\psi_p\left(x, 2^k r\right)}.$$

Thus, we get

$$\left\| |x - \cdot|^\alpha \chi_{\mathbb{R}^n \setminus \tilde{B}(x,r)} \right\|_{L^{p(\cdot)}(\mathbb{R}^n)}$$
$$\leq \sum_{k=1}^\infty \left\| |x - \cdot|^\alpha \chi_{\tilde{B}_k(x,r)} \right\|_{L^{p(\cdot)}(\mathbb{R}^n)}$$
$$\lesssim \sum_{k=1}^\infty \left(2^k r\right)^\alpha \left\| \chi_{\tilde{B}_k(x,2^k r)} \right\|_{L^{p(\cdot)}(\mathbb{R}^n)}$$
$$\lesssim \sum_{k=1}^\infty \left(2^k r\right)^{\alpha + \psi_p\left(x, 2^k r\right)} \tag{5.29}$$
$$\lesssim \frac{\displaystyle\int_r^\infty t^{\alpha + \psi_p(x,t)} \frac{dt}{t}}{\ln 2}$$
$$= \frac{1}{\ln 2} \cdot \begin{cases} \dfrac{r^{\alpha + \frac{n}{p(x)}}}{\alpha + \frac{n}{p(x)}} & , \quad 0 < r < 1 \\[3ex] \dfrac{1}{\alpha + \frac{n}{p(x)}} + \dfrac{r^{\alpha + \frac{n}{p(\infty)}} - 1}{\alpha + \frac{n}{p(\infty)}} & , \quad r > 1 \end{cases}.$$

As a result, (5.29) implies (5.28).

## 5.3   Main Results

The main results of this work are stated as follows.

***Theorem 5.2***

***(Spanne type result)*** *(our main result) Let $x \in \mathbb{R}^n$, $0 < \alpha < n$, $T_{\Omega,\alpha}$ be defined as in (5.9) and $p(\cdot), q(\cdot) \in \mathcal{P}_\infty^{\log}(\mathbb{R}^n)$ such that $1 < p_- \leq p_+ < \frac{n}{\alpha}$, $\frac{1}{p(x)} - \frac{1}{q(x)} = \frac{\alpha}{n}$. Let $\Omega \in L^s(S^{n-1})$, $1 < s \leq \infty$ with $1 < \frac{s}{s-1} < p^- \leq p(\cdot) < \frac{n}{\alpha}$.*

*If the functions $w_1(x,r)$ and $w_2(x,r)$ satisfy (5.19) as well as the following Zygmund condition*

$$\int_r^\infty \frac{w_1^{\frac{1}{p(x)}}(x,t)}{t^{1-\alpha(x)}} dt \lesssim w_2^{\frac{1}{q(x)}}(x,r) \tag{5.30}$$

*and additionally these functions satisfy the conditions (5.21)-(5.22),*

$$c_\delta := \int_\delta^\infty \sup_{x \in \mathbb{R}^n} \frac{w_1^{\frac{1}{p(x)}}(x,t)}{t^{1-\alpha(x)}} dt < \infty, \qquad \delta > 0 \tag{5.31}$$

*then the operator $T_{\Omega,\alpha}$ is $\left(VL^{p(\cdot),w_1}(\mathbb{R}^n) \to VL^{q(\cdot),w_2}(\mathbb{R}^n)\right)$-bounded. Moreover,*

$$\|T_{\Omega,\alpha}f\|_{VL^{q(\cdot),w_2}(\mathbb{R}^n)} \lesssim \|f\|_{VL^{p(\cdot),w_1}(\mathbb{R}^n)}, \tag{5.32}$$

**Proof 5.6** For all $f \in L_{loc}^{p(\cdot)}(\mathbb{R}^n)$ and arbitrary $x_0 \in \mathbb{R}^n$, we assume that

$$\int_1^\infty \|f\|_{L^{p(\cdot)}(\tilde{B}(x_0,t))} \frac{dt}{t^{\psi_q(x_0,t)+1}} < \infty. \tag{5.33}$$

Then for $\frac{s}{s-1} < p^- \leq p(\cdot) < \frac{n}{\alpha(\cdot)}$ and any ball $\tilde{B}(x_0,r)$, since inequality

$$\|T_{\Omega,\alpha}f\|_{L^{q(\cdot)}(\tilde{B}(x_0,r))} \lesssim r^{\psi_q(x_0,r)} \int_{2r}^\infty \|f\|_{L^{p(\cdot)}(\tilde{B}(x_0,t))} \frac{dt}{t^{\psi_q(x_0,t)+1}} \tag{5.34}$$

is the key of the proof of (5.32), we first prove (5.34).

For any $x \in \mathbb{R}^n$, we write as

$$f(y) = f_1(y) + f_2(y), \tag{5.35}$$

where $f_1(y) = f(y)\chi_{\tilde{B}(x_0,2r)}(y)$, $r > 0$ such that

$$T_{\Omega,\alpha}f(y) = T_{\Omega,\alpha}f_1(y) + T_{\Omega,\alpha}f_2(y).$$

By using the sublinearity of $T_{\Omega,\alpha}$, we get

$$\|T_{\Omega,\alpha}f\|_{L^{q(\cdot)}(\tilde{B}(x_0,r))} \leq \|T_{\Omega,\alpha}f_1\|_{L^{q(\cdot)}(\tilde{B}(x_0,r))} + \|T_{\Omega,\alpha}f_2\|_{L^{q(\cdot)}(\tilde{B}(x_0,r))}.$$

Now, let us estimate $\|T_{\Omega,\alpha}f_1\|_{L^{q(\cdot)}(\tilde{B}(x_0,r))}$ and $\|T_{\Omega,\alpha}f_2\|_{L^{q(\cdot)}(\tilde{B}(x_0,r))}$, respectively.

Noting that $\frac{q(\cdot)(n-\alpha)}{n} \in \mathcal{B}(\mathbb{R}^n)$, then by (5.24), we obtain that

$$\|T_{\Omega,\alpha}f_1\|_{L^{q(\cdot)}(\tilde{B}(x_0,r))} \leq \|T_{\Omega,\alpha}f_1\|_{L^{q(\cdot)}(\mathbb{R}^n)} \lesssim \|f_1\|_{L^{p(\cdot)}(\mathbb{R}^n)} = \|f\|_{L^{p(\cdot)}(\tilde{B}(x_0,2r))}$$

$$\approx r^{\psi_q(x_0,r)} \|f\|_{L^{p(\cdot)}(\tilde{B}(x_0,2r))} \int\limits_{2r}^{\infty} \frac{dt}{t^{\psi_q(x_0,t)+1}}$$

$$\leq r^{\psi_q(x_0,r)} \int\limits_{r}^{\infty} \|f\|_{L^{p(\cdot)}(\tilde{B}(x_0,t))} \frac{dt}{t^{\psi_q(x_0,t)+1}},$$

where in the last inequality, we have used the following fact:

$$\|f\|_{L^{p(\cdot)}(\tilde{B}(x_0,2r))} \leq \|f\|_{L^{p(\cdot)}(\tilde{B}(x_0,t))}, \text{ for } t > 2r.$$

Now, turn to estimate $\|T_{\Omega,\alpha}f_2\|_{L^{q(\cdot)}(\tilde{B}(x_0,r))}$. Note that, if $x \in \tilde{B}(x_0,2r)$ and $y \in \mathbb{R}^n \setminus \tilde{B}(x_0,2r)$, then $|x_0 - y| \approx |x - y|$. Let $v(x) \in \mathcal{P}_\infty^{\log}(\mathbb{R}^n)$ such that $\frac{1}{p(\cdot)} + \frac{1}{s} + \frac{1}{v(\cdot)} = 1$, $x \in \mathbb{R}^n$. By (5.18), (5.3), (5.8) and (5.33), we obtain

$$|T_{\Omega,\alpha}f_2| \leq \int\limits_{\mathbb{R}^n \setminus \tilde{B}(x_0,2r)} \frac{|\Omega(x-y)|\,|f(y)|}{|x-y|^{n-\alpha}} dy$$

$$\lesssim \int\limits_{\mathbb{R}^n \setminus \tilde{B}(x_0,2r)} |\Omega(x-y)|\,|f(y)| \left( \int\limits_{|x-y|}^{\infty} \frac{dt}{t^{n+1+\alpha}} \right) dy$$

$$= \int\limits_{2r}^{\infty} \frac{1}{t^{n+1-\alpha}} \|f\|_{L^{p(\cdot)}(\tilde{B}(x_0,t))} \|\Omega(x-\cdot)\|_{L^s(\tilde{B}(x_0,t))} \left\| \chi_{\mathbb{R}^n \cap \tilde{B}(x_0,t)} \right\|_{L^{v(\cdot)}(\mathbb{R}^n)} dt$$

$$\lesssim \int\limits_{2r}^{\infty} \frac{1}{t^{n+1-\alpha}} \|f\|_{L^{p(\cdot)}(\tilde{B}(x_0,t))} \|\Omega(x-\cdot)\|_{L^s(\tilde{B}(x_0,t))} t^{\psi_v(x_0,t)}$$

$$\lesssim \int\limits_{2r}^{\infty} \frac{1}{t^{n+1-\alpha}} \|f\|_{L^{p(\cdot)}(\tilde{B}(x_0,t))} \|\Omega\|_{L^s(S^{n-1})} |\tilde{B}(x_0,2t)|^{\frac{1}{s}} t^{\psi_v(x_0,t)} dt$$

for $x \in \tilde{B}(x_0,r)$. Thus, we get

$$|T_{\Omega,\alpha}f_2| \lesssim \int\limits_{2r}^{\infty} \|f\|_{L^{p(\cdot)}(\tilde{B}(x_0,t))} \frac{dt}{t^{\psi_q(x_0,t)+1}} < \infty. \tag{5.36}$$

Since $\mathbb{R}^n = \bigcup\limits_{t>0} \tilde{B}(x_0,t)$, from (5.36) we get $T_{\Omega,\alpha}f_2(x)$ converges absolutely for any $f_2 \in L^{p(\cdot)}_{loc}(\mathbb{R}^n)$ and almost every $x \in \mathbb{R}^n$. Thus, we get $T_{\Omega,\alpha}f(x)$ converges absolutely for any $f \in L_{loc}^{p(\cdot)}(\mathbb{R}^n)$ with condition (5.33) and almost every $x \in \mathbb{R}^n$. Therefore, by (5.7) we get

$$\left\|T_{\Omega,\alpha}f_2\right\|_{L^{q(\cdot)}\left(\tilde{B}(x_0,r)\right)} \lesssim \left\|\chi_{\tilde{B}(x,r)}\right\|_{L^{p(\cdot)}(\mathbb{R}^n)} \int_{2r}^{\infty} \|f\|_{L^{p(\cdot)}\left(\tilde{B}(x_0,t)\right)} \frac{dt}{t^{\psi_q(x_0,t)+1}}$$

$$\lesssim r^{\psi_q(x_0,r)} \int_{2r}^{\infty} \|f\|_{L^{p(\cdot)}\left(\tilde{B}(x_0,t)\right)} \frac{dt}{t^{\psi_q(x_0,t)+1}}.$$

Combining all the estimates for $\left\|T_{\Omega,\alpha}f_1\right\|_{L^{q(\cdot)}\left(\tilde{B}(x_0,r)\right)}$ and $\left\|T_{\Omega,\alpha}f_2\right\|_{L^{q(\cdot)}\left(\tilde{B}(x_0,r)\right)}$, we get (5.34).

At last, by (5.34) and (5.30) for every $f \in L^{p(\cdot),w}(\mathbb{R}^n)$ we obtain that

$$\left\|T_{\Omega,\alpha}f\right\|_{VL^{q(\cdot),w_2}(\mathbb{R}^n)} = \sup_{x\in\mathbb{R}^n,r>0} \frac{1}{r^{\psi_q(x,r)} w_2(x,r)^{\frac{1}{q(x)}}} \left\|T_{\Omega,\alpha}f\right\|_{L^{q(\cdot)}\left(\tilde{B}(x,r)\right)}$$

$$\lesssim \sup_{x\in\mathbb{R}^n,r>0} \frac{1}{w_2(x,r)^{\frac{1}{q(x)}}} \int_{2r}^{\infty} \|f\|_{L^{p(\cdot)}\left(\tilde{B}(x,t)\right)} \frac{dt}{t^{\psi_q(x,t)+1}}$$

$$\lesssim \|f\|_{VL^{p(\cdot),w_1}(\mathbb{R}^n)} \sup_{x\in\mathbb{R}^n,r>0} \frac{1}{w_2(x,r)^{\frac{1}{q(x)}}} \int_{r}^{\infty} \frac{w_1^{\frac{1}{p(x)}}(x,t)}{t^{1-\alpha}} dt$$

$$\lesssim \|f\|_{VL^{p(\cdot),w_1}(\mathbb{R}^n)}.$$

At last, we need to prove that

$$\lim_{r\to 0} \sup_{x\in\mathbb{R}^n} \frac{1}{r^{\psi_q(x,r)} w_2(x,r)^{\frac{1}{q(x)}}} \left\|T_{\Omega,\alpha}f\right\|_{L^{q(\cdot)}\left(\tilde{B}(x,r)\right)} \lesssim \lim_{r\to 0} \sup_{x\in\mathbb{R}^n} \frac{1}{r^{\psi_p(x,r)} w_1(x,r)^{\frac{1}{p(x)}}} \|f\|_{L^{p(\cdot)}\left(\tilde{B}(x,r)\right)} = 0.$$

But, because the proof of above inequality is similar to Theorem 3.1. in [18], we omit the details, which completes the proof of (5.32).

As a result, the proof of Theorem 5.2 is completed.

### Corollary 5.2

*Obviously, under the conditions of Theorem 5.2, if the operators $M_{\Omega,\alpha}$ and $I_{\Omega,\alpha}$ are $\left(L^{p(\cdot)}(\mathbb{R}^n) \to L^{q(\cdot)}(\mathbb{R}^n)\right)$-bounded and satisfy (5.9), the result in Theorem 5.2 still holds.*

For $\alpha = 0$ in Theorem 5.2, we get the following new result:

### Corollary 5.3

*Let $\Omega$, $p(x)$ be the same as in Theorem 5.2. Let $T_\Omega$ be defined as in (5.10). For all $f \in L_{loc}^{p(\cdot)}(\mathbb{R}^n)$ and arbitrary $x_0 \in \mathbb{R}^n$, we assume that*

$$\int_1^\infty \|f\|_{L^{p(\cdot)}(\tilde{B}(x_0,t))} \frac{dt}{t^{\psi_p(x_0,t)+1}} < \infty.$$

*Then, for $\frac{s}{s-1} < p^- \leq p(\cdot) \leq p_+ < \infty$, the following pointwise estimate*

$$\|T_\Omega f\|_{L^{p(\cdot)}(\tilde{B}(x_0,r))} \lesssim r^{\psi_p(x_0,r)} \int_{2r}^\infty \|f\|_{L^{p(\cdot)}(\tilde{B}(x_0,t))} \frac{dt}{t^{\psi_p(x_0,t)+1}}$$

*holds for any ball $\tilde{B}(x_0,r)$ and for all $f \in L_{loc}^{p(\cdot)}(\mathbb{R}^n)$.*
*If the function $w(x,r)$ satisfies (5.19) as well as the following Zygmund condition*

$$\int_r^\infty \frac{w^{\frac{1}{p(x)}}(x,t)}{t} dt \lesssim w^{\frac{1}{p(x)}}(x,r)$$

*and additionally this function satisfies the conditions (5.21)-(5.22),*

$$c_\delta := \int_\delta^\infty \sup_{x \in \mathbb{R}^n} \frac{w^{\frac{1}{p(x)}}(x,t)}{t} dt < \infty, \qquad \delta > 0$$

*then the operator $T_\Omega$ is bounded on $VL^{p(\cdot),w}(\mathbb{R}^n)$. Moreover,*

$$\|T_\Omega f\|_{VL^{p(\cdot),w}(\mathbb{R}^n)} \lesssim \|f\|_{VL^{p(\cdot),w}(\mathbb{R}^n)}.$$

### Corollary 5.4

*Obviously, under the conditions of Corollary 5.3, since the operators $M_\Omega$ and $T_\Omega$ satisfy (5.10), the result in Corollary 5.3 still holds.*

### Theorem 5.3

*(**Adams type result**) (our main result) Let $\Omega$, $p(x)$, $q(x)$, $\alpha$ be the same as in Theorem 5.2. Let $T_{\Omega,\alpha}$ be defined as in (5.9). Then, for $\frac{s}{s-1} < p^- \leq p(\cdot) < \frac{n}{\alpha}$, the following pointwise estimate*

$$|T_{\Omega,\alpha} f(x)| \lesssim r^\alpha M_\Omega f(x) + \int_r^\infty \|f\|_{L^{p(\cdot)}(\tilde{B}(x,t))} \frac{dt}{t^{\psi_p(x,t)-\alpha+1}} \tag{5.37}$$

*holds for any ball $\tilde{B}(x,r)$ and for all $f \in L_{loc}^{p(\cdot)}(\mathbb{R}^n)$.*

*The function $w(x,t)$ satisfies (5.19), (5.21)-(5.22) as well as the following conditions:*

$$\int_r^\infty \frac{w^{\frac{1}{p(x)}}(x,t)}{t}\,dt \lesssim w^{\frac{1}{p(x)}}(x,r),$$

$$\int_r^\infty \frac{w^{\frac{1}{p(x)}}(x,t)}{t^{1-\alpha}}\,dt \lesssim r^{-\frac{\alpha p(x)}{q(x)-p(x)}}, \tag{5.38}$$

*where $p(x) < q(x)$. Then the operator $T_{\Omega,\alpha}$ is $\left(VL_{\Pi}^{p(\cdot),w^{\frac{1}{p(\cdot)}}}(\mathbb{R}^n) \to VL_{\Pi}^{q(\cdot),w^{\frac{1}{q(\cdot)}}}(\mathbb{R}^n)\right)$-bounded. Moreover,*

$$\|T_{\Omega,\alpha}f\|_{VL^{q(\cdot),w^{\frac{1}{q(\cdot)}}}(\mathbb{R}^n)} \lesssim \|f\|_{VL^{p(\cdot),w^{\frac{1}{p(\cdot)}}}(\mathbb{R}^n)}. \tag{5.39}$$

**Proof 5.7**   As in the proof of Theorem 5.2, we represent the function $f$ in the form (5.35) and have

$$T_{\Omega,\alpha}f(x) = T_{\Omega,\alpha}f_1(x) + T_{\Omega,\alpha}f_2(x).$$

For $T_{\Omega,\alpha}f_1(x)$, we must prove the following pointwise estimate:

$$|T_{\Omega,\alpha}f_1(x)| \lesssim r^\alpha M_\Omega f(x). \tag{5.40}$$

For this, we first have to prove the following:

$$|T_{\Omega,\alpha}f_1(x)| := \left| \int_{\|x-y\|<r} \frac{\Omega(x-y)}{|x-y|^{n-\alpha}} f(y)dy \right| \lesssim \frac{2^n r^\alpha}{2^\alpha - 1} M_\Omega f(x). \tag{5.41}$$

Indeed, for $f(x) \geq 0$ we have

$$T_{\Omega,\alpha}f_1(x) = \sum_{j=0}^\infty \int_{2^{-j-1}r \leq |x-y| < 2^{-j}r} \frac{\Omega(x-y)}{|x-y|^{n-\alpha}} f(y)dy$$

$$\leq \sum_{j=0}^\infty \frac{1}{(2^{-j-1}r)^{n-\alpha}} \int_{|x-y|<2^{-j}r} \Omega(x-y)f(y)dy$$

$$\leq 2^{n-\alpha} M_\Omega f(x) \sum_{j=0}^\infty \frac{|B(x,2^{-j}r)|}{(2^{-j}r)^{n-\alpha}}.$$

Hence by $|B(x,2^{-j}r)| \lesssim (2^{-j}r)^n$, we obtain

$$T_{\Omega,\alpha}f_1(x) \lesssim 2^{n-\alpha} r^\alpha M_\Omega f(x) \sum_{j=0}^\infty (2^{-j\alpha}),$$

which gives the estimate (5.41). Then by (5.41), we get (5.40).

For $T_{\Omega,\alpha}f_2(x)$, similar to the proof of (5.36), by (5.3), (5.18) and (5.8), we get

$$|T_{\Omega,\alpha}f_2(x)| \lesssim \int_r^\infty \|f\|_{L^{p(\cdot)}(\tilde{B}(x,t))} \frac{dt}{t^{\psi_p(x,t)-\alpha+1}} \tag{5.42}$$

and by (5.40) and (5.42) complete the proof of (5.37).

At last, by (5.37) and (5.38), we obtain

$$|T_{\Omega,\alpha}f(x)| \lesssim r^\alpha M_\Omega f(x) + r^{-\frac{\alpha p(x)}{q(x)-p(x)}} \|f\|_{VL^{p(\cdot),w}(\mathbb{R}^n)}.$$

Then, choosing $r = \left(\dfrac{\|f\|_{VL^{p(\cdot),w}(\mathbb{R}^n)}}{M_\Omega f(x)}\right)^{\frac{q(x)-p(x)}{\alpha p(x)}}$ for every $x \in \mathbb{R}^n$ supposing that $f$ is not equal 0, thus we have

$$|T_{\Omega,\alpha}f(x)| \lesssim (M_\Omega f(x))^{\frac{p(x)}{q(x)}} \|f\|_{VL^{p(\cdot),w}(\mathbb{R}^n)}^{1-\frac{p(x)}{q(x)}}. \tag{5.43}$$

Finally, by (5.43) and Corollary 5.4 for every $f \in L^{p(\cdot),w}(\mathbb{R}^n)$ we obtain that

$$\begin{aligned}
\|T_{\Omega,\alpha}f\|_{VL^{q(\cdot),w^{\frac{1}{q(\cdot)}}}(\mathbb{R}^n)} &= \sup_{x\in\mathbb{R}^n,r>0} \frac{1}{r^{\psi_q(x,r)}w(x,r)^{\frac{1}{q(x)}}} \|T_{\Omega,\alpha}f\|_{L^{q(\cdot)}(\tilde{B}(x,r))} \\
&\lesssim \|f\|_{VL^{p(\cdot),w}(\mathbb{R}^n)}^{1-\frac{p(x)}{q(x)}} \sup_{x\in\mathbb{R}^n,r>0} \frac{1}{r^{\psi_q(x,r)}w(x,r)^{\frac{1}{q(x)}}} \|M_\Omega f\|_{L^{p(\cdot)}(\tilde{B}(x,r))}^{\frac{p(x)}{q(x)}} \\
&\lesssim \|f\|_{VL^{p(\cdot),w}(\mathbb{R}^n)}^{1-\frac{p(x)}{q(x)}} \left(\sup_{x\in\mathbb{R}^n,r>0} \frac{1}{r^{\psi_p(x,r)}w(x,r)^{\frac{1}{p(x)}}} \|M_\Omega f\|_{L^{p(\cdot)}(\tilde{B}(x,r))}\right)^{\frac{p(x)}{q(x)}} \\
&\lesssim \|f\|_{VL^{p(\cdot),w}(\mathbb{R}^n)}^{1-\frac{p(x)}{q(x)}} \|M_\Omega f\|_{VL^{p(\cdot),w^{\frac{1}{p(\cdot)}}}(\mathbb{R}^n)}^{\frac{p(x)}{q(x)}} \\
&\lesssim \|f\|_{VL^{p(\cdot),w^{\frac{1}{p(\cdot)}}}(\mathbb{R}^n)}
\end{aligned}$$

if $p(x) < q(x)$.

Hence, we need to prove that

$$\lim_{r\to 0}\sup_{x\in\mathbb{R}^n} \frac{1}{r^{\psi_q(x,r)}w_2(x,r)^{\frac{1}{q(x)}}} \|T_{\Omega,\alpha}f\|_{L^{q(\cdot)}(\tilde{B}(x,r))} \lesssim \lim_{r\to 0}\sup_{x\in\mathbb{R}^n} \frac{1}{r^{\psi_p(x,r)}w_1(x,r)^{\frac{1}{p(x)}}} \|f\|_{L^{p(\cdot)}(\tilde{B}(x,r))} = 0.$$

But, because the proof of above inequality is similar to Theorem 3.1. in [18], we omit the details, which completes the proof of (5.39).

As a result, the proof of Theorem 5.3 is completed.

***Corollary 5.5***

*Obviously, under the conditions of Theorem 5.3, if the operators $M_{\Omega,\alpha}$ and $I_{\Omega,\alpha}$ are $\left(L^{p(\cdot)}\left(\mathbb{R}^n\right) \to L^{q(\cdot)}\left(\mathbb{R}^n\right)\right)$-bounded and satisfy (5.9), the result in Theorem 5.3 still holds.*

**Definition 5.13**    Let $\lambda(x)$ be a measurable function on $\mathbb{R}^n$ with values in $[0,n]$.

Then, the variable exponent vanishing Morrey space $VL^{p(\cdot),\lambda(\cdot)} \equiv VL^{p(\cdot),\lambda(\cdot)}\left(\mathbb{R}^n\right)$ is defined by

$$VL^{p(\cdot),\lambda(\cdot)} \equiv VL^{p(\cdot),\lambda(\cdot)}\left(\mathbb{R}^n\right) = \left\{ \begin{array}{c} f \in L^{p(\cdot),\lambda(\cdot)}\left(\mathbb{R}^n\right): \\ \|f\|_{VL^{p(\cdot),\lambda(\cdot)}} = \lim_{r\to 0} \sup_{\substack{x\in\mathbb{R}^n \\ 0<t<r}} t^{-\frac{\lambda(x)}{p(x)}} \left\|f\chi_{\tilde{B}(x,t)}\right\|_{L^{p(\cdot)}\left(\mathbb{R}^n\right)} = 0 \end{array} \right\}.$$

***Theorem 5.4***

*(**Spanne type result**) Let $\lambda(x)$ be a measurable function on $\mathbb{R}^n$ with values in $[0,n]$. Let $\Omega$, $p(x)$, $q(x)$ and $\alpha$ be the same as in Theorem 5.2 such that $1 < p_- \leq p(\cdot) < \frac{n-\lambda(\cdot)}{\alpha}$, $\frac{1}{p(x)} - \frac{1}{q(x)} = \frac{\alpha}{n}$ and $\frac{\lambda(\cdot)}{p(\cdot)} = \frac{\mu(\cdot)}{q(\cdot)}$. Let $T_{\Omega,\alpha}$ be defined as in (5.9). If the functions $w_1(x,r)$ and $w_2(x,r)$ satisfy (5.19) as well as the following Zygmund condition*

$$\int_r^\infty w_1^{\frac{1}{p(x)}}(x,t)\frac{dt}{t^{\Psi_q(x,r)+1}} \lesssim \frac{w_2^{\frac{1}{q(x)}}(x,r)}{r^{\Psi_q(x,r)}} \tag{5.44}$$

*and these functions satisfy the conditions (5.21)-(5.22) and*

$$c_\delta := \int_\delta^\infty \sup_{x\in\mathbb{R}^n} w_1^{\frac{1}{p(x)}}(x,t)\frac{dt}{t^{\Psi_q(x,r)+1}} < \infty, \qquad \delta > 0$$

*then for $\frac{s}{s-1} < p^-$ or $(p')_+ \leq s$, the operator $T_{\Omega,\alpha}$ is $\left(VL^{p(\cdot),\lambda(\cdot)}\left(\mathbb{R}^n\right) \to VL^{q(\cdot),\mu(\cdot)}\left(\mathbb{R}^n\right)\right)$-bounded. Moreover,*

$$\|T_{\Omega,\alpha}f\|_{VL^{q(\cdot),\mu(\cdot)}\left(\mathbb{R}^n\right)} \lesssim \|f\|_{VL^{p(\cdot),\lambda(\cdot)}\left(\mathbb{R}^n\right)}. \tag{5.45}$$

**Proof 5.8**   Let $\frac{s}{s-1} < p^-$ or $(p')_+ \le s$. By using $w_1(x,r) = r^{\frac{\lambda(\cdot)}{p(\cdot)}}$ and $w_2(x,r) = r^{\frac{\mu(\cdot)}{q(\cdot)}}$ in the proof of (5.34) and condition (5.44), it follows that

$$
\begin{aligned}
\|T_{\Omega,\alpha}f\|_{VL^{q(\cdot),\mu(\cdot)}(\mathbb{R}^n)} &= \sup_{x\in\mathbb{R}^n,r>0} r^{-\frac{\mu(x)}{q(x)}} \|T_{\Omega,\alpha}f\|_{L^{q(\cdot)}(\tilde{B}(x_0,r))} \\
&\lesssim \sup_{x\in\mathbb{R}^n,r>0} r^{-\frac{\mu(x)}{q(x)}} r^{\Psi_q(x_0,r)} \int_{2r}^{\infty} \|f\|_{L^{p(\cdot)}(\tilde{B}(x_0,t))} \frac{1}{t^{\Psi_q(x_0,t)+1}} \frac{t^{\frac{\lambda(x)}{p(x)}}}{t^{\frac{\lambda(x)}{p(x)}}} dt \\
&\lesssim \|f\|_{VL^{p(\cdot),\lambda(\cdot)}(\mathbb{R}^n)} \sup_{x\in\mathbb{R}^n,r>0} r^{-\frac{\mu(x)}{q(x)}} r^{\Psi_q(x_0,r)} \int_{2r}^{\infty} \frac{t^{\frac{\lambda(x)}{p(x)}}}{t^{\Psi_q(x_0,t)+1}} dt \\
&\lesssim \|f\|_{VL^{p(\cdot),\lambda(\cdot)}(\mathbb{R}^n)}.
\end{aligned}
$$

Thus, we need to prove that

$$
\lim_{\substack{r\to 0}} \sup_{\substack{x\in\mathbb{R}^n \\ 0<t<r}} t^{-\frac{\mu(x)}{q(x)}} \left\|T_{\Omega,\alpha}f\chi_{\tilde{B}(x,t)}\right\|_{L^{q(\cdot)}(\mathbb{R}^n)} = 0 \lesssim \lim_{\substack{r\to 0}} \sup_{\substack{x\in\mathbb{R}^n \\ 0<t<r}} t^{-\frac{\lambda(x)}{p(x)}} \left\|f\chi_{\tilde{B}(x,t)}\right\|_{L^{p(\cdot)}(\mathbb{R}^n)} = 0.
$$

But, because the proof of above inequality is similar to Theorem 3.1. in [18] (by taking $w(x,r) = r^{-\Psi_p(x,r)+\frac{\lambda(x)}{p(x)}}$ in the definition of the variable exponent generalized Morrey space $L^{p(\cdot),w}(\mathbb{R}^n)$), we omit the details, which completes the proof of (5.45).

As a result, the proof of Theorem 5.4 is completed.

### *Corollary 5.6*
*Obviously, under the conditions of Theorem 5.4, since the operators $M_{\Omega,\alpha}$ and $I_{\Omega,\alpha}$ satisfy (5.9), then the result in Theorem 5.4 still holds.*

For $\alpha = 0$ in Theorem 5.4, we get the following new result:

### *Corollary 5.7*
*Let $\Omega$, $p(x)$ be the same as in Theorem 5.2. Let $T_\Omega$ be defined as in (5.10). For all $f \in L^{p(\cdot)}_{loc}(\mathbb{R}^n)$ and arbitrary $x_0 \in \mathbb{R}^n$, we assume that*

$$
\int_1^{\infty} \|f\|_{L^{p(\cdot)}(\tilde{B}(x_0,t))} \frac{dt}{t^{\Psi_p(x_0,t)+1}} < \infty.
$$

*Then, for $\frac{s}{s-1} < p^- \le p(\cdot) \le p_+ < \infty$, the following pointwise estimate*

$$
\|T_\Omega f\|_{L^{p(\cdot)}(\tilde{B}(x_0,r))} \lesssim r^{\Psi_p(x_0,r)} \int_{2r}^{\infty} \|f\|_{L^{p(\cdot)}(\tilde{B}(x_0,t))} \frac{dt}{t^{\Psi_p(x_0,t)+1}}
$$

*holds for any ball $\tilde{B}(x_0,r)$ and for all $f \in L^{p(\cdot)}_{loc}(\mathbb{R}^n)$.*

*If the functions $w_1(x,r)$ and $w_2(x,r)$ satisfy (5.19) as well as the following Zygmund condition*

$$\int_r^\infty w_1^{\frac{1}{p(x)}}(x,t)\frac{dt}{t^{\Psi_p(x,r)+1}} \lesssim \frac{w_2^{\frac{1}{p(x)}}(x,r)}{r^{\Psi_p(x,r)}}$$

*and these functions satisfy the conditions (5.21)–(5.22) and*

$$c_\delta := \int_\delta^\infty \sup_{x\in\mathbb{R}^n} w_1^{\frac{1}{p(x)}}(x,t)\frac{dt}{t^{\Psi_p(x,r)+1}} < \infty, \qquad \delta > 0$$

*then for $\frac{s}{s-1} < p^-$ or $(p')_+ \leq s$ the operator $T_\Omega$ is bounded on $VL^{p(\cdot),\lambda(\cdot)}(\mathbb{R}^n)$. Moreover,*

$$\|T_\Omega f\|_{VL^{p(\cdot),\lambda(\cdot)}(\mathbb{R}^n)} \lesssim \|f\|_{VL^{p(\cdot),\lambda(\cdot)}(\mathbb{R}^n)}.$$

### Corollary 5.8

*Obviously, under the conditions of Corollary 5.7, since the operators $M_\Omega$ and $T_\Omega$ satisfy (5.10), the result in Corollary 5.7 still holds.*

### Lemma 5.8

*Let $\Omega$, $p(x),q(x)$, $\alpha$ be the same as in Theorem 5.2. Define $q(x)$ by $\frac{1}{q(x)} = \frac{1}{p(x)} - \frac{\alpha}{n-\lambda(x)}$. Let $T_{\Omega,\alpha}$ be defined as in (5.9). Let also the following conditions hold:*

$$\lambda(x) \geq 0, \qquad \operatorname*{esssup}_{x\in\mathbb{R}^n}[\lambda(x) + \alpha p(x)] < n.$$

*Then for $\frac{s}{s-1} < p^-$ or $(p_-)' \leq s$, the operator $I_{\Omega,\alpha}$ is $\left(VL^{p(\cdot),\lambda(\cdot)}(\mathbb{R}^n) \to VL^{q(\cdot),\lambda(\cdot)}(\mathbb{R}^n)\right)$-bounded. Moreover,*

$$\|I_{\Omega,\alpha}f\|_{VL^{q(\cdot),\lambda(\cdot)}(\mathbb{R}^n)} \lesssim \|f\|_{VL^{p(\cdot),\lambda(\cdot)}(\mathbb{R}^n)}.$$

**Proof 5.9**  By the embedding property in Lemma 7 in [3], we only need to prove that the operator $I_{\Omega,\alpha}$ is bounded in $L^{p(\cdot),\lambda(\cdot)}(\mathbb{R}^n)$.

**Hedberg's trick:**

$$I_{\Omega,\alpha}f(x) = \int_{B(x,2r)} \frac{\Omega(x-y)}{|x-y|^{n-\alpha}}f(y)dy + \int_{B^C(x,2r)} \frac{\Omega(x-y)}{|x-y|^{n-\alpha}}f(y)dy$$

$$= \mathcal{F}(x,r) + \mathcal{G}(x,r). \tag{5.46}$$

We may assume that $\|f\|_{L^{p(\cdot),\lambda(\cdot)}(E)} \leq 1$. For $\mathcal{F}(x,r)$, by (5.41):

$$|\mathcal{F}(x,r)| \lesssim r^{\alpha(x)}M_\Omega f(x).$$

For $\mathcal{G}(x,r)$, from Lemma 5.7 and the procedure of Theorem 3 in [3], we may show that

$$|\mathcal{G}(x,r)| \lesssim r^{\alpha(x) - \frac{n-\lambda(x)}{p(x)}}.$$

Then, from (5.46) we get

$$I_{\Omega,\alpha(\cdot)}f(x) \lesssim \left[ r^{\alpha(x)} M_\Omega f(x) + r^{\alpha(x) - \frac{n-\lambda(x)}{p(x)}} \right]. \tag{5.47}$$

As usual in Hedberg approach, we choose

$$r = [M_\Omega f(x)]^{-\frac{p(x)}{n-\lambda(x)}}.$$

Substituting this into the (5.47), we get

$$\left| I_{\Omega,\alpha(\cdot)}f(x) \right| \lesssim (M_\Omega f(x))^{\frac{p(x)}{q(x)}},$$

here we need the (5.8). Therefore, by Theorem 5.1 in [34] we know that

$$\int_{\tilde{B}(x,r)} \left| I_{\Omega,\alpha(\cdot)}f(y) \right|^{q(y)} dy \lesssim \int_{\tilde{B}(x,r)} |M_\Omega f(y)|^{p(y)} dy \lesssim r^{\lambda(x)},$$

which completes the proof of Lemma 5.8.


### Theorem 5.5
*(Adams type result) Let $\Omega$, $p(x)$, $q(x)$, $\alpha$ be the same as in Theorem 5.2. Let $T_{\Omega,\alpha}$ be defined as in (5.9). Let also the following conditions hold:*

$$\lambda(x) \geq 0, \qquad \underset{x \in \mathbb{R}^n}{\mathrm{esssup}}\, [\lambda(x) + \alpha p(x)] < n.$$

*Define $q(x)$, $\mu(x)$ by $\frac{1}{q(\cdot)} = \frac{1}{p(\cdot)} - \frac{\alpha}{n}$, $\frac{n-\mu(\cdot)}{q(\cdot)} = \frac{n-\lambda(\cdot)}{p(\cdot)} - \alpha$, respectively. Then, for $\frac{s}{s-1} < p^-$ or $(p_-)' \leq s$, the operator $T_{\Omega,\alpha}$ is $\left( L^{p(\cdot),\lambda(\cdot)}(\mathbb{R}^n) \to L^{q(\cdot),\mu(\cdot)}(\mathbb{R}^n) \right)$-bounded. Moreover,*

$$\|T_{\Omega,\alpha}f\|_{L^{q(\cdot),\mu(\cdot)}(\mathbb{R}^n)} \lesssim \|f\|_{L^{p(\cdot),\lambda(\cdot)}(\mathbb{R}^n)}, \tag{5.48}$$

*where*

$$1 \leq q(\cdot) \leq \frac{p(\cdot)(n-\lambda(\cdot))}{n-\lambda(\cdot)-\alpha p(\cdot)}.$$

**Proof 5.10**   Since

$$\frac{p(\cdot)(n-\lambda(\cdot))}{n-\lambda(\cdot)-\alpha p(\cdot)} < \frac{np(\cdot)}{n-\alpha p(\cdot)},$$

from Lemma 5.8 and (5.24) we obtain

$$\begin{aligned}
\|T_{\Omega,\alpha}f\|_{L^{q(\cdot),\mu(\cdot)}(\mathbb{R}^n)} &= \sup_{x\in\mathbb{R}^n, r>0} r^{-\frac{\mu(x)}{q(x)}} \left\|T_{\Omega,\alpha}f\chi_{\tilde{B}(x,r)}\right\|_{L^{q(\cdot)}(\mathbb{R}^n)} \\
&\lesssim \sup_{x\in\mathbb{R}^n, r>0} r^{-\frac{\lambda(x)}{p(x)}} \left\|f\chi_{\tilde{B}(x,r)}\right\|_{L^{p(\cdot)}(\mathbb{R}^n)} \\
&= \|f\|_{L^{p(\cdot),\lambda(\cdot)}(\mathbb{R}^n)},
\end{aligned}$$

which completes the proof of (5.48). Clearly, Theorem 5.5 holds.

### *Corollary 5.9*

*Obviously, under the conditions of Theorem 5.5, if the operators $M_{\Omega,\alpha}$ and $I_{\Omega,\alpha}$ are*

$\left(L^{p(\cdot)}(\mathbb{R}^n) \to L^{q(\cdot)}(\mathbb{R}^n)\right)$*-bounded and satisfy (5.9), the result in Theorem 5.5 still holds.*

## Acknowledgement

This study has been given as the plenary talk by the author at the "The 23rd International Pure Mathematics Conference 2023 (23rd IPMC 2023)", 27–28 August 2023, Online, Islamabad, Pakistan.

# References


[1] Adams, D.R. 1975. A note on Riesz potentials. Duke Math. J. 42(4): 765–778.

[2] Adams, D.R. 2015. Morrey spaces. Lecture Notes in Applied and Numerical Harmonic Analysis. Birkhäuser/Springer, Cham.

[3] Almeida, A., Hasanov, J.J. and Samko, S.G. 2008. Maximal and potential operators in variable exponent Morrey spaces. Georgian Math. J. 15(2): 195–208.

[4] Antontsev, S.N. and Rodrigues, J.F. 2006. On stationary thermorheological viscous flows. Ann. Univ. Ferrara, Sez. VII, Sci. Mat. 52(1): 19–36.

[5] Capone, C., Cruz-Uribe, D. and Fiorenza, A. 2007. The fractional maximal operator and fractional integrals on variable $L^p$ spaces. Rev. Mat. Iberoam. 23: 743–770.

[6] Chen, Y., Levine, S. and Rao, M. 2006. Variable exponent, linear growth functionals in image restoration. SIAM J. Appl. Math. 66(4): 1383–1406.

[7] Chiarenza, F. and Frasca, M. 1987. Morrey spaces and Hardy-Littlewood maximal function. Rend. Math. 7(7): 273–279.

[8] Cruz-Uribe, D., Fiorenza, A., Martell, J. and Pérez, C. 2006. The boundedness of classical operators on variable $L^p$ spaces. Ann. Acad. Sci Fenn. Math. 31(2): 239–264.

[9] Cruz-Uribe, D. and Fiorenza, A. 2013. Variable Lebesgue spaces. Foundations and Harmonic Analysis. Birkhäuser/Springer, New York.

[10] Diening, L., Harjulehto, P., Hästö, P. and Růžička, M. 2011. Lebesgue and Sobolev space with variable exponents. Springer Lecture Notes, Vol. 2017, Berlin: Springer-Verlag.

[11] Diening, L. and Rüžička, M. 2003. Calderón-Zygmund operators on generalized Lebesgue Spaces $L^{p(x)}(\Omega)$ and problems related to fluid dynamics. J. Reine Angew. Math. 563: 197–220.

[12] Ding, Y. and Lu, S.Z. 1998. Weighted norm inequalities for fractional integral operators with rough kernel. Canad. J. Math. 50(1): 29–39.

[13] Fan, X.L. and Zhao, D. 2001. On the spaces $L^{p(x)}(\Omega)$ and $W^{m,p(x)}(\Omega)$. J. Math. Anal. Appl. 263(2): 424—446.

[14] Giaquinta, M. 1983. Multiple integrals in the calculus of variations and nonlinear elliptic systems. Annals of Mathematics Studies, 105. Princeton University Press, Princeton, NJ.

[15] Guliyev, V.S., Hasanov, J.J. and Samko, S.G. 2010. Boundedness of the maximal, potential nd singular operators in the generalized variable exponent Morrey spaces. Math. Scand. 107(2): 285–304.

[16] Guliyev, V.S. and Samko, S.G. 2013. Maximal, potential and singular operators in the generalized variable exponent Morrey spaces on unbounded sets. J. Math. Sci. 193(2): 228–248.

[17] Gürbüz, F. 2018. Marcinkiewicz integrals with rough kernel associated with Schrödinger operators and commutators on generalized vanishing local Morrey spaces. Tbilisi Math. J. 11(3): 133–156.

[18] Gürbüz, F. 2020a. Sublinear operators with rough kernel generated by fractional integrals and commutators on generalized vanishing local Morrey spaces. TWMS J. App. Eng. Math. 10(Special Issue): 73–84.

[19] Gürbüz, F. 2020b. The boundedness of a class of fractional type rough higher order commutators on vanishing generalized weighted Morrey spaces. TWMS J. App. Eng. Math. 10(Special Issue): 97–104.

[20] Gürbüz, F., Ding, S., Han, H. and Long, P. 2021a. Norm inequalities on variable exponent vanishing Morrey type spaces for the rough singular type integral operators. Int. J. Nonlinear Sci. Numer. Simul. 22(6): 721–739.

[21] Gürbüz, F., Ding, S., Han, H. and Long, P. 2021. Characterizations of rough fractional-type integral operators on variable exponent vanishing Morrey-type spaces. pp. 95–123. *In:* Dutta, H. (ed.). Topics in Contemporary Mathematical Analysis and Applications. CRC Press.

[22] Gürbüz, F. 2021b. About rough commutators with variable kernel of fractional type on vanishing generalized weighted Morrey spaces. Rom. J. Math. Comput. Sci. 11(1): 1–9.

[23] Gürbüz, F. 2021c. Product generalized local Morrey spaces and commutators of multi-sublinear operators generated by multilinear Calderon-Zygmund operators and local Campanato functions. Filomat 35(9): 2849–2868.

[24] Harjulehto, P., Hästö, P., Lê, Ú.V. and Nuortio, M. 2010. Overview of differential equations with nonstandard growth. Nonlinear. Anal. 72(12): 4551–4574.

[25] Ho, K.-P. 2013. The fractional integral operators on Morrey spaces with variable exponent on unbounded domains. Math. Inequal. Appl. 16(2): 363–373.

[26] Ho, K.-P. 2017. Fractional integral operators with homogeneous kernels on Morrey spaces with variable exponents. J. Math. Soc. Japan 69(3): 1059–1077.

[27] Kováčik, O. and Rákosník, J. 1991. On spaces $L^{p(x)}$ and $W^{k,p(x)}$. Czechoslovak Math. J. 41(11): 592–618.

[28] Kurata, K., Nishigaki, S. and Sugano, S. 2000. Boundedness of integral operators on generalized Morrey spaces and its application to Schrōdinger operators. Proc. Am. Math. Soc. 128(4): 1125–1134.

[29] Long, P. and Han, H. 2016. Characterizations of some operators on the vanishing generalized Morrey spaces with variable exponent. J. Math. Anal. Appl. 437(1): 419–430.

[30] Mastylo, M., Sawano, Y. and Tanaka, H. 2018. Morrey type pace and its Köthe dual space. Bulletin of the Malaysian Math. Sci. Soc. 41(3): 1181–1198.

[31] Morrey, C.B. 1938. On the solutions of quasi-linear elliptic partial differential equations. Trans. Amer. Math. Soc. 43: 126–166.

[32] Orlicz, W. 1931. Über konjugierte Exponentenfolgen. Studia Math. 3: 200—211.

[33] Peetre, J. 1969. On the theory of $L_{p,\lambda}$. J. Func. Anal. 4: 71–87.

[34] Rafeiro, H. and Samko, S.G. 2016. On maximal and potential operators with rough kernels in variable exponent spaces. Rend. Lincei Mat. Appl. 27(3): 309–325.

[35] Růžička, M. 2000. Electrorheological fluids: Modeling and mathematical theory. Lecture Notes in Mathematics, 1748. Springer-Verlag, Berlin.

[36] Softova, L. 2006. Singular integrals and commutators in generalized Morrey spaces. Acta Math. Sin., Engl. Ser. 22(3): 757–766.

[37] Taylor, M.E. 2000. Tools for PDE: Pseudodifferential Operators, Paradifferential Operators, and Layer Potentials, Volume 81 of Math. Surveys and Monogr. AMS, Providence, R.I.

[38] Vitanza, C. 1990. Functions with vanishing Morrey norm and elliptic partial differential equations, 147–150. The Proceedings of Methods of Real Analysis and Partial Differential Equations, Capri, Springer.

[39] Wu, J. 2015. Boundedness for Riesz-type potential operators on Herz-Morrey spaces with variable exponent. Math. Inequal. Appl. 18(2): 471–484.

[40] Wunderli, T. 2010. On time flows of minimizers of general convex functionals of linear growth with variable exponent in BV space and stability of pseudosolutions. J. Math. Anal. Appl. 364(2): 591–598.

[41] Yang, D.C. and Yang, S. 2011. New characterizations of weighted Morrey-Campanato spaces. Taiwanese J. Math. 15(1): 141–163.

# Chapter 6

# Boundedness of the Intrinsic Square Function on Herz Spaces with Variable Exponents

*Liwei Wang*

## 6.1 Introduction

For $0 < \beta \leq 1$, the family $\mathcal{C}_\beta$ consists of those functions $\phi : \mathbb{R}^n \to \mathbb{R}$ such that

$$\operatorname{supp}\phi \subset \{x \in \mathbb{R}^n : |x| \leq 1\}, \tag{6.1}$$

$$\int_{\mathbb{R}^n} \phi(x)dx = 0, \tag{6.2}$$

$$|\phi(x_1) - \phi(x_2)| \leq |x_1 - x_2|^\beta, \quad \forall x_1, x_2 \in \mathbb{R}^n. \tag{6.3}$$

School of Mathematics-Physics and Finance, Anhui Polytechnic University, Wuhu, 241000, China.
Email: wangliwei@ahpu.edu.cn

For any $f \in L^1_{loc}(\mathbb{R}^n)$ and $(y,t) \in \mathbb{R}^{n+1}_+ := \mathbb{R}^n \times (0,\infty)$, we set

$$A_\beta(f)(y,t) = \sup_{\phi \in \mathcal{C}_\beta} |f * \phi_t(y)| = \sup_{\phi \in \mathcal{C}_\beta} \left| \int_{\mathbb{R}^n} \phi_t(y-z)f(z)dz \right|. \qquad (6.4)$$

Then the intrinsic square function of $f$ (of order $\beta$) is defined by

$$S_\beta(f)(x) = \left( \int\!\!\int_{\Gamma(x)} (A_\beta(f)(y,t))^2 \frac{dydt}{t^{n+1}} \right)^{\frac{1}{2}}, \qquad (6.5)$$

where $\phi_t(x) = \frac{1}{t^n}\phi(\frac{x}{t})$ and $\Gamma(x) = \{(y,t) \in \mathbb{R}^{n+1}_+ : |x-y| < t\}$.

Wilson [35, 36] first introduced the intrinsic square function and proved that if $\omega \in A_p$ (the class of Muckenhoupt weights), then the intrinsic square function $S_\beta(f)$ is bounded on $L^p_\omega(\mathbb{R}^n)$ for any $1 < p < \infty$. Subsequently, Lerner [20] showed sharp $L^p_\omega(\mathbb{R}^n)$ norm inequalities for the $S_\beta(f)$ in terms of the $A_p$ characteristic constant of $\omega$ for all $1 < p < \infty$. Recently, many researchers have studied the boundedness of such intrinsic square function on various other function spaces, we refer to [3, 17, 29, 34] and their references.

A locally integrable function $b$ is said to belong to space $\mathrm{BMO}(\mathbb{R}^n)$, if it satisfies

$$\|b\|_* := \sup_B \frac{1}{|B|} \int_B |b(y) - b_B| \, dy < \infty,$$

where $B$ denotes the ball centered at $x \in \mathbb{R}^n$ and radius of $r > 0$, and $b_B$ denotes the average of $b$ on $B$, that is, $b_B := |B|^{-1} \int_B b(t)dt$. For $b \in \mathrm{BMO}(\mathbb{R}^n)$, $[b, S_\beta]$, the commutator of the intrinsic square function, is then defined by

$$[b, S_\beta](f)(x) = \left( \int\!\!\int_{\Gamma(x)} \sup_{\phi \in \mathcal{C}_\beta} \left| \int_{\mathbb{R}^n} (b(x) - b(z))\phi_t(y-z)f(z)dz \right|^2 \frac{dydt}{t^{n+1}} \right)^{\frac{1}{2}}.$$

$$(6.6)$$

Wang [29] proved that the commutator $[b, S_\beta]$ is bounded on the weighted Lebesgue space $L^p_\omega(\mathbb{R}^n)$ if $\omega \in A_p$ for some $1 < p < \infty$, and is also bounded on the weighted Morrey space $L^{p,\kappa}_\omega(\mathbb{R}^n)$ for $0 < \kappa < 1$. Hu and Wang [13] studied its boundedness on the weighted Herz spaces. More applications of such intrinsic square function were also given by Guliyev et al. [7] and Zhuo et al. [39].

In recent years, function spaces with variable exponent have been intensively studied by many authors. The growing interest in such spaces comes not only from theoretical purposes, but also from applications to fluid dynamics [26], image restoration [2] and PDE with non-standard growth conditions [11]. The Lebesgue spaces with variable exponent were first investigated by Orlicz [24] and were systematically studied by Kováčik and Rákosník [19]. Since then, various other function spaces have been studied in the variable exponent setting, we refer to the monographs [4, 6, 18] and the recent papers [9, 10, 12, 22, 27, 28, 31, 32, 37].

The classical Herz spaces play a key role in harmonic analysis and partial differential equations. For instance, they are good substitutes of the ordinary Hardy spaces when considering the boundedness of non-translation invariant singular integral operators. They also appear in the characterization of multiplier on Hardy spaces and in the regularity theory for elliptic and parabolic equations in divergence form, see [21, 25]. Herz spaces $\dot{K}^{\alpha}_{p(\cdot),q}(\mathbb{R}^n)$ and $K^{\alpha}_{p(\cdot),q}(\mathbb{R}^n)$ with variable exponent $p$ were introduced by Izuki [14, 15]. Almeida and Direhem [1] futher defined the Herz spaces $\dot{K}^{\alpha(\cdot)}_{p(\cdot),q}(\mathbb{R}^n)$ and $K^{\alpha(\cdot)}_{p(\cdot),q}(\mathbb{R}^n)$, where $\alpha$ and $p$ are variable. The most generalized Herz spaces $\dot{K}^{\alpha(\cdot)}_{p(\cdot),q(\cdot)}(\mathbb{R}^n)$ and $K^{\alpha(\cdot)}_{p(\cdot),q(\cdot)}(\mathbb{R}^n)$ in discrete and continual settings were introduced by Izuki and Noi [16] and Samko [27], respectively, where the exponent $q$ is variable as well, and they proved the boundedness of a wide class of sublinear operators on such spaces, which includes maximal, potential and Calderón-Zygmund operators.

Wang [30] proved the boundedness of the intrinsic square function $S_\beta$ on weighted Herz spaces under proper assumptions on each exponent and weight. Izuki and Noi [17] extended the above Wang's results to the variable exponent case. The boundedness of the intrinsic square function $S_\beta$ and its commutator $[b, S_\beta]$ on Herz spaces $\dot{K}^{\alpha(\cdot)}_{p(\cdot),q}(\mathbb{R}^n)$ was futher discussed in [8, 33, 34]. Motivated by the work of [8, 16, 33, 34], in the present paper, we will study the boundedness properties for $S_\beta$ and $[b, S_\beta]$ on Herz spaces $\dot{K}^{\alpha(\cdot)}_{p(\cdot),q(\cdot)}(\mathbb{R}^n)$ in discrete setting, where all the three main indices are variable.

Throughout this paper, we denote by $\mathbb{Z}$ the set of all integral numbers. Given a measurable set $E \subset \mathbb{R}^n$, $|E|$ is the Lebesgue measure of $E$ and $\chi_E$ means the characteristic function of $E$. We define $B := B(x, r) = \{y \in \mathbb{R}^n : |x - y| < r\}$. For $k \in \mathbb{Z}$, we set $B_k := B(0, 2^k)$, $C_k = B_k \setminus B_{k-1}$, $\chi_k = \chi_{C_k}$ and $f_k = f\chi_k$. $p'(\cdot)$ denotes the conjugate exponent defined by $1/p(\cdot) + 1/p'(\cdot) = 1$. The letter $C$ stands for a positive constant, which may vary from line to line. The expression $f \lesssim g$ means $f \leqslant Cg$, and $f \approx g$ means $f \lesssim g \lesssim f$.

## 6.2 Preliminaries

We begin with a brief review of the Lebesgue space with variable exponent $L^{p(\cdot)}(\mathbb{R}^n)$, further details can be found in [4, 5, 6, 18, 23].

Let $p(\cdot) : \mathbb{R}^n \to [1, \infty)$ be a measurable function. We suppose that

$$1 \leq p_- \leq p(x) \leq p_+ < \infty,$$

where $p_- := \operatorname{ess\,inf}_{x \in \mathbb{R}^n} p(x)$, $p_+ := \operatorname{ess\,sup}_{x \in \mathbb{R}^n} p(x)$. The Lebesgue space with variable exponent $L^{p(\cdot)}(\mathbb{R}^n)$ is the set of all measurable functions $f$ on $\mathbb{R}^n$ such that

$$I_{p(\cdot)}(f) := \int_{\mathbb{R}^n} |f(x)|^{p(x)} \mathrm{d}x < \infty.$$

This is a Banach space equipped with the norm

$$\|f\|_{L^{p(\cdot)}(\mathbb{R}^n)} = \inf\{\lambda > 0 : I_{p(\cdot)}(f/\lambda) \le 1\}.$$

Let $\Omega \subset \mathbb{R}^n$ be an open set, the space $L_{\mathrm{loc}}^{p(\cdot)}(\Omega)$ is defined by

$$L_{\mathrm{loc}}^{p(\cdot)}(\Omega) = \{f : f \in L^{p(\cdot)}(K) \text{ for all compact subsets } K \subset \Omega\}.$$

We denote by $\mathcal{P}(\mathbb{R}^n)$ the set of all measurable functions $p(\cdot) : \mathbb{R}^n \to (1, \infty)$ with $1 < p_- \le p(x) \le p_+ < \infty$. For $p(\cdot) \in \mathcal{P}(\mathbb{R}^n)$, the generalized Hölder inequality holds in the form

$$\int_{\mathbb{R}^n} |f(x)g(x)|\,dx \le c_p \|f\|_{L^{p(\cdot)}(\mathbb{R}^n)} \|g\|_{L^{p'(\cdot)}(\mathbb{R}^n)}, \tag{6.7}$$

where $c_p = 1 + \frac{1}{p_-} - \frac{1}{p_+}$, see [19, Theorem 2.1].

The set $\mathcal{B}(\mathbb{R}^n)$ is defined by

$$\mathcal{B}(\mathbb{R}^n) := \{p(\cdot) \in \mathcal{P}(\mathbb{R}^n) : \ M \text{ is bounded on } L^{p(\cdot)}(\mathbb{R}^n)\},$$

where $M$ denotes the Hardy-Littlewood maximal operator defined on locally integrable functions by

$$Mf(x) = \sup_{x \in \mathbb{R}^n, r > 0} r^{-n} \int_{B(x,r)} |f(y)|\,dy.$$

The mixed Lebesgue sequence spaces and Herz spaces with variable exponents are defined respectively as follows:

**Definition 6.1**     Let $p(\cdot), q(\cdot) \in \mathcal{P}(\mathbb{R}^n)$. The mixed Lebesgue sequence space $\ell^{q(\cdot)}(L^{p(\cdot)})$ consists of all sequences $\{f_j\}_{j=-\infty}^{\infty}$ of measurable functions on $\mathbb{R}^n$ such that

$$\|\{f_j\}_{j=-\infty}^{\infty}\|_{\ell^{q(\cdot)}(L^{p(\cdot)})} := \inf\left\{\mu > 0 : \rho_{\ell^{q(\cdot)}(L^{p(\cdot)})}\left(\left\{\frac{f_j}{\mu}\right\}_{j=-\infty}^{\infty}\right) \le 1\right\} < \infty,$$

where

$$\rho_{\ell^{q(\cdot)}(L^{p(\cdot)})}(\{f_j\}_{j=-\infty}^{\infty}) := \sum_{j=-\infty}^{\infty} \inf\left\{\lambda_j : \int_{\mathbb{R}^n} \left(\frac{|f_j(x)|}{\lambda_j^{\frac{1}{q(x)}}}\right)^{p(x)} dx \le 1\right\}.$$

If $q^+ < \infty$, then we can write

$$\rho_{\ell^{q(\cdot)}(L^{p(\cdot)})}(\{f_j\}_{j=-\infty}^{\infty}) = \sum_{j=-\infty}^{\infty} \left\| |f_j|^{q(\cdot)} \right\|_{L^{\frac{p(\cdot)}{q(\cdot)}}}.$$

**Definition 6.2** Let $p(\cdot), q(\cdot) \in \mathcal{P}(\mathbb{R}^n)$ and $\alpha(\cdot) : \mathbb{R}^n \to \mathbb{R}$ such that $-\infty < \alpha_- \leq \alpha_+ < \infty$. The homogeneous Herz space $\dot{K}^{\alpha(\cdot)}_{p(\cdot),q(\cdot)}(\mathbb{R}^n)$ is defined by

$$\dot{K}^{\alpha(\cdot)}_{p(\cdot),q(\cdot)}(\mathbb{R}^n) := \left\{ f \in L^{p(\cdot)}_{\mathrm{loc}}(\mathbb{R}^n \setminus \{0\}) : \|f\|_{\dot{K}^{\alpha(\cdot)}_{p(\cdot),q(\cdot)}(\mathbb{R}^n)} < \infty \right\},$$

where

$$\|f\|_{\dot{K}^{\alpha(\cdot)}_{p(\cdot),q(\cdot)}(\mathbb{R}^n)} := \left\| (2^{k\alpha(\cdot)} |f\chi_k|)_{k=-\infty}^{\infty} \right\|_{\ell^{q(\cdot)}(L^{p(\cdot)})}$$

$$= \inf\left\{ \lambda > 0 : \sum_{k=-\infty}^{\infty} \left\| \left( \frac{2^{k\alpha(\cdot)} |f\chi_k|}{\lambda} \right)^{q(\cdot)} \right\|_{L^{\frac{p(\cdot)}{q(\cdot)}}} \leq 1 \right\}.$$

The non-homogeneous Herz space $K^{\alpha(\cdot)}_{p(\cdot),q(\cdot)}(\mathbb{R}^n)$ is defined by

$$K^{\alpha(\cdot)}_{p(\cdot),q(\cdot)}(\mathbb{R}^n) := \left\{ f \in L^{p(\cdot)}_{\mathrm{loc}}(\mathbb{R}^n) : \|f\|_{K^{\alpha(\cdot)}_{p(\cdot),q(\cdot)}(\mathbb{R}^n)} < \infty \right\},$$

where

$$\|f\|_{K^{\alpha(\cdot)}_{p(\cdot),q(\cdot)}(\mathbb{R}^n)} := \left\| (2^{k\alpha(\cdot)} |f\chi_k|)_{k=0}^{\infty} \right\|_{\ell^{q(\cdot)}(L^{p(\cdot)})}$$

$$= \inf\left\{ \lambda > 0 : \left\| \left( \frac{|f\chi_{B_0}|}{\lambda} \right)^{q(\cdot)} \right\|_{L^{\frac{p(\cdot)}{q(\cdot)}}} + \sum_{k=1}^{\infty} \left\| \left( \frac{2^{k\alpha(\cdot)} |f\chi_k|}{\lambda} \right)^{q(\cdot)} \right\|_{L^{\frac{p(\cdot)}{q(\cdot)}}} \leq 1 \right\}.$$

The following are the key lemmas which are needed in the proofs of the main results.

### Lemma 6.1
*(see [16]) Let $p(\cdot)$ be a measurable function on $\mathbb{R}^n$ with $0 < \alpha \leq p(\cdot) \leq \beta$ and $q(\cdot) \in \mathcal{P}(\mathbb{R}^n)$. If $p(\cdot)q(\cdot) \in \mathcal{P}(\mathbb{R}^n)$ and $g(\cdot) \in L^{p(\cdot)q(\cdot)}(\mathbb{R}^n)$, then we have*

$$\min\{\|g(\cdot)\|^{\beta}_{L^{p(\cdot)q(\cdot)}}, \|g(\cdot)\|^{\alpha}_{L^{p(\cdot)q(\cdot)}}\} \leq \||g(\cdot)|^{p(\cdot)}\|_{L^{q(\cdot)}}$$

$$\leq \max\{\|g(\cdot)\|^{\beta}_{L^{p(\cdot)q(\cdot)}}, \|g(\cdot)\|^{\alpha}_{L^{p(\cdot)q(\cdot)}}\}.$$

### Lemma 6.2
*(see [1]) Let $\alpha(\cdot) \in L^{\infty}(\mathbb{R}^n)$ and $r_1 > 0$. If $\alpha(\cdot)$ is log-Hölder continuous both at the origin and at infinity, then we have*

$$r_1^{\alpha(x)} \lesssim r_2^{\alpha(y)} \times \begin{cases} \left( \frac{r_1}{r_2} \right)^{\alpha_+}, & 0 < r_2 \leq r_1/2, \\ 1, & r_1/2 < r_2 \leq 2r_1, \\ \left( \frac{r_1}{r_2} \right)^{\alpha_-}, & r_2 > 2r_1, \end{cases}$$

*for any $x \in B(0, r_1) \setminus B(0, r_1/2)$ and $y \in B(0, r_2) \setminus B(0, r_2/2)$.*

### Lemma 6.3

*(see [38]) If $a_k > 0$ and $1 < p_k < \infty$ with $k \in \mathbb{Z}$, then we have*

$$\sum_{k=-\infty}^{\infty} a_k^{p_k} \leq \left( \sum_{k=-\infty}^{\infty} a_k \right)^{p_*},$$

*where*

$$p_* := \begin{cases} \min\limits_{k \in \mathbb{Z}} p_k, & \sum\limits_{k=-\infty}^{\infty} a_k \leq 1, \\[2ex] \max\limits_{k \in \mathbb{Z}} p_k, & \sum\limits_{k=-\infty}^{\infty} a_k \geq 1. \end{cases}$$

### Lemma 6.4

*(see [14]) If $p(\cdot) \in \mathcal{B}(\mathbb{R}^n)$, then we have for all measurable subsets $S \subset B$,*

$$\frac{\|\chi_S\|_{L^{p(\cdot)}(\mathbb{R}^n)}}{\|\chi_B\|_{L^{p(\cdot)}(\mathbb{R}^n)}} \lesssim \left( \frac{|S|}{|B|} \right)^{\delta_1}, \quad \frac{\|\chi_S\|_{L^{p'(\cdot)}(\mathbb{R}^n)}}{\|\chi_B\|_{L^{p'(\cdot)}(\mathbb{R}^n)}} \lesssim \left( \frac{|S|}{|B|} \right)^{\delta_2},$$

*where $\delta_1$ and $\delta_2$ are constants with $0 < \delta_1, \delta_2 < 1$.*

### Lemma 6.5

*(see [14]) If $p(\cdot) \in \mathcal{B}(\mathbb{R}^n)$, then we have*

$$\frac{1}{|B|} \|\chi_B\|_{L^{p(\cdot)}(\mathbb{R}^n)} \|\chi_B\|_{L^{p'(\cdot)}(\mathbb{R}^n)} \lesssim 1.$$

### Lemma 6.6

*(see [15]) Let $b \in \mathrm{BMO}(\mathbb{R}^n)$ and $k > j$ $(k, j \in \mathbb{N})$. If $p(\cdot) \in \mathcal{B}(\mathbb{R}^n)$, then we have*

$$\sup_{B \subset \mathbb{R}^n} \frac{1}{\|\chi_B\|_{L^{p(\cdot)}(\mathbb{R}^n)}} \|(b - b_B)\chi_B\|_{L^{p(\cdot)}(\mathbb{R}^n)} \approx \|b\|_*,$$

$$\|(b - b_{B_j})\chi_{B_k}\|_{L^{p(\cdot)}(\mathbb{R}^n)} \lesssim (k - j)\|b\|_* \|\chi_{B_k}\|_{L^{p(\cdot)}(\mathbb{R}^n)}.$$

## 6.3   Main Results and their Proofs

In this section, we prove that the intrinsic square function $S_\beta$ and it commutator $[b, S_\beta]$ are bounded on Herz spaces with variable exponents. Our main results can be stated as follows.

### Theorem 6.1

*Suppose that $p(\cdot) \in \mathcal{B}(\mathbb{R}^n)$, $q_1(\cdot), q_2(\cdot) \in \mathcal{P}(\mathbb{R}^n)$ and $(q_1)_+ \leq (q_2)_+$. Let $\alpha(\cdot) \in L^\infty(\mathbb{R}^n)$ be log-Hölder continuous both at the origin and at infinity, such that*

$$-n\delta_1 < \alpha_- \leq \alpha_+ < n\delta_2,$$

*where $0 < \delta_1, \delta_2 < 1$ are the constants appearing in Lemma 6.4. Then we have that for all $f \in \dot{K}^{\alpha(\cdot)}_{p(\cdot),q_1(\cdot)}(\mathbb{R}^n)$,*

$$\|S_\beta(f)\|_{\dot{K}^{\alpha(\cdot)}_{p(\cdot),q_2(\cdot)}(\mathbb{R}^n)} \leq C\|f\|_{\dot{K}^{\alpha(\cdot)}_{p(\cdot),q_1(\cdot)}(\mathbb{R}^n)}.$$

**Theorem 6.2**

*Suppose that $b \in \mathrm{BMO}(\mathbb{R}^n)$, $p(\cdot) \in \mathcal{B}(\mathbb{R}^n)$, $q_1(\cdot), q_2(\cdot) \in \mathcal{P}(\mathbb{R}^n)$ and $(q_1)_+ \leq (q_2)_+$. Let $\alpha(\cdot) \in L^\infty(\mathbb{R}^n)$ be log-Hölder continuous both at the origin and at infinity, such that*

$$-n\delta_1 < \alpha_- \leq \alpha_+ < n\delta_2,$$

*where $0 < \delta_1, \delta_2 < 1$ are the constants appearing in Lemma 6.4. Then we have that for all $f \in \dot{K}^{\alpha(\cdot)}_{p(\cdot),q_1(\cdot)}(\mathbb{R}^n)$,*

$$\|[b,S_\beta](f)\|_{\dot{K}^{\alpha(\cdot)}_{p(\cdot),q_2(\cdot)}(\mathbb{R}^n)} \leq C\|b\|_*\|f\|_{\dot{K}^{\alpha(\cdot)}_{p(\cdot),q_1(\cdot)}(\mathbb{R}^n)}.$$

**Remark 6.1**   When $q_1 = q_2 = q$ is a positive constant, the corresponding statements to Theorems 6.1 and 6.2 were proved by Wang and Shu [34] and Wang [33], respectively.   ■

**Proof of Theorem 6.1.**  Without loss of generality, we may assume that $f \geq 0$. For simplicity, we denote $L^{p(\cdot)}(\mathbb{R}^n)$ by $L^{p(\cdot)}$. By definition 6.2, we have

$$\|S_\beta(f)\|_{\dot{K}^{\alpha(\cdot)}_{p(\cdot),q_2(\cdot)}(\mathbb{R}^n)} = \inf\left\{\lambda > 0 : \sum_{k=-\infty}^{\infty}\left\|\left(\frac{2^{k\alpha(\cdot)}|S_\beta(f)|\chi_k}{\lambda}\right)^{q_2(\cdot)}\right\|_{L^{\frac{p(\cdot)}{q_2(\cdot)}}} \leq 1\right\}.$$

For any $k \in \mathbb{Z}$, we consider the norm $\left\|\left(\frac{2^{k\alpha(\cdot)}|S_\beta(f)|\chi_k}{\lambda}\right)^{q_2(\cdot)}\right\|_{L^{\frac{p(\cdot)}{q_2(\cdot)}}}$. Let $f_j := f\chi_j$ for $j \in \mathbb{Z}$, then we have $f = \sum_{j=-\infty}^{\infty} f_j$. Let

$$\lambda_1 := \left\|\left\{2^{k\alpha(\cdot)}\sum_{j=-\infty}^{k-2}|S_\beta(f_j)|\chi_k\right\}_{k\in\mathbb{Z}}\right\|_{l^{q_2(\cdot)}(L^{p(\cdot)})},$$

$$\lambda_2 := \left\|\left\{2^{k\alpha(\cdot)}\sum_{j=k-1}^{k+1}|S_\beta(f_j)|\chi_k\right\}_{k\in\mathbb{Z}}\right\|_{l^{q_2(\cdot)}(L^{p(\cdot)})},$$

$$\lambda_3 := \left\|\left\{2^{k\alpha(\cdot)}\sum_{j=k+2}^{\infty}|S_\beta(f_j)|\chi_k\right\}_{k\in\mathbb{Z}}\right\|_{l^{q_2(\cdot)}(L^{p(\cdot)})}.$$

Then we have

$$\left\|\left(\frac{2^{k\alpha(\cdot)}|S_\beta(f)|\chi_k}{\lambda}\right)^{q_2(\cdot)}\right\|_{L^{\frac{p(\cdot)}{q_2(\cdot)}}} \leq \left\|\left(\frac{2^{k\alpha(\cdot)}\sum_{j=-\infty}^{\infty}|S_\beta(f_j)|\chi_k}{\lambda}\right)^{q_2(\cdot)}\right\|_{L^{\frac{p(\cdot)}{q_2(\cdot)}}}$$

$$\leq \left\|\left(\frac{2^{k\alpha(\cdot)}\sum_{j=-\infty}^{k-2}|S_\beta(f_j)|\chi_k}{\lambda_1}\right)^{q_2(\cdot)}\right\|_{L^{\frac{p(\cdot)}{q_2(\cdot)}}}$$

$$+ \left\|\left(\frac{2^{k\alpha(\cdot)}\sum_{j=k-1}^{k+1}|S_\beta(f_j)|\chi_k}{\lambda_2}\right)^{q_2(\cdot)}\right\|_{L^{\frac{p(\cdot)}{q_2(\cdot)}}}$$

$$+ \left\|\left(\frac{2^{k\alpha(\cdot)}\sum_{j=k+2}^{\infty}|S_\beta(f_j)|\chi_k}{\lambda_3}\right)^{q_2(\cdot)}\right\|_{L^{\frac{p(\cdot)}{q_2(\cdot)}}},$$

where

$$\lambda := \lambda_1 + \lambda_2 + \lambda_3.$$

From the definitions of $\lambda_1, \lambda_2$ and $\lambda_3$, it follows that

$$\sum_{k=-\infty}^{\infty}\left\|\left(\frac{2^{k\alpha(\cdot)}|S_\beta(f)|\chi_k}{\lambda}\right)^{q_2(\cdot)}\right\|_{L^{\frac{p(\cdot)}{q_2(\cdot)}}} \lesssim 1,$$

which implies that

$$\|S_\beta(f)\|_{\dot{K}^{\alpha(\cdot)}_{p(\cdot),q_2(\cdot)}(\mathbb{R}^n)} \lesssim \lambda_1 + \lambda_2 + \lambda_3.$$

Hence, it suffices to prove

$$\lambda_i \lesssim \|f\|_{\dot{K}^{\alpha(\cdot)}_{p(\cdot),q_1(\cdot)}(\mathbb{R}^n)}, \quad i = 1, 2, 3.$$

Let

$$u = \|f\|_{\dot{K}^{\alpha(\cdot)}_{p(\cdot),q_1(\cdot)}(\mathbb{R}^n)}.$$

First, we estimate $\lambda_2$. For each $k \in \mathbb{Z}$, we define

$$(q_{2*})_k := \begin{cases} (q_2)_+, & \left\|\left(\dfrac{2^{k\alpha(\cdot)}\sum_{j=k-1}^{k+1}|S_\beta(f_j)|\chi_k}{u}\right)^{q_2(\cdot)}\right\|_{L^{\frac{p(\cdot)}{q_2(\cdot)}}} \geq 1, \\ (q_2)_-, & \text{otherwise.} \end{cases}$$

By Lemma 6.1 and Lemma 6.2, we have

$$\sum_{k=-\infty}^{\infty}\left\|\left(\frac{2^{k\alpha(\cdot)}\sum_{j=k-1}^{k+1}|S_\beta(f_j)|\chi_k}{u}\right)^{q_2(\cdot)}\right\|_{L^{\frac{p(\cdot)}{q_2(\cdot)}}}$$

$$\lesssim \sum_{k=-\infty}^{\infty}\left\|\frac{2^{k\alpha(\cdot)}\sum_{j=k-1}^{k+1}|S_\beta(f_j)|\chi_k}{u}\right\|_{L^{p(\cdot)}}^{(q_{2*})_k}$$

$$\lesssim \sum_{k=-\infty}^{\infty}\left\|\frac{\sum_{j=k-1}^{k+1}|S_\beta(2^{j\alpha}f_j)|\chi_k}{u}\right\|_{L^{p(\cdot)}}^{(q_{2*})_k}$$

$$\lesssim \sum_{k=-\infty}^{\infty}\left(\sum_{j=k-1}^{k+1}\left\|\frac{|S_\beta(2^{j\alpha}f_j)|\chi_k}{u}\right\|_{L^{p(\cdot)}}\right)^{(q_{2*})_k}$$

$$\lesssim \sum_{k=-\infty}^{\infty}\left(\sum_{j=k-1}^{k+1}\left\|\frac{S_\beta(2^{j\alpha}f_j)}{u}\right\|_{L^{p(\cdot)}}\right)^{(q_{2*})_k}.$$

We use the boundedness of $S_\beta$ on $L^{p(\cdot)}(\mathbb{R}^n)$ (see [34]) and obtain

$$\sum_{k=-\infty}^{\infty}\left\|\left(\frac{2^{k\alpha(\cdot)}\sum_{j=k-1}^{k+1}|S_\beta(f_j)|\chi_k}{u}\right)^{q_2(\cdot)}\right\|_{L^{\frac{p(\cdot)}{q_2(\cdot)}}}$$

$$\lesssim \sum_{k=-\infty}^{\infty}\left(\sum_{j=k-1}^{k+1}\left\|\frac{2^{j\alpha(\cdot)}f_j}{u}\right\|_{L^{p(\cdot)}}\right)^{(q_{2*})_k}$$

$$\lesssim \sum_{k=-\infty}^{\infty}\left\|\frac{2^{k\alpha(\cdot)}f_k}{u}\right\|_{L^{p(\cdot)}}^{(q_{2*})_k}$$

$$\lesssim \sum_{k=-\infty}^{\infty}\left\|\left(\frac{2^{k\alpha(\cdot)}f_k}{u}\right)^{q_1(\cdot)}\right\|_{L^{p(\cdot)/q_1(\cdot)}}^{(q_{2*})_k/(q_1)_+}$$

$$\lesssim \left(\sum_{k=-\infty}^{\infty}\left\|\left(\frac{2^{k\alpha(\cdot)}f_k}{u}\right)^{q_1(\cdot)}\right\|_{L^{p(\cdot)/q_1(\cdot)}}\right)^{q_*}$$

$$\lesssim 1,$$

here and afterhere $q_* \geq 1$ is a constant as defined in Lemma 6.3. Hence we conclude that

$$\lambda_2 \lesssim \|f\|_{\dot{K}_{p(\cdot),q_1(\cdot)}^{\alpha(\cdot)}(\mathbb{R}^n)}.$$

For $\lambda_1$, noting that if $x \in C_k$, $(y,t) \in \Gamma(x)$, $z \in C_j \bigcap\{z : |y-z| \leq t\}$ and $j \leq k-2$, then we have

$$t \geq \frac{1}{2}(|x-y|+|y-z|) \geq \frac{1}{2}|x-z| \geq \frac{1}{2}(|x|-|z|) \geq \frac{1}{4}|x|.$$

This together with the generalized Hölder inequality yields

$$
\begin{aligned}
|S_\beta(f_j)(x)| &= \left( \int\!\!\int_{\Gamma(x)} \sup_{\phi \in \mathcal{C}_\beta} \left| \int_{\mathbb{R}^n} \phi_t(y-z) f_j(z)\,dz \right|^2 \frac{dy\,dt}{t^{n+1}} \right)^{\frac{1}{2}} \\
&\lesssim \left( \int_{\frac{|x|}{4}}^{\infty} \int_{|x-y|<t} \left| \frac{1}{t^n} \int_{C_j \cap \{z:|y-z|\le t\}} f_j(z)\,dz \right|^2 \frac{dy\,dt}{t^{n+1}} \right)^{\frac{1}{2}} \qquad (6.8) \\
&\lesssim \left( \int_{C_j} |f_j(z)|\,dz \right) \left( \int_{\frac{|x|}{4}}^{\infty} \frac{dt}{t^{2n+1}} \right)^{\frac{1}{2}} \\
&\lesssim 2^{-kn} \|f_j\|_{L^{p(\cdot)}(\mathbb{R}^n)} \|\chi_j\|_{L^{p'(\cdot)}(\mathbb{R}^n)}.
\end{aligned}
$$

Thus, by Lemma 6.1, Lemma 6.2, and (6.8), we have

$$
\begin{aligned}
& \left\| \left( \frac{2^{k\alpha(\cdot)} \sum_{j=-\infty}^{k-2} |S_\beta(f_j)|\chi_k}{u} \right)^{q_2(\cdot)} \right\|_{L^{\frac{p(\cdot)}{q_2(\cdot)}}} \\
&\lesssim \left\| \frac{2^{k\alpha(\cdot)} \sum_{j=-\infty}^{k-2} |S_\beta(f_j)|\chi_k}{u} \right\|_{L^{p(\cdot)}}^{(q_{2**})_k} \\
&\lesssim \left\| \frac{\sum_{j=-\infty}^{k-2} 2^{(k-j)\alpha_+} |S_\beta(2^{j\alpha} f_j)|\chi_k}{u} \right\|_{L^{p(\cdot)}}^{(q_{2**})_k} \\
&\lesssim \left\| \frac{\sum_{j=-\infty}^{k-2} 2^{(k-j)\alpha_+} 2^{-kn} \|2^{j\alpha(\cdot)} f_j\|_{L^{p(\cdot)}} \|\chi_{B_j}\|_{L^{p'(\cdot)}} \chi_k}{u} \right\|_{L^{p(\cdot)}}^{(q_{2**})_k} \\
&\lesssim \left( \sum_{j=-\infty}^{k-2} 2^{(k-j)\alpha_+} 2^{-kn} \left\| \frac{2^{j\alpha} f_j}{u} \right\|_{L^{p(\cdot)}} \|\chi_{B_j}\|_{L^{p'(\cdot)}} \|\chi_{B_k}\|_{L^{p(\cdot)}} \right)^{(q_{2**})_k},
\end{aligned}
$$

where

$$
(q_{2**})_k := \begin{cases} (q_2)_+, & \left\| \left( \frac{2^{k\alpha(\cdot)} \sum_{j=-\infty}^{k-2} |S_\beta(f_j)|\chi_k}{u} \right)^{q_2(\cdot)} \right\|_{L^{\frac{p(\cdot)}{q_2(\cdot)}}} \ge 1, \\ (q_2)_-, & \text{otherwise.} \end{cases}
$$

By Lemma 6.4 and Lemma 6.5, we have

$$
\begin{aligned}
\|\chi_{B_j}\|_{L^{p'(\cdot)}} \|\chi_{B_k}\|_{L^{p(\cdot)}} &\lesssim 2^{kn} \|\chi_{B_j}\|_{L^{p'(\cdot)}} \|\chi_{B_k}\|_{L^{p'(\cdot)}}^{-1} \\
&\lesssim 2^{kn} \|\chi_{B_j}\|_{L^{p'(\cdot)}} \|\chi_{B_j}\|_{L^{p'(\cdot)}}^{-1} \frac{\|\chi_{B_j}\|_{L^{p'(\cdot)}}}{\|\chi_{B_k}\|_{L^{p'(\cdot)}}} \\
&\lesssim 2^{kn} 2^{(j-k)n\delta_2}.
\end{aligned}
$$

Therefore, we get

$$
\sum_{k=-\infty}^{\infty} \left\| \left( \frac{2^{k\alpha(\cdot)} \sum_{j=-\infty}^{k-2} |S_\beta(f_j)| \chi_k}{u} \right)^{q_2(\cdot)} \right\|_{L^{\frac{p(\cdot)}{q_2(\cdot)}}}
$$

$$
\lesssim \sum_{k=-\infty}^{\infty} \left( \sum_{j=-\infty}^{k-2} 2^{(k-j)(\alpha_+ - n\delta_2)} \left\| \frac{2^{j\alpha(\cdot)} f_j}{u} \right\|_{L^{p(\cdot)}} \right)^{(q_{2**})_k}
$$

$$
\lesssim \sum_{k=-\infty}^{\infty} \left( \sum_{j=-\infty}^{k-2} 2^{(k-j)(\alpha_+ - n\delta_2)} \left\| \left( \frac{2^{j\alpha(\cdot)} f_j}{u} \right)^{q_1(\cdot)} \right\|_{L^{p(\cdot)/q_1(\cdot)}}^{1/(q_1)_+} \right)^{(q_{2**})_k} .
$$

When $(q_1)_+ \leq 1$, by Lemma 6.1 and Lemma 6.3, we have

$$
\sum_{k=-\infty}^{\infty} \left\| \left( \frac{2^{k\alpha(\cdot)} \sum_{j=-\infty}^{k-2} |S_\beta(f_j)| \chi_k}{u} \right)^{q_2(\cdot)} \right\|_{L^{\frac{p(\cdot)}{q_2(\cdot)}}}
$$

$$
\lesssim \left( \sum_{k=-\infty}^{\infty} \sum_{j=-\infty}^{k-2} 2^{(k-j)(\alpha_+ - n\delta_2)(q_1)_+} \left\| \left( \frac{2^{j\alpha(\cdot)} f_j}{u} \right)^{q_1(\cdot)} \right\|_{L^{p(\cdot)/q_1(\cdot)}} \right)^{q_*}
$$

$$
\lesssim \left( \sum_{j=-\infty}^{\infty} \sum_{k=j+2}^{\infty} 2^{(k-j)(\alpha_+ - n\delta_2)(q_1)_+} \left\| \left( \frac{2^{j\alpha(\cdot)} f_j}{u} \right)^{q_1(\cdot)} \right\|_{L^{p(\cdot)/q_1(\cdot)}} \right)^{q_*}
$$

$$
\lesssim \left( \sum_{j=-\infty}^{\infty} \left\| \left( \frac{2^{j\alpha(\cdot)} f_j}{u} \right)^{q_1(\cdot)} \right\|_{L^{p(\cdot)/q_1(\cdot)}} \right)^{q_*}
$$

$$
\lesssim 1.
$$

When $(q_1)_+ > 1$, by Hölder's inequality and Lemma 6.3, we have

$$
\sum_{k=-\infty}^{\infty} \left\| \left( \frac{2^{k\alpha(\cdot)} \sum_{j=-\infty}^{k-2} |S_\beta(f_j)| \chi_k}{u} \right)^{q_2(\cdot)} \right\|_{L^{\frac{p(\cdot)}{q_2(\cdot)}}}
$$

$$
\lesssim \sum_{k=-\infty}^{\infty} \left\{ \left( \sum_{j=-\infty}^{k-2} 2^{(k-j)(\alpha_+ - n\delta_2)(q_1)_+/2} \left\| \left( \frac{2^{j\alpha(\cdot)} f_j}{u} \right)^{q_1(\cdot)} \right\|_{L^{p(\cdot)/q_1(\cdot)}} \right)^{1/q_{1+}} \right.
$$

$$
\left. \times \left( \sum_{j=-\infty}^{k-2} 2^{(k-j)(\alpha_+ - n\delta_2)((q_1)_+)'/2} \right)^{1/((q_1)_+)'} \right\}^{(q_{2**})_k}
$$

$$
\lesssim \left( \sum_{k=-\infty}^{\infty} \sum_{j=-\infty}^{k-2} 2^{(k-j)(\alpha_+ - n\delta_2)(q_1)_+/2} \left\| \left( \frac{2^{j\alpha(\cdot)} f_j}{u} \right)^{q_1(\cdot)} \right\|_{L^{p(\cdot)/q_1(\cdot)}} \right)^{q_*}
$$

$$\lesssim \left( \sum_{j=-\infty}^{\infty} \sum_{k=j+2}^{\infty} 2^{(k-j)(\alpha_+ - n\delta_2)(q_1)_+/2} \left\| \left( \frac{2^{j\alpha(\cdot)} f_j}{u} \right)^{q_1(\cdot)} \right\|_{L^{p(\cdot)/q_1(\cdot)}} \right)^{q_*}$$

$$\lesssim \left( \sum_{j=-\infty}^{\infty} \left\| \left( \frac{2^{j\alpha(\cdot)} f_j}{u} \right)^{q_1(\cdot)} \right\|_{L^{p(\cdot)/q_1(\cdot)}} \right)^{q_*}$$

$$\lesssim 1.$$

Hence we have

$$\lambda_1 \lesssim \|f\|_{\dot{K}^{\alpha(\cdot)}_{p(\cdot), q_1(\cdot)}(\mathbb{R}^n)}.$$

For $\lambda_3$, noting that if $x \in C_k$, $(y,t) \in \Gamma(x)$, $z \in C_j \bigcap \{z : |y-z| \le t\}$ and $j \ge k+2$, we have

$$t \ge \frac{1}{2}(|x-y| + |y-z|) \ge \frac{1}{2}|x-z| \ge \frac{1}{2}(|z|-|x|) \ge \frac{1}{4}|z|.$$

Then, by Hölder's inequality, we get

$$|S_\beta(f_j)(x)| = \left( \int\!\!\int_{\Gamma(x)} \sup_{\phi \in \mathcal{C}_\beta} \left| \int_{\mathbb{R}^n} \phi_t(y-z) f_j(z) dz \right|^2 \frac{dydt}{t^{n+1}} \right)^{\frac{1}{2}}$$

$$\lesssim \left( \int_{\frac{|z|}{4}}^{\infty} \int_{|x-y|<t} \left| t^{-n} \int_{C_j \bigcap \{z:|y-z| \le t\}} f_j(z) dz \right|^2 \frac{dydt}{t^{n+1}} \right)^{\frac{1}{2}} \tag{6.9}$$

$$\lesssim \left( \int_{C_j} |f_j(z)| dz \right) \left( \int_{\frac{|z|}{4}}^{\infty} \frac{dt}{t^{2n+1}} \right)^{\frac{1}{2}}$$

$$\lesssim 2^{-jn} \|f_j\|_{L^{p(\cdot)}(\mathbb{R}^n)} \|\chi_j\|_{L^{p'(\cdot)}(\mathbb{R}^n)}.$$

From (6.9), Lemma 6.1 and Lemma 6.2, it follows that

$$\left\| \left( \frac{2^{k\alpha(\cdot)} \sum_{j=k+2}^{\infty} |S_\beta(f_j)| \chi_k}{u} \right)^{q_2(\cdot)} \right\|_{L^{\frac{p(\cdot)}{q_2(\cdot)}}}$$

$$\lesssim \left\| \frac{2^{k\alpha(\cdot)} \sum_{j=k+2}^{\infty} |S_\beta(f_j)| \chi_k}{u} \right\|_{L^{p(\cdot)}}^{(q_{2***})_k}$$

$$\lesssim \left\| \frac{\sum_{j=k+2}^{\infty} 2^{(k-j)\alpha_-} |S_\beta(2^{j\alpha} f_j)| \chi_k}{u} \right\|_{L^{p(\cdot)}}^{(q_{2***})_k}$$

$$\lesssim \left\| \frac{\sum_{j=k+2}^{\infty} 2^{(k-j)\alpha_-} 2^{-jn} \|2^{j\alpha(\cdot)} f_j\|_{L^{p(\cdot)}} \|\chi_{B_j}\|_{L^{p'(\cdot)}} \chi_k}{u} \right\|_{L^{p(\cdot)}}^{(q_{2***})_k}$$

$$\lesssim \left( \sum_{j=k+2}^{\infty} 2^{(k-j)\alpha_-} 2^{-jn} \left\| \frac{2^{j\alpha(\cdot)} f_j}{u} \right\|_{L^{p(\cdot)}} \|\chi_{B_j}\|_{L^{p'(\cdot)}} \|\chi_{B_k}\|_{L^{p(\cdot)}} \right)^{(q_{2***})_k},$$

where

$$(q_{2***})_k := \begin{cases} (q_2)_+, & \left\| \left( \dfrac{2^{k\alpha(\cdot)} \sum_{j=k+2}^{\infty} |S_\beta(f_j)|\chi_k}{u} \right)^{q_2(\cdot)} \right\|_{L^{\frac{p(\cdot)}{q_2(\cdot)}}} \geq 1, \\ (q_2)_-, & \text{otherwise.} \end{cases}$$

By Lemma 6.4 and Lemma 6.5, we get

$$\|\chi_{B_j}\|_{L^{p'(\cdot)}}\|\chi_{B_k}\|_{L^{p(\cdot)}} \lesssim 2^{jn}\|\chi_{B_j}\|_{L^{p(\cdot)}}^{-1}\|\chi_{B_k}\|_{L^{p(\cdot)}}$$
$$\lesssim 2^{jn}\|\chi_{B_k}\|_{L^{p(\cdot)}}\|\chi_{B_k}\|_{L^{p(\cdot)}}^{-1} \frac{\|\chi_{B_k}\|_{L^{p(\cdot)}}}{\|\chi_{B_j}\|_{L^{p(\cdot)}}}$$
$$\lesssim 2^{jn}2^{(k-j)n\delta_1}.$$

Then we have

$$\sum_{k=-\infty}^{\infty} \left\| \left( \frac{2^{k\alpha(\cdot)} \sum_{j=k+2}^{\infty} |S_\beta(f_j)|\chi_k}{u} \right)^{q_2(\cdot)} \right\|_{L^{\frac{p(\cdot)}{q_2(\cdot)}}}$$
$$\lesssim \sum_{k=-\infty}^{\infty} \left( \sum_{j=k+2}^{\infty} 2^{(k-j)(\alpha_-+n\delta_1)} \left\| \frac{2^{j\alpha(\cdot)}f_j}{u} \right\|_{L^{p(\cdot)}} \right)^{(q_{2***})_k}$$
$$\lesssim \sum_{k=-\infty}^{\infty} \left( \sum_{j=k+2}^{\infty} 2^{(k-j)(\alpha_-+n\delta_1)} \left\| \left( \frac{2^{j\alpha(\cdot)}f_j}{u} \right)^{q_1(\cdot)} \right\|_{L^{p(\cdot)/q_1(\cdot)}}^{1/(q_1)_+} \right)^{(q_{2***})_k}.$$

When $(q_1)_+ \leq 1$, by Lemma 6.1 and Lemma 6.3, we have

$$\sum_{k=-\infty}^{\infty} \left\| \left( \frac{2^{k\alpha(\cdot)} \sum_{j=k+2}^{\infty} |S_\beta(f_j)|\chi_k}{u} \right)^{q_2(\cdot)} \right\|_{L^{\frac{p(\cdot)}{q_2(\cdot)}}}$$
$$\lesssim \sum_{k=-\infty}^{\infty} \left( \sum_{j=k+2}^{\infty} 2^{(k-j)(\alpha_-+n\delta_1)} \left\| \left( \frac{2^{j\alpha(\cdot)}f_j}{u} \right)^{q_1(\cdot)} \right\|_{L^{p(\cdot)/q_1(\cdot)}} \right)^{(q_{2***})_k/(q_1)_+}$$
$$\lesssim \left( \sum_{k=-\infty}^{\infty} \sum_{j=k+2}^{\infty} 2^{(k-j)(\alpha_-+n\delta_1)(q_1)_+} \left\| \left( \frac{2^{j\alpha(\cdot)}f_j}{u} \right)^{q_1(\cdot)} \right\|_{L^{p(\cdot)/q_1(\cdot)}} \right)^{q_*}$$
$$\lesssim \left( \sum_{j=-\infty}^{\infty} \sum_{k=-\infty}^{j-2} 2^{(k-j)(\alpha_-+n\delta_1)(q_1)_+} \left\| \left( \frac{2^{j\alpha(\cdot)}f_j}{u} \right)^{q_1(\cdot)} \right\|_{L^{p(\cdot)/q_1(\cdot)}} \right)^{q_*}$$
$$\lesssim \left( \sum_{j=-\infty}^{\infty} \left\| \left( \frac{2^{j\alpha(\cdot)}f_j}{u} \right)^{q_1(\cdot)} \right\|_{L^{p(\cdot)/q_1(\cdot)}} \right)^{q_*}$$
$$\lesssim 1.$$

When $(q_1)_+ > 1$, by Hölder's inequality and Lemma 6.3, we have

$$\sum_{k=-\infty}^{\infty} \left\| \left( \frac{2^{k\alpha(\cdot)} \sum_{j=k+2}^{\infty} |S_\beta(f_j)| \chi_k}{u} \right)^{q_2(\cdot)} \right\|_{L^{\frac{p(\cdot)}{q_2(\cdot)}}}$$

$$\lesssim \sum_{k=-\infty}^{\infty} \left\{ \left( \sum_{j=k+2}^{\infty} 2^{(k-j)(\alpha_-+n\delta_1)(q_1)_+/2} \left\| \left( \frac{2^{j\alpha(\cdot)} f_j}{u} \right)^{q_1(\cdot)} \right\|_{L^{p(\cdot)/q_1(\cdot)}} \right)^{1/q_{1+}} \right.$$

$$\times \left. \left( \sum_{j=k+2}^{\infty} 2^{(k-j)(\alpha_-+n\delta_1)((q_1)_+)'/2} \right)^{1/((q_1)_+)'} \right\}^{(q_{2***})k}$$

$$\lesssim \left( \sum_{k=-\infty}^{\infty} \sum_{j=k+2}^{\infty} 2^{(k-j)(\alpha_-+n\delta_1)(q_1)_+/2} \left\| \left( \frac{2^{j\alpha(\cdot)} f_j}{u} \right)^{q_1(\cdot)} \right\|_{L^{p(\cdot)/q_1(\cdot)}} \right)^{q_*}$$

$$\lesssim \left( \sum_{j=-\infty}^{\infty} \sum_{k=-\infty}^{j-2} 2^{(k-j)(\alpha_-+n\delta_1)(q_1)_+/2} \left\| \left( \frac{2^{j\alpha(\cdot)} f_j}{u} \right)^{q_1(\cdot)} \right\|_{L^{p(\cdot)/q_1(\cdot)}} \right)^{q_*}$$

$$\lesssim \left( \sum_{j=-\infty}^{\infty} \left\| \left( \frac{2^{j\alpha(\cdot)} f_j}{u} \right)^{q_1(\cdot)} \right\|_{L^{p(\cdot)/q_1(\cdot)}} \right)^{q_*}$$

$$\lesssim 1.$$

Hence we conclude that

$$\lambda_3 \lesssim \|f\|_{\dot{K}^{\alpha(\cdot)}_{p(\cdot),q_1(\cdot)}(\mathbb{R}^n)}.$$

Combining the estimates of $\lambda_1, \lambda_2$ and $\lambda_3$, we complete the proof of Theorem 6.1.

**Proof of Theorem 6.2.** Without loss of generality, we may assume that $f \geq 0$. In view of Definition 6.2, we have

$$\|[b, S_\beta](f)\|_{\dot{K}^{\alpha(\cdot)}_{p(\cdot),q_2(\cdot)}(\mathbb{R}^n)} = \inf \left\{ \eta > 0 : \sum_{k=-\infty}^{\infty} \left\| \left( \frac{2^{k\alpha(\cdot)} |[b, S_\beta](f)| \chi_k}{\eta} \right)^{q_2(\cdot)} \right\|_{L^{\frac{p(\cdot)}{q_2(\cdot)}}} \leq 1 \right\}.$$

For any $k \in \mathbb{Z}$, we consider the norm $\left\| \left( \frac{2^{k\alpha(\cdot)} |[b, S_\beta](f)| \chi_k}{\eta} \right)^{q_2(\cdot)} \right\|_{L^{\frac{p(\cdot)}{q_2(\cdot)}}}$. Let $f_j := f\chi_j$ for $j \in \mathbb{Z}$. then we have $f = \sum_{j=-\infty}^{\infty} f_j$. Let

$$\eta_1 := \left\| \left\{ 2^{k\alpha(\cdot)} \sum_{j=-\infty}^{k-2} |[b, S_\beta](f_j)| \chi_k \right\}_{k\in\mathbb{Z}} \right\|_{l^{q_2(\cdot)}(L^{p(\cdot)})},$$

$$\eta_2 := \left\| \left\{ 2^{k\alpha(\cdot)} \sum_{j=k-1}^{k+1} |[b, S_\beta](f_j)| \chi_k \right\}_{k\in\mathbb{Z}} \right\|_{l^{q_2(\cdot)}(L^{p(\cdot)})},$$

$$\eta_3 := \left\| \left\{ 2^{k\alpha(\cdot)} \sum_{j=k+2}^{\infty} |[b, S_\beta](f_j)| \chi_k \right\}_{k \in \mathbb{Z}} \right\|_{l^{q_2(\cdot)}(L^{p(\cdot)})}.$$

The Minkowski's inequality implies that

$$\left\| \left( \frac{2^{k\alpha(\cdot)}|[b, S_\beta](f)|\chi_k}{\eta} \right)^{q_2(\cdot)} \right\|_{L^{\frac{p(\cdot)}{q_2(\cdot)}}} \leq \left\| \left( \frac{2^{k\alpha(\cdot)} \sum_{j=-\infty}^{\infty} |[b, S_\beta](f_j)|\chi_k}{\eta} \right)^{q_2(\cdot)} \right\|_{L^{\frac{p(\cdot)}{q_2(\cdot)}}}$$

$$\leq \left\| \left( \frac{2^{k\alpha(\cdot)} \sum_{j=-\infty}^{k-2} |[b, S_\beta](f_j)|\chi_k}{\eta_1} \right)^{q_2(\cdot)} \right\|_{L^{\frac{p(\cdot)}{q_2(\cdot)}}}$$

$$+ \left\| \left( \frac{2^{k\alpha(\cdot)} \sum_{j=k-1}^{k+1} |[b, S_\beta](f_j)|\chi_k}{\eta_2} \right)^{q_2(\cdot)} \right\|_{L^{\frac{p(\cdot)}{q_2(\cdot)}}}$$

$$+ \left\| \left( \frac{2^{k\alpha(\cdot)} \sum_{j=k+2}^{\infty} |[b, S_\beta](f_j)|\chi_k}{\eta_3} \right)^{q_2(\cdot)} \right\|_{L^{\frac{p(\cdot)}{q_2(\cdot)}}},$$

where
$$\eta := \eta_1 + \eta_2 + \eta_3.$$

From the definitions of $\eta_1, \eta_2$ and $\eta_3$, we get

$$\sum_{k=-\infty}^{\infty} \left\| \left( \frac{2^{k\alpha(\cdot)}|[b, S_\beta](f)|\chi_k}{\eta} \right)^{q_2(\cdot)} \right\|_{L^{\frac{p(\cdot)}{q_2(\cdot)}}} \lesssim 1,$$

which implies that

$$\|[b, S_\beta](f)\|_{\dot{K}^{\alpha(\cdot)}_{p(\cdot),q_2(\cdot)}(\mathbb{R}^n)} \lesssim \eta_1 + \eta_2 + \eta_3.$$

Hence, it suffices to prove

$$\eta_i \lesssim \|f\|_{\dot{K}^{\alpha(\cdot)}_{p(\cdot),q_1(\cdot)}(\mathbb{R}^n)}, \quad i = 1, 2, 3.$$

Let

$$v = \|f\|_{\dot{K}^{\alpha(\cdot)}_{p(\cdot),q_1(\cdot)}(\mathbb{R}^n)}.$$

First, we estimate $\eta_2$. For each $k \in \mathbb{Z}$, we define

$$(q_{2*})_k := \begin{cases} (q_2)_+, & \left\| \left( \frac{2^{k\alpha(\cdot)} \sum_{j=k-1}^{k+1} |[b, S_\beta](f_j)|\chi_k}{v\|b\|_*} \right)^{q_2(\cdot)} \right\|_{L^{\frac{p(\cdot)}{q_2(\cdot)}}} \geq 1, \\[2ex] (q_2)_-, & \text{otherwise.} \end{cases}$$

By Lemma 6.1 and Lemma 6.2, we have

$$\sum_{k=-\infty}^{\infty} \left\| \left( \frac{2^{k\alpha(\cdot)} \sum_{j=k-1}^{k+1} |[b,S_\beta](f_j)| \chi_k}{v \|b\|_*} \right)^{q_2(\cdot)} \right\|_{L^{\frac{p(\cdot)}{q_2(\cdot)}}}$$

$$\lesssim \sum_{k=-\infty}^{\infty} \left\| \frac{2^{k\alpha(\cdot)} \sum_{j=k-1}^{k+1} |[b,S_\beta](f_j)| \chi_k}{v \|b\|_*} \right\|_{L^{p(\cdot)}}^{(q_{2*})_k}$$

$$\lesssim \sum_{k=-\infty}^{\infty} \left\| \frac{\sum_{j=k-1}^{k+1} |[b,S_\beta](2^{j\alpha} f_j)| \chi_k}{v \|b\|_*} \right\|_{L^{p(\cdot)}}^{(q_{2*})_k}$$

$$\lesssim \sum_{k=-\infty}^{\infty} \left( \sum_{j=k-1}^{k+1} \left\| \frac{|[b,S_\beta](2^{j\alpha} f_j)| \chi_k}{v \|b\|_*} \right\|_{L^{p(\cdot)}} \right)^{(q_{2*})_k}$$

$$\lesssim \sum_{k=-\infty}^{\infty} \left( \sum_{j=k-1}^{k+1} \left\| \frac{[b,S_\beta](2^{j\alpha} f_j)}{v \|b\|_*} \right\|_{L^{p(\cdot)}} \right)^{(q_{2*})_k}.$$

After applying the boundedness of $[b,S_\beta]$ on $L^{p(\cdot)}(\mathbb{R}^n)$ (see [33]) and Lemma 6.3, we get

$$\sum_{k=-\infty}^{\infty} \left\| \left( \frac{2^{k\alpha(\cdot)} \sum_{j=k-1}^{k+1} |[b,S_\beta](f_j)| \chi_k}{v \|b\|_*} \right)^{q_2(\cdot)} \right\|_{L^{\frac{p(\cdot)}{q_2(\cdot)}}}$$

$$\lesssim \sum_{k=-\infty}^{\infty} \left( \sum_{j=k-1}^{k+1} \left\| \frac{2^{j\alpha(\cdot)} f_j}{v} \right\|_{L^{p(\cdot)}} \right)^{(q_{2*})_k}$$

$$\lesssim \sum_{k=-\infty}^{\infty} \left\| \frac{2^{k\alpha(\cdot)} f_k}{v} \right\|_{L^{p(\cdot)}}^{(q_{2*})_k}$$

$$\lesssim \sum_{k=-\infty}^{\infty} \left\| \left( \frac{2^{k\alpha(\cdot)} f_k}{v} \right)^{q_1(\cdot)} \right\|_{L^{p(\cdot)/q_1(\cdot)}}^{(q_{2*})_k/(q_1)_+}$$

$$\lesssim \left( \sum_{k=-\infty}^{\infty} \left\| \left( \frac{2^{k\alpha(\cdot)} f_k}{v} \right)^{q_1(\cdot)} \right\|_{L^{p(\cdot)/q_1(\cdot)}} \right)^{q_*}$$

$$\lesssim 1.$$

Hence we conclude that

$$\eta_2 \lesssim \|f\|_{\dot{K}_{p(\cdot),q_1(\cdot)}^{\alpha(\cdot)}(\mathbb{R}^n)}.$$

For $\eta_1$, noting that $x \in C_k$, $(y,t) \in \Gamma(x)$, $z \in C_j \bigcap \{z : |y-z| \le t\}$ and $j \le k-2$, we have

$$t \ge \frac{1}{2}(|x-y| + |y-z|) \ge \frac{1}{2}|x-z| \ge \frac{1}{2}(|x| - |z|) \ge \frac{1}{4}|x|.$$

Thus, we get

$$|[b,S_\beta](f_j)(x)|$$

$$= \left( \int \int_{\Gamma(x)} \sup_{\phi \in \mathcal{C}_\beta} \left| \int_{\mathbb{R}^n} (b(x)-b(z))\phi_t(y-z)f_j(z)dz \right|^2 \frac{dydt}{t^{n+1}} \right)^{\frac{1}{2}}$$

$$\lesssim \left( \int_{\frac{|x|}{4}}^{\infty} \int_{|x-y|<t} \left| t^{-n} \int_{C_j \cap \{z:|y-z|\leq t\}} (b(x)-b(z))f_j(z)dz \right|^2 \frac{dydt}{t^{n+1}} \right)^{\frac{1}{2}}$$

$$\lesssim \left( \int_{C_j} |b(x)-b(z)||f_j(z)|dz \right) \left( \int_{\frac{|x|}{4}}^{\infty} \frac{dt}{t^{2n+1}} \right)^{\frac{1}{2}}$$

$$\lesssim 2^{-kn} \int_{C_j} |b(x)-b(z)||f_j(z)|dz$$

$$\lesssim 2^{-kn} \left( |b(x)-b_{B_j}| \int_{C_j} |f_j(z)|dz + \int_{C_j} |b_{B_j}-b(z)||f_j(z)|dz \right)$$

$$\lesssim 2^{-kn} \|f_j\|_{L^{p(\cdot)}(\mathbb{R}^n)} \left( |b(x)-b_{B_j}|\|\chi_j\|_{L^{p'(\cdot)}(\mathbb{R}^n)} + \|(b_{B_j}-b)\chi_j\|_{L^{p'(\cdot)}(\mathbb{R}^n)} \right).$$

From this and Lemmas 6.4–6.6, we get

$$\|[b,S_\beta](f_j)\chi_k\|_{L^{p(\cdot)}(\mathbb{R}^n)}$$

$$\lesssim 2^{-kn}\|f_j\|_{L^{p(\cdot)}(\mathbb{R}^n)} \left( \|(b-b_{B_j})\chi_k\|_{L^{p(\cdot)}(\mathbb{R}^n)}\|\chi_j\|_{L^{p'(\cdot)}(\mathbb{R}^n)} \right.$$

$$+ \left. \|(b_{B_j}-b)\chi_j\|_{L^{p'(\cdot)}(\mathbb{R}^n)}\|\chi_k\|_{L^{p(\cdot)}(\mathbb{R}^n)} \right)$$

$$\lesssim \|b\|_* 2^{-kn}\|f_j\|_{L^{p(\cdot)}(\mathbb{R}^n)} \left( (k-j)\|\chi_{B_k}\|_{L^{p(\cdot)}(\mathbb{R}^n)}\|\chi_j\|_{L^{p'(\cdot)}(\mathbb{R}^n)} \right. \tag{6.10}$$

$$+ \left. \|\chi_{B_j}\|_{L^{p'(\cdot)}(\mathbb{R}^n)}\|\chi_k\|_{L^{p(\cdot)}(\mathbb{R}^n)} \right)$$

$$\lesssim (k-j)\|b\|_*\|f_j\|_{L^{p(\cdot)}(\mathbb{R}^n)} 2^{-kn}\|\chi_{B_k}\|_{L^{p(\cdot)}(\mathbb{R}^n)}\|\chi_{B_j}\|_{L^{p'(\cdot)}(\mathbb{R}^n)}$$

$$\lesssim (k-j)2^{(j-k)n\delta_2}\|b\|_*\|f_j\|_{L^{p(\cdot)}(\mathbb{R}^n)}.$$

We define

$$(q_{2**})_k := \begin{cases} (q_2)_+, & \left\| \left( \dfrac{2^{k\alpha(\cdot)}\sum_{j=-\infty}^{k-2}|[b,S_\beta](f_j)|\chi_k}{v\|b\|_*} \right)^{q_2(\cdot)} \right\|_{L^{\frac{p(\cdot)}{q_2(\cdot)}}} \geq 1, \\ (q_2)_-, & \text{otherwise.} \end{cases}$$

By Lemma 6.1, Lemma 6.2 and (6.10), we have

$$\sum_{k=-\infty}^{\infty}\left\|\left(\frac{2^{k\alpha(\cdot)}\sum_{j=-\infty}^{k-2}|[b,S_\beta](f_j)|\chi_k}{v\|b\|_*}\right)^{q_2(\cdot)}\right\|_{L^{\frac{p(\cdot)}{q_2(\cdot)}}}$$

$$\lesssim \sum_{k=-\infty}^{\infty}\left\|\frac{2^{k\alpha(\cdot)}\sum_{j=-\infty}^{k-2}|[b,S_\beta](f_j)|\chi_k}{v\|b\|_*}\right\|_{L^{p(\cdot)}}^{(q_{2**})_k}$$

$$\lesssim \sum_{k=-\infty}^{\infty}\left\|\frac{\sum_{j=-\infty}^{k-2}2^{(k-j)\alpha_+}|[b,S_\beta](2^{j\alpha}f_j)|\chi_k}{v\|b\|_*}\right\|_{L^{p(\cdot)}}^{(q_{2**})_k}$$

$$\lesssim \sum_{k=-\infty}^{\infty}\left(\sum_{j=-\infty}^{k-2}2^{(k-j)\alpha_+}\left\|\frac{|[b,S_\beta](2^{j\alpha}f_j)|\chi_k}{v\|b\|_*}\right\|_{L^{p(\cdot)}}\right)^{(q_{2**})_k}$$

$$\lesssim \sum_{k=-\infty}^{\infty}\left(\sum_{j=-\infty}^{k-2}(k-j)2^{(k-j)(\alpha_+-n\delta_2)}\left\|\frac{2^{j\alpha}f_j}{v}\right\|_{L^{p(\cdot)}}\right)^{(q_{2**})_k}$$

$$\lesssim \sum_{k=-\infty}^{\infty}\left(\sum_{j=-\infty}^{k-2}(k-j)2^{(k-j)(\alpha_+-n\delta_2)}\left\|\left(\frac{2^{j\alpha(\cdot)}f_j}{v}\right)^{q_1(\cdot)}\right\|_{L^{p(\cdot)/q_1(\cdot)}}^{1/(q_1)_+}\right)^{(q_{2**})_k}.$$

When $(q_1)_+ \leq 1$, by Lemma 6.1 and Lemma 6.3, we have

$$\sum_{k=-\infty}^{\infty}\left\|\left(\frac{2^{k\alpha(\cdot)}\sum_{j=-\infty}^{k-2}|[b,S_\beta](f_j)|\chi_k}{v\|b\|_*}\right)^{q_2(\cdot)}\right\|_{L^{\frac{p(\cdot)}{q_2(\cdot)}}}$$

$$\lesssim \sum_{k=-\infty}^{\infty}\left(\sum_{j=-\infty}^{k-2}(k-j)2^{(k-j)(\alpha_+-n\delta_2)}\left\|\left(\frac{2^{j\alpha(\cdot)}f_j}{v}\right)^{q_1(\cdot)}\right\|_{L^{p(\cdot)/q_1(\cdot)}}\right)^{(q_{2**})_k/(q_1)_+}$$

$$\lesssim \left(\sum_{k=-\infty}^{\infty}\sum_{j=-\infty}^{k-2}(k-j)^{(q_1)_+}2^{(k-j)(\alpha_+-n\delta_2)(q_1)_+}\left\|\left(\frac{2^{j\alpha(\cdot)}f_j}{v}\right)^{q_1(\cdot)}\right\|_{L^{p(\cdot)/q_1(\cdot)}}\right)^{q_*}$$

$$\lesssim \left(\sum_{j=-\infty}^{\infty}\sum_{k=j+2}^{\infty}(k-j)^{(q_1)_+}2^{(k-j)(\alpha_+-n\delta_2)(q_1)_+}\left\|\left(\frac{2^{j\alpha(\cdot)}f_j}{v}\right)^{q_1(\cdot)}\right\|_{L^{p(\cdot)/q_1(\cdot)}}\right)^{q_*}$$

$$\lesssim \left(\sum_{j=-\infty}^{\infty}\left\|\left(\frac{2^{j\alpha(\cdot)}f_j}{v}\right)^{q_1(\cdot)}\right\|_{L^{p(\cdot)/q_1(\cdot)}}\right)^{q_*}$$

$$\lesssim 1.$$

When $(q_1)_+ > 1$, by applying the Hölder's inequality and Lemma 6.3, give

$$\sum_{k=-\infty}^{\infty}\left\|\left(\frac{2^{k\alpha(\cdot)}\sum_{j=-\infty}^{k-2}|[b,S_\beta](f_j)|\chi_k}{v\|b\|_*}\right)^{q_2(\cdot)}\right\|_{L^{\frac{p(\cdot)}{q_2(\cdot)}}}$$

$$\lesssim \sum_{k=-\infty}^{\infty}\left\{\left(\sum_{j=-\infty}^{k-2}2^{(k-j)(\alpha_+-n\delta_2)(q_1)_+/2}\left\|\left(\frac{2^{j\alpha(\cdot)}f_j}{\nu}\right)^{q_1(\cdot)}\right\|_{L^{p(\cdot)/q_1(\cdot)}}\right)^{1/q_{1+}}\right.$$

$$\left.\times\left(\sum_{j=-\infty}^{k-2}(k-j)^{((q_1)_+)'}2^{(k-j)(\alpha_+-n\delta_2)((q_1)_+)'/2}\right)^{1/((q_1)_+)'}\right\}^{(q_{2**})_k}$$

$$\lesssim\left(\sum_{k=-\infty}^{\infty}\sum_{j=-\infty}^{k-2}2^{(k-j)(\alpha_+-n\delta_2)(q_1)_+/2}\left\|\left(\frac{2^{j\alpha(\cdot)}f_j}{\nu}\right)^{q_1(\cdot)}\right\|_{L^{p(\cdot)/q_1(\cdot)}}\right)^{q_*}$$

$$\lesssim\left(\sum_{j=-\infty}^{\infty}\sum_{k=j+2}^{\infty}2^{(k-j)(\alpha_+-n\delta_2)(q_1)_+/2}\left\|\left(\frac{2^{j\alpha(\cdot)}f_j}{\nu}\right)^{q_1(\cdot)}\right\|_{L^{p(\cdot)/q_1(\cdot)}}\right)^{q_*}$$

$$\lesssim\left(\sum_{j=-\infty}^{\infty}\left\|\left(\frac{2^{j\alpha(\cdot)}f_j}{\nu}\right)^{q_1(\cdot)}\right\|_{L^{p(\cdot)/q_1(\cdot)}}\right)^{q_*}$$

$$\lesssim 1.$$

Thus we have

$$\eta_1\lesssim\|f\|_{\dot{K}^{\alpha(\cdot)}_{p(\cdot),q_1(\cdot)}(\mathbb{R}^n)}.$$

For $\eta_3$, noting that if $x\in C_k$, $(y,t)\in\Gamma(x)$, $z\in C_j\bigcap\{z:|y-z|\leq t\}$ and $j\geq k+2$, then we have

$$t\geq\frac{1}{2}(|x-y|+|y-z|)\geq\frac{1}{2}|x-z|\geq\frac{1}{2}(|z|-|x|)\geq\frac{1}{4}|z|.$$

Hölder's inequality implies that

$$|[b,S_\beta](f_j)(x)|$$

$$=\left(\int\int_{\Gamma(x)}\sup_{\phi\in\mathcal{C}_\beta}\left|\int_{\mathbb{R}^n}(b(x)-b(z))\phi_t(y-z)f_j(z)dz\right|^2\frac{dydt}{t^{n+1}}\right)^{\frac{1}{2}}$$

$$\lesssim\left(\int_{\frac{|z|}{4}}^{\infty}\int_{|x-y|<t}\left|t^{-n}\int_{C_j\bigcap\{z:|y-z|\leq t\}}(b(x)-b(z))f_j(z)dz\right|^2\frac{dydt}{t^{n+1}}\right)^{\frac{1}{2}}$$

$$\lesssim\left(\int_{C_j}|b(x)-b(z)||f_j(z)|dz\right)\left(\int_{\frac{|z|}{4}}^{\infty}\frac{dt}{t^{2n+1}}\right)^{\frac{1}{2}}$$

$$\lesssim 2^{-jn}\int_{C_j}|b(x)-b(z)||f_j(z)|dz$$

$$\lesssim 2^{-jn}\left(|b(x)-b_{B_k}|\int_{C_j}||f_j(z)|dz+\int_{C_j}|b_{B_k}-b(z)||f_j(z)|dz\right)$$

$$\lesssim 2^{-jn}\|f_j\|_{L^{p(\cdot)}(\mathbb{R}^n)}\left(|b(x)-b_{B_k}|\|\chi_j\|_{L^{p'(\cdot)}(\mathbb{R}^n)}+\|(b_{B_k}-b)\chi_j\|_{L^{p'(\cdot)}(\mathbb{R}^n)}\right)$$

Applying Lemma 6.4, Lemma 6.5 and Lemma 6.6, we have

$$
\begin{aligned}
&\|[b, S_\beta](f_j)\chi_k\|_{L^{p(\cdot)}(\mathbb{R}^n)} \\
&\lesssim 2^{-jn}\|f_j\|_{L^{p(\cdot)}(\mathbb{R}^n)}\Big( \|(b - b_{B_k})\chi_k\|_{L^{p(\cdot)}(\mathbb{R}^n)}\|\chi_j\|_{L^{p'(\cdot)}(\mathbb{R}^n)} \\
&\quad + \|(b_{B_k} - b)\chi_j\|_{L^{p'(\cdot)}(\mathbb{R}^n)}\|\chi_k\|_{L^{p(\cdot)}(\mathbb{R}^n)}\Big) \\
&\lesssim \|b\|_* 2^{-jn}\|f_j\|_{L^{p(\cdot)}(\mathbb{R}^n)}\Big( (j-k)\|\chi_{B_k}\|_{L^{p(\cdot)}(\mathbb{R}^n)}\|\chi_{B_j}\|_{L^{p'(\cdot)}(\mathbb{R}^n)} \\
&\quad + \|\chi_j\|_{L^{p'(\cdot)}(\mathbb{R}^n)}\|\chi_{B_k}\|_{L^{p(\cdot)}(\mathbb{R}^n)}\Big) \\
&\lesssim (j-k)\|b\|_* 2^{-jn}\|f_j\|_{L^{p(\cdot)}(\mathbb{R}^n)}\|\chi_{B_k}\|_{L^{p(\cdot)}(\mathbb{R}^n)}\|\chi_{B_j}\|_{L^{p'(\cdot)}(\mathbb{R}^n)} \\
&\lesssim (j-k)\|b\|_* 2^{(k-j)n\delta_1}\|f_j\|_{L^{p(\cdot)}(\mathbb{R}^n)}.
\end{aligned}
\tag{6.11}
$$

This together with Lemma 6.1 and Lemma 6.2 gives

$$
\begin{aligned}
&\sum_{k=-\infty}^{\infty}\left\|\left(\frac{2^{k\alpha(\cdot)}\sum_{j=k+2}^{\infty}|[b, S_\beta](f_j)|\chi_k}{v\|b\|_*}\right)^{q_2(\cdot)}\right\|_{L^{\frac{p(\cdot)}{q_2(\cdot)}}} \\
&\lesssim \sum_{k=-\infty}^{\infty}\left\|\frac{2^{k\alpha(\cdot)}\sum_{j=k+2}^{\infty}|[b, S_\beta](f_j)|\chi_k}{v\|b\|_*}\right\|_{L^{p(\cdot)}}^{(q_{2***})_k} \\
&\lesssim \sum_{k=-\infty}^{\infty}\left\|\frac{\sum_{j=k+2}^{\infty}2^{(k-j)\alpha_-}|[b, S_\beta](2^{j\alpha}f_j)|\chi_k}{v\|b\|_*}\right\|_{L^{p(\cdot)}}^{(q_{2***})_k} \\
&\lesssim \sum_{k=-\infty}^{\infty}\left(\sum_{j=k+2}^{\infty}2^{(k-j)\alpha_-}\left\|\frac{|[b, S_\beta](2^{j\alpha}f_j)|\chi_k}{v\|b\|_*}\right\|_{L^{p(\cdot)}}\right)^{(q_{2***})_k} \\
&\lesssim \sum_{k=-\infty}^{\infty}\left(\sum_{j=k+2}^{\infty}(j-k)2^{(k-j)(\alpha_- + n\delta_1)}\left\|\frac{2^{j\alpha(\cdot)}f_j}{v}\right\|_{L^{p(\cdot)}}\right)^{(q_{2***})_k} \\
&\lesssim \sum_{k=-\infty}^{\infty}\left(\sum_{j=k+2}^{\infty}(j-k)2^{(k-j)(\alpha_- + n\delta_1)}\left\|\left(\frac{2^{j\alpha(\cdot)}f_j}{v}\right)^{q_1(\cdot)}\right\|_{L^{p(\cdot)/q_1(\cdot)}}^{1/(q_1)_+}\right)^{(q_{2***})_k},
\end{aligned}
$$

where

$$
(q_{2***})_k := \begin{cases} (q_2)_+, & \left\|\left(\dfrac{2^{k\alpha(\cdot)}\sum_{j=k+2}^{\infty}|[b, S_\beta](f_j)|\chi_k}{v\|b\|_*}\right)^{q_2(\cdot)}\right\|_{L^{\frac{p(\cdot)}{q_2(\cdot)}}} \geq 1, \\ (q_2)_-, & \text{otherwise.} \end{cases}
$$

When $(q_1)_+ \leq 1$, by Lemma 6.1 and Lemma 6.3, we have

$$\sum_{k=-\infty}^{\infty} \left\| \left( \frac{2^{k\alpha(\cdot)} \sum_{j=k+2}^{\infty} |[b,S_\beta](f_j)|\chi_k}{v\|b\|_*} \right)^{q_2(\cdot)} \right\|_{L^{\frac{p(\cdot)}{q_2(\cdot)}}}$$

$$\lesssim \sum_{k=-\infty}^{\infty} \left( \sum_{j=k+2}^{\infty} (j-k)2^{(k-j)(\alpha_-+n\delta_1)} \left\| \left( \frac{2^{j\alpha(\cdot)}f_j}{v} \right)^{q_1(\cdot)} \right\|_{L^{p(\cdot)/q_1(\cdot)}} \right)^{(q_{2***})_k/(q_1)_+}$$

$$\lesssim \left( \sum_{k=-\infty}^{\infty} \sum_{j=k+2}^{\infty} (j-k)^{(q_1)_+} 2^{(k-j)(\alpha_-+n\delta_1)(q_1)_+} \left\| \left( \frac{2^{j\alpha(\cdot)}f_j}{v} \right)^{q_1(\cdot)} \right\|_{L^{p(\cdot)/q_1(\cdot)}} \right)^{q_*}$$

$$\lesssim \left( \sum_{j=-\infty}^{\infty} \sum_{k=-\infty}^{j-2} (j-k)^{(q_1)_+} 2^{(k-j)(\alpha_-+n\delta_1)(q_1)_+} \left\| \left( \frac{2^{j\alpha(\cdot)}f_j}{v} \right)^{q_1(\cdot)} \right\|_{L^{p(\cdot)/q_1(\cdot)}} \right)^{q_*}$$

$$\lesssim \left( \sum_{j=-\infty}^{\infty} \left\| \left( \frac{2^{j\alpha(\cdot)}f_j}{v} \right)^{q_1(\cdot)} \right\|_{L^{p(\cdot)/q_1(\cdot)}} \right)^{q_*}$$

$$\lesssim 1.$$

When $(q_1)_+ > 1$, by Hölder's inequality and Lemma 6.3, we have

$$\sum_{k=-\infty}^{\infty} \left\| \left( \frac{2^{k\alpha(\cdot)} \sum_{j=k+2}^{\infty} |[b,S_\beta](f_j)|\chi_k}{v\|b\|_*} \right)^{q_2(\cdot)} \right\|_{L^{\frac{p(\cdot)}{q_2(\cdot)}}}$$

$$\lesssim \sum_{k=-\infty}^{\infty} \left\{ \left( \sum_{j=k+2}^{\infty} 2^{(k-j)(\alpha_-+n\delta_1)(q_1)_+/2} \left\| \left( \frac{2^{j\alpha(\cdot)}f_j}{v} \right)^{q_1(\cdot)} \right\|_{L^{p(\cdot)/q_1(\cdot)}} \right)^{1/q_{1+}} \right.$$

$$\left. \times \left( \sum_{j=k+2}^{\infty} (j-k)^{((q_1)_+)'} 2^{(k-j)(\alpha_-+n\delta_1)((q_1)_+)'/2} \right)^{1/((q_1)_+)'} \right\}^{(q_{2***})_k}$$

$$\lesssim \left( \sum_{k=-\infty}^{\infty} \sum_{j=k+2}^{\infty} 2^{(k-j)(\alpha_-+n\delta_1)(q_1)_+/2} \left\| \left( \frac{2^{j\alpha(\cdot)}f_j}{v} \right)^{q_1(\cdot)} \right\|_{L^{p(\cdot)/q_1(\cdot)}} \right)^{q_*}$$

$$\lesssim \left( \sum_{j=-\infty}^{\infty} \sum_{k=-\infty}^{j-2} 2^{(k-j)(\alpha_-+n\delta_1)(q_1)_+/2} \left\| \left( \frac{2^{j\alpha(\cdot)}f_j}{v} \right)^{q_1(\cdot)} \right\|_{L^{p(\cdot)/q_1(\cdot)}} \right)^{q_*}$$

$$\lesssim \left( \sum_{j=-\infty}^{\infty} \left\| \left( \frac{2^{j\alpha(\cdot)}f_j}{v} \right)^{q_1(\cdot)} \right\|_{L^{p(\cdot)/q_1(\cdot)}} \right)^{q_*}$$

$$\lesssim 1.$$

Hence we derive the estimate

$$\eta_3 \lesssim \|f\|_{\dot{K}^{\alpha(\cdot)}_{p(\cdot),q_1(\cdot)}(\mathbb{R}^n)}.$$

Combining the estimates of $\eta_1, \eta_2$ and $\eta_3$, we complete the proof of Theorem 6.2.

## 6.4　Acknowledgement

This work was supported by the Natural Science Foundation of Anhui Higher Education Institutions (KJ2021A1050).

# References


[1] Almeida, A. and Drihem, D. 2012. Maximal, potential and singular type operators on Herz spaces with variable exponents. J. Math. Anal. Appl. 394: 781–795.

[2] Chen, Y., Levine, S. and Rao, R. 2006. Variable exponent, linear growth functionals in image restoration. SIAM J. Appl. Math. 66: 1383–1406.

[3] Chen, J., Wang, L. and Liu, K. 2022. Boundedness of the intrinsic square function on grand variable Herz spaces. J. Math. Res. Appl. 42: 611–627.

[4] Cruz-uribe, D. and Fiorenza, A. 2013. Variable Lebesgue spaces: Foundations and harmonic analysis. Applied and Numerical Harmonic Analysis, Birkhäuser, Basel.

[5] Diening, L. 2004. Maximal function on generalized Lebesgue spaces $L^{p(\cdot)}$. Math. Inequal. Appl. 7: 245–253.

[6] Diening, L., Harjulehto, P., Hästö, P. and Růžička, M. 2011. Lebesgue and Sobolev spaces with variable exponents. Volume 2017 of Lecture Notes in Mathematics, Springer, Heidelberg.

[7] Guliyev, V., Omarova, M. and Sawano, Y. 2015. Boundedness of intrinsic square functions and their commutators on generalized weighted Orlicz-Morrey spaces. Banach J. Math. Anal. 9: 44–62.

[8] Gürbüz, F. 2022. Some estimates for intrinsic square functions and commutators on homogeneous Herz spaces with variable exponents. Eğitim Publishing, Editor: Gürbüz, Ferit, 1st Edition, April 2022, Volume:1, pp. 40, ISBN: 978-625-8108-88-0 (In Turkish).

[9] *Gürbüz*, F., Ding, S., Han, H. and Long, P. 2021. Characterizations of rough fractional-type integral Operators on variable exponent vanishing Morrey-type spaces. pp. 95–123. *In*: Dutta, H. (ed.). Topics in Contemporary Mathematical Analysis and Applications. CRC Press.

[10] *Gürbüz*, F., Ding, S., Han, H. and Long, P. 2021. Norm inequalities on variable exponent vanishing Morrey type spaces for the rough singular type integral operators. Int. J. Nonlinear Sci. Numer. Simul. 22: 721–739.

[11] Harjulehto, P., Hästö, P., Lê, Ú.V. and Nuortio, M. 2010. Overview of differential equations with non-standard growth. Nonlinear Anal. 72: 4551–4574.

[12] Ho, K.P. 2017. Intrinsic square functions on Morrey and Block spaces with variable exponents. Bull. Malays. Math. Sci. Soc. 40: 995–1010.

[13] Hu, Y. and Wang, Y. 2016. The commutators of intrinsic square functions on weighted Herz spaces. Bull. Malays. Math. Sci. Soc. 39: 1421–1437.

[14] Izuki, M. 2009. Herz and amalgam spaces with variable exponent, the Haar wavelets and greediness of the wavelet system. East J. Approx. 15: 87–109.

[15] Izuki, M. 2010. Commutators of fractional integrals on Lebesgue and Herz spaces with variable exponent. Rend. Circ. Mat. Palermo. 59: 461–472.

[16] Izuki, M. and Noi, T. 2011. Boundedness of some integral operators and commutators on generalized Herz spaces with variable exponents. OCAMI Preprint Ser, 11–15.

[17] Izuki, M. and Noi, T. 2017. An intrinsic square function on weighted Herz spaces with variable exponent. J. Math. Inequal. 11: 799–816.

[18] Kokilashvili, V., Meskhi, A., Rafeiro, H. and Samko, S. 2016. Integral operators in non-standard function spaces. Vol. I–II, Birkhäuser/Springer.

[19] Kováčik, O. and Rákosník, J. 1991. On spaces $L^{p(x)}$ and $W^{k,p(x)}$. Czechoslovak Math. J. 41: 592–618.

[20] Lerner, A.K. 2011. Sharp weighted norm inequalities for Littlewood-Paley operators and singular integrals. Adv. Math. 226: 3912–3926.

[21] Lu, S., Yang, D. and Hu, G. 2008. Herz Type Spaces and their Applications. Science Press, Beijing.

[22] Nakai, E. and Sawano, Y. 2012. Hardy spaces with variable exponents and generalized Campanato spaces. J. Funct. Anal. 262: 3665–3748.

[23] Nekvinda, A. 2004. Hardy-littlewood maximal operator in $L^{p(x)}(\mathbb{R}^n)$. Math. Ineq. Appl. 7: 255–265.

[24] Orlicz, W. 1931. Uber konjugierte Exponentenfolgen. Studia Math. 3: 200–211.

[25] Ragusa, M. 2009. Homogeneous Herz spaces and regularity results. Nonlinear Anal. 71: 1909–1914.

[26] Růžička, M. 2000. Electrorheological Fluids: Modeling and Mathematical Theory. Springer-Verlag, Berlin.

[27] Samko, S. 2013. Variable exponent Herz spaces. Mediterr. J. Math. 10: 2007–2025.

[28] Tan, J. 2023. Real-variable theory of local variable Hardy spaces. Acta Math. Sin. (Engl. Ser.) 39: 1229–1262.

[29] Wang, H. 2012. Intrinsic square functions on the weighted Morrey spaces. J. Math. Anal. Appl. 396: 302–314.

[30] Wang, H. 2014. The boundedness of intrinsic square functions on the weighted Herz spaces. J. Funct. Spaces, Art. ID 274521, 14 pp.

[31] Wang, H. and Liu, Z. 2012. The Herz-type Hardy spaces with variable exponent and their applications. Taiwanese. J. Math. 16: 1363–1389.

[32] Wang, H. and Liu, Z. 2020. Boundedness of singular integral operators on weak Herz type spaces with variable exponent. Ann. Funct. Anal. 11: 1363–1389.

[33] Wang, L. 2018. Boundedness of the commutator of the intrinsic square function in variable exponent spaces. J. Korean Math. Soc. 55: 939–962.

[34] Wang, L. and Shu, L. 2018. Boundedness of the intrinsic square functions on variable exponent Herz and Herz-Hardy spaces. Acta Math. Scientia Ser. A 38A: 716–727.

[35] Wilson, M. 2007. The intrinsic square function. Rev. Mat. Iberoamericana 23: 771–791.

[36] Wilson, M. 2007. Weighted Littlewood-Paley Theory and Exponential-Square Integrability. *In*: Lecture Notes in Math., Vol. 1924, Springer-Verlag.

[37] Yang, D., Zhuo, C. and Yuan, W. 2015. Besov-type spaces with variable smoothness and integrability. J. Funct. Anal. 269: 1840–1898.

[38] Yu, X. and Liu, Z. 2021. Boundedness of some integral operators and commutators on homogeneous Herz spaces with three variable exponents. Front. Math. China 16: 211–237.

[39] Zhuo, C., Yang, D. and Liang, Y. 2016. Intrinsic square function characterizations of Hardy spaces with variable exponents. Bull. Malays. Math. Sci. Soc. 39: 1541–1577.

# Chapter 7

# $q$-Deformed and $\lambda$-Parametrized Hyperbolic Tangent Function Relied Complex Valued Trigonometric and Hyperbolic Neural Network High Order Approximations

*George A. Anastassiou*

## 7.1 Introduction

The author in [1] and [2], see Chapters 2–5, was the first to establish neural network approximation to continuous functions with rates by very specifically de-

Department of Mathematical Sciences, University of Memphis, Memphis, TN 38152, U.S.A.
Email: ganastss@memphis.edu1

fined neural network operators of Cardaliaguet-Euvrard and "Squashing" types, by employing the modulus of continuity of the engaged function or its high order derivative, and producing very tight Jackson type inequalities. He treats both the univariate and multivariate cases. The defining these operators "bell-shaped" and "squashing" functions are assumed to be of compact support.

Again the author inspired by [12], continued his studies on neural networks approximation by introducing and using the proper quasi-interpolation operators of sigmoidal and hyperbolic tangent type which resulted into [3]–[7], by treating both the univariate and multivariate cases.

Let $h$ be a general sigmoid activation function with $h(0) = 0$, and $y = \pm 1$ the horizontal asymptotes. Of course $h$ is strictly increasing over $\mathbb{R}$. Let the parameter $0 < r < 1$ and $x > 0$. Then clearly $-x < x$ and $-x < -rx < rx < x$, furthermore it holds $h(-x) < h(-rx) < h(rx) < h(x)$. Consequently the sigmoid $y = h(rx)$ has a graph inside the graph of $y = h(x)$, of course with the same asymptotes $y = \pm 1$. Therefore $h(rx)$ has derivatives (gradients) at more points $x$ than $h(x)$ has different than zero or not as close to zero, thus killing less number of neurons! And of course $h(rx)$ is more distant from $y = \pm 1$, than $h(x)$ it is. A highly desired fact in Neural Networks theory.

Different activation functions allow for different non-linearities which might work better for solving a specific function. So the need to use neural networks with various activation functions is vivid. Thus, performing neural network approximations using different activation functions is not only necessary but fully justified.

Also brain asymmetry has been observed in animals and humans in terms of structure, function and behaviour. This lateralization is thought to reflect evolutionary, hereditary, developmental, experiential and pathological factors. Therefore it is natural to consider for our study deformed neural network activation functions and operators. So this chapter is a specific study under this philosophy of approaching reality as close as possible.

Consequently the author here performs $q$-deformed and $\lambda$-parametrized hyperbolic tangent function activated high order neural network approximations to continuous functions over compact intervals of the real line with complex values. All convergences are with rates expressed via the modulus of continuity of the involved functions high order derivatives, deriving by very tight Jackson type inequalities.

The basis of our higher order approximations are some newly discovered by the author trigonometric and hyperbolic type Taylor's formulae.

Our compact intervals are not necessarily symmetric to the origin. In preparation to prove our results we describe important properties of the basic density function defining our operators which is induced by a $q$-deformed and $\lambda$-parametrized hyperbolic tangent function, which is a sigmoid activation function.

Feed-forward $X$-valued neural networks (FNNs) with one hidden layer, the only type of networks we deal with in this article, are mathematically expressed as

$$N_n(x) = \sum_{j=0}^{n} c_j \sigma\left(\langle a_j \cdot x \rangle + b_j\right), \quad x \in \mathbb{R}^s, \ s \in \mathbb{N},$$

where for $0 \le j \le n$, $b_j \in \mathbb{R}$ are the thresholds, $a_j \in \mathbb{R}^s$ are the connection weights, $c_j \in \mathbb{C}$ are the coefficients, $\langle a_j \cdot x \rangle$ is the inner product of $a_j$ and $x$, and $\sigma$ is the activation function of the network. About neural networks in general read [13], [14], [15].

## 7.2  About $q$-Deformed and $\lambda$-parametrized Hyperbolic Tangent Function $g_{q,\lambda}$

Here all this background comes from [10, Ch. 17].

We use $g_{q,\lambda}$, see (1), and exhibit that it is a sigmoid function and we will present several of its properties related to the approximation by neural network operators.

So, let us consider the function

$$g_{q,\lambda}(x) := \frac{e^{\lambda x} - q e^{-\lambda x}}{e^{\lambda x} + q e^{-\lambda x}}, \quad \lambda, q > 0, \ x \in \mathbb{R}. \tag{1}$$

We have that

$$g_{q,\lambda}(0) = \frac{1-q}{1+q}.$$

We notice also that

$$g_{q,\lambda}(-x) = \frac{e^{-\lambda x} - q e^{\lambda x}}{e^{-\lambda x} + q e^{\lambda x}} = \frac{\frac{1}{q} e^{-\lambda x} - e^{\lambda x}}{\frac{1}{q} e^{-\lambda x} + e^{\lambda x}} = -\frac{\left(e^{\lambda x} - \frac{1}{q} e^{-\lambda x}\right)}{e^{\lambda x} + \frac{1}{q} e^{-\lambda x}} = -g_{\frac{1}{q},\lambda}(x). \tag{2}$$

That is

$$g_{q,\lambda}(-x) = -g_{\frac{1}{q},\lambda}(x), \quad \forall\, x \in \mathbb{R}, \tag{3}$$

and

$$g_{\frac{1}{q},\lambda}(x) = -g_{q,\lambda}(-x),$$

hence

$$g'_{\frac{1}{q},\lambda}(x) = g'_{q,\lambda}(-x). \tag{4}$$

It is

$$g_{q,\lambda}(x) = \frac{e^{2\lambda x} - q}{e^{2\lambda x} + q} = \frac{1 - \frac{q}{e^{2\lambda x}}}{1 + \frac{q}{e^{2\lambda x}}} \underset{(x \to +\infty)}{\to} 1,$$

i.e.,

$$g_{q,\lambda}(+\infty) = 1, \tag{5}$$

Furthermore

$$g_{q,\lambda}(x) = \frac{e^{2\lambda x} - q}{e^{2\lambda x} + q} \underset{(x \to -\infty)}{\to} \frac{-q}{q} = -1,$$

i.e.,

$$g_{q,\lambda}(-\infty) = -1. \tag{6}$$

We find that

$$g'_{q,\lambda}(x) = \frac{4q\lambda e^{2\lambda x}}{\left(e^{2\lambda x} + q\right)^2} > 0, \tag{7}$$

therefore $g_{q,\lambda}$ is striclty increasing.

Next we obtain ($x \in \mathbb{R}$)

$$g''_{q,\lambda}(x) = 8q\lambda^2 e^{2\lambda x}\left(\frac{q - e^{2\lambda x}}{\left(e^{2\lambda x} + q\right)^3}\right) \in C(\mathbb{R}). \tag{8}$$

We observe that

$$q - e^{2\lambda x} \gtrless 0 \Leftrightarrow q \gtrless e^{2\lambda x} \Leftrightarrow \ln q \gtrless 2\lambda x \Leftrightarrow x \lessgtr \frac{\ln q}{2\lambda}.$$

So, in case of $x < \frac{\ln q}{2\lambda}$, we have that $g_{q,\lambda}$ is strictly concave up, with $g''_{q,\lambda}\left(\frac{\ln q}{2\lambda}\right) = 0$.

And in case of $x > \frac{\ln q}{2\lambda}$, we have that $g_{q,\lambda}$ is strictly concave down.

Clearly, $g_{q,\lambda}$ is a shifted sigmoid function with $g_{q,\lambda}(0) = \frac{1-q}{1+q}$, and $g_{q,\lambda}(-x) = -g_{q^{-1},\lambda}(x)$, (a semi-odd function), see also [9].

By $1 > -1$, $x + 1 > x - 1$, we consider the activation function

$$M_{q,\lambda}(x) := \frac{1}{4}\left(g_{q,\lambda}(x+1) - g_{q,\lambda}(x-1)\right) > 0, \tag{9}$$

$\forall x \in \mathbb{R}$; $q, \lambda > 0$. Notice that $M_{q,\lambda}(\pm\infty) = 0$, so the $x$-axis is horizontal asymptote.

We have that

$$M_{q,\lambda}(-x) = \frac{1}{4}\left(g_{q,\lambda}(-x+1) - g_{q,\lambda}(-x-1)\right) =$$

$$\frac{1}{4}\left(g_{q,\lambda}(-(x-1)) - g_{q,\lambda}(-(x+1))\right) =$$

$$\frac{1}{4}\left(-g_{\frac{1}{q},\lambda}(x-1) + g_{\frac{1}{q},\lambda}(x+1)\right) = \tag{10}$$

$$\frac{1}{4}\left(g_{\frac{1}{q},\lambda}(x+1) - g_{\frac{1}{q},\lambda}(x-1)\right) = M_{\frac{1}{q},\lambda}(x), \quad \forall\, x \in \mathbb{R}.$$

Thus

$$M_{q,\lambda}(-x) = M_{\frac{1}{q},\lambda}(x), \quad \forall\, x \in \mathbb{R};\; q,\lambda > 0, \tag{11}$$

a deformed symmetry.

Next, we have that

$$M'_{q,\lambda}(x) = \frac{1}{4}\left(g'_{q,\lambda}(x+1) - g'_{q,\lambda}(x-1)\right), \quad \forall\, x \in \mathbb{R}. \tag{12}$$

Let $x < \frac{\ln q}{2\lambda} - 1$, then $x - 1 < x + 1 < \frac{\ln q}{2\lambda}$ and $g'_{q,\lambda}(x+1) > g'_{q,\lambda}(x-1)$ (by $g_{q,\lambda}$ being strictly concave up for $x < \frac{\ln q}{2\lambda}$), that is $M'_{q,\lambda}(x) > 0$. Hence $M_{q,\lambda}$ is striclty increasing over $\left(-\infty, \frac{\ln q}{2\lambda} - 1\right)$.

Let now $x - 1 > \frac{\ln q}{2\lambda}$, then $x + 1 > x - 1 > \frac{\ln q}{2\lambda}$, and $g'_{q,\lambda}(x+1) < g'_{q,\lambda}(x-1)$, that is $M'_{q,\lambda}(x) < 0$.

Therefore $M_{q,\lambda}$ is strictly decreasing over $\left(\frac{\ln q}{2\lambda} + 1, +\infty\right)$.

Let us next consider, $\frac{\ln q}{2\lambda} - 1 \le x \le \frac{\ln q}{2\lambda} + 1$. We have that

$$M''_{q,\lambda}(x) = \frac{1}{4}\left(g''_{q,\lambda}(x+1) - g''_{q,\lambda}(x-1)\right) =$$

$$2q\lambda^2\left[e^{2\lambda(x+1)}\left(\frac{q - e^{2\lambda(x+1)}}{\left(e^{2\lambda(x+1)} + q\right)^3}\right) - e^{2\lambda(x-1)}\left(\frac{q - e^{2\lambda(x-1)}}{\left(e^{2\lambda(x-1)} + q\right)^3}\right)\right]. \tag{13}$$

By $\frac{\ln q}{2\lambda} - 1 \le x \Leftrightarrow \frac{\ln q}{2\lambda} \le x + 1 \Leftrightarrow \ln q \le 2\lambda(x+1) \Leftrightarrow q \le e^{2\lambda(x+1)} \Leftrightarrow q - e^{2\lambda(x+1)} \le 0$.

By $x \le \frac{\ln q}{2\lambda} + 1 \Leftrightarrow x - 1 \le \frac{\ln q}{2\lambda} \Leftrightarrow 2\lambda(x-1) \le \ln q \Leftrightarrow e^{2\lambda(x-1)} \le q \Leftrightarrow q - e^{2\lambda\beta(x-1)} \ge 0$.

Clearly by (13) we get that $M''_{q,\lambda}(x) \le 0$, for $x \in \left[\frac{\ln q}{2\lambda} - 1, \frac{\ln q}{2\lambda} + 1\right]$.

More precisely $M_{q,\lambda}$ is concave down over $\left[\frac{\ln q}{2\lambda} - 1, \frac{\ln q}{2\lambda} + 1\right]$, and strictly concave down over $\left(\frac{\ln q}{2\lambda} - 1, \frac{\ln q}{2\lambda} + 1\right)$.

Consequently $M_{q,\lambda}$ has a bell-type shape over $\mathbb{R}$.

Of course it holds $M''_{q,\lambda}\left(\frac{\ln q}{2\lambda}\right) < 0$.

At $x = \frac{\ln q}{2\lambda}$, we have

$$M'_{q,\lambda}(x) = \frac{1}{4}\left(g'_{q,\lambda}(x+1) - g'_{q,\lambda}(x-1)\right) =$$

$$q\lambda\left(\frac{e^{2\lambda(x+1)}}{\left(e^{2\lambda(x+1)} + q\right)^2} - \frac{e^{2\lambda(x-1)}}{\left(e^{2\lambda(x-1)} + q\right)^2}\right). \tag{14}$$

Thus

$$M'_{q,\lambda}\left(\frac{\ln q}{2\lambda}\right) = q\lambda\left(\frac{e^{2\lambda\left(\frac{\ln q}{2\lambda}+1\right)}}{\left(e^{2\lambda\left(\frac{\ln q}{2\lambda}+1\right)}+q\right)^2} - \frac{e^{2\lambda\left(\frac{\ln q}{2\lambda}-1\right)}}{\left(e^{2\lambda\left(\frac{\ln q}{2\lambda}-1\right)}+q\right)^2}\right) =$$

$$q\lambda\left(\frac{qe^{2\lambda}}{\left(qe^{2\lambda}+q\right)^2} - \frac{qe^{-2\lambda}}{\left(qe^{-2\lambda}+q\right)^2}\right) =$$

$$\lambda\left(\frac{e^{2\lambda}}{\left(e^{2\lambda}+1\right)^2} - \frac{e^{-2\lambda}}{\left(e^{-2\lambda}+1\right)^2}\right) = \tag{15}$$

$$\lambda\left(\frac{e^{2\lambda}\left(e^{-2\lambda}+1\right)^2 - e^{-2\lambda}\left(e^{2\lambda}+1\right)^2}{\left(e^{2\lambda}+1\right)^2\left(e^{-2\lambda}+1\right)^2}\right) = 0.$$

That is, $\frac{\ln q}{2\lambda}$ is the only critical number of $M_{q,\lambda}$ over $\mathbb{R}$. Hence at $x = \frac{\ln q}{2\lambda}$, $M_{q,\lambda}$ achieves its global maximum, which is

$$M_{q,\lambda}\left(\frac{\ln q}{2\lambda}\right) = \frac{1}{4}\left[g_{q,\lambda}\left(\frac{\ln q}{2\lambda}+1\right) - g_{q,\lambda}\left(\frac{\ln q}{2\lambda}-1\right)\right] =$$

$$\frac{1}{4}\left[\left(\frac{e^{\lambda\left(\frac{\ln q}{2\lambda}+1\right)} - qe^{-\lambda\left(\frac{\ln q}{2\lambda}+1\right)}}{e^{\lambda\left(\frac{\ln q}{2\lambda}+1\right)} + qe^{-\lambda\left(\frac{\ln q}{2\lambda}+1\right)}}\right) - \left(\frac{e^{\lambda\left(\frac{\ln q}{2\lambda}-1\right)} - qe^{-\lambda\left(\frac{\ln q}{2\lambda}-1\right)}}{e^{\lambda\left(\frac{\ln q}{2\lambda}-1\right)} + qe^{-\lambda\left(\frac{\ln q}{2\lambda}-1\right)}}\right)\right] =$$

$$\frac{1}{4}\left[\left(\frac{\sqrt{q}e^{\lambda} - qq^{-\frac{1}{2}}e^{-\lambda}}{\sqrt{q}e^{\lambda} + qq^{-\frac{1}{2}}e^{-\lambda}}\right) - \left(\frac{\sqrt{q}e^{-\lambda} - qq^{-\frac{1}{2}}e^{\lambda}}{\sqrt{q}e^{-\lambda} + qq^{-\frac{1}{2}}e^{\lambda}}\right)\right] = \tag{16}$$

$$\frac{1}{4}\left[\left(\frac{e^{\lambda} - e^{-\lambda}}{e^{\lambda} + e^{-\lambda}}\right) - \left(\frac{e^{-\lambda} - e^{\lambda}}{e^{-\lambda} + e^{\lambda}}\right)\right] =$$

$$\frac{1}{4}\left[\frac{2\left(e^{\lambda} - e^{-\lambda}\right)}{e^{\lambda} + e^{-\lambda}}\right] = \frac{1}{2}\left(\frac{e^{\lambda} - e^{-\lambda}}{e^{\lambda} + e^{-\lambda}}\right) = \frac{\tanh(\lambda)}{2}. \tag{17}$$

**Conclusion:** The maximum value of $M_{q,\lambda}$ is

$$M_{q,\lambda}\left(\frac{\ln q}{2\lambda}\right) = \frac{\tanh(\lambda)}{2}, \quad \lambda > 0. \tag{18}$$

We give

**Theorem 7.1**

*([10, Ch. 17]) We have that*

$$\sum_{i=-\infty}^{\infty} M_{q,\lambda}(x-i) = 1, \quad \forall\, x \in \mathbb{R}, \ \forall\, \lambda, q > 0. \tag{19}$$

*Thus*

$$\sum_{i=-\infty}^{\infty} M_{q,\lambda}\left(nx-i\right) = 1,\ \forall\, n \in \mathbb{N},\ \forall\, x \in \mathbb{R}. \tag{20}$$

*Similarly, it holds*

$$\sum_{i=-\infty}^{\infty} M_{\frac{1}{q},\lambda}\left(x-i\right) = 1,\ \forall\, x \in \mathbb{R}. \tag{21}$$

*But* $M_{\frac{1}{q},\lambda}\left(x-i\right) \overset{(11)}{=} M_{q,\lambda}\left(i-x\right),\ \forall\, x \in \mathbb{R}.$
    *Hence*

$$\sum_{i=-\infty}^{\infty} M_{q,\lambda}\left(i-x\right) = 1,\ \forall\, x \in \mathbb{R}, \tag{22}$$

*and*

$$\sum_{i=-\infty}^{\infty} M_{q,\lambda}\left(i+x\right) = 1,\ \forall\, x \in \mathbb{R}. \tag{23}$$

It follows

**Theorem 7.2**
*([10, Ch. 17]) It holds*

$$\int_{-\infty}^{\infty} M_{q,\lambda}\left(x\right) dx = 1,\quad \lambda, q > 0. \tag{24}$$

So that $M_{q,\lambda}$ is a density function on $\mathbb{R}$; $\lambda, q > 0$.
We need the following result

**Theorem 7.3**
*([10, Ch. 17]) Let $0 < \alpha < 1$, and $n \in \mathbb{N}$ with $n^{1-\alpha} > 2$; $q, \lambda > 0$. Then*

$$\sum_{\substack{k=-\infty \\ :\, |nx-k| \ge n^{1-\alpha}}}^{\infty} M_{q,\lambda}\left(nx-k\right) < \max\left\{q, \frac{1}{q}\right\} e^{4\lambda} e^{-2\lambda n^{(1-\alpha)}} = T e^{-2\lambda n^{(1-\alpha)}},$$

$$\tag{25}$$

*where* $T := \max\left\{q, \frac{1}{q}\right\} e^{4\lambda}.$

Let $\lceil \cdot \rceil$ the ceiling of the number, and $\lfloor \cdot \rfloor$ the integral part of the number.

**Theorem 7.4**

*([10, Ch. 17]) Let $x \in [a,b] \subset \mathbb{R}$ and $n \in \mathbb{N}$ so that $\lceil na \rceil \leq \lfloor nb \rfloor$. For $q > 0$, $\lambda > 0$, we consider the number $\lambda_q > z_0 > 0$ with $M_{q,\lambda}(z_0) = M_{q,\lambda}(0)$ and $\lambda_q > 1$. Then*

$$\frac{1}{\displaystyle\sum_{k=\lceil na \rceil}^{\lfloor nb \rfloor} M_{q,\lambda}(nx-k)} < max\left\{\frac{1}{M_{q,\lambda}(\lambda_q)}, \frac{1}{M_{\frac{1}{q},\lambda}\left(\lambda_{\frac{1}{q}}\right)}\right\} =: \Delta(q). \tag{26}$$

We make

**Remark 7.1** ([10, Ch. 17]) (i) We have that

$$\lim_{n \to +\infty} \sum_{k=\lceil na \rceil}^{\lfloor nb \rfloor} M_{q,\lambda}(nx-k) \neq 1, \quad \text{for at least some } x \in [a,b], \tag{27}$$

where $\lambda, q > 0$.

(ii) Let $[a,b] \subset \mathbb{R}$. For large $n$ we always have $\lceil na \rceil \leq \lfloor nb \rfloor$. Also $a \leq \frac{k}{n} \leq b$, iff $\lceil na \rceil \leq k \leq \lfloor nb \rfloor$. In general it holds

$$\sum_{k=\lceil na \rceil}^{\lfloor nb \rfloor} M_{q,\lambda}(nx-k) \leq 1. \tag{28}$$

■

Let $(\mathbb{C}, |\cdot|)$ be the Banach space of the complex numbers over the reals.

**Definition 7.1** Let $f \in C([a,b], \mathbb{C})$ and $n \in \mathbb{N} : \lceil na \rceil \leq \lfloor nb \rfloor$. We introduce and define the $\mathbb{C}$-valued linear neural network operators

$$H_n(f,x) := \frac{\displaystyle\sum_{k=\lceil na \rceil}^{\lfloor nb \rfloor} f\left(\frac{k}{n}\right) M_{q,\lambda}(nx-k)}{\displaystyle\sum_{k=\lceil na \rceil}^{\lfloor nb \rfloor} M_{q,\lambda}(nx-k)}, \quad x \in [a,b]; \ q > 0, q \neq 1. \tag{29}$$

For large enough $n$ we always obtain $\lceil na \rceil \leq \lfloor nb \rfloor$. Also $a \leq \frac{k}{n} \leq b$, iff $\lceil na \rceil \leq k \leq \lfloor nb \rfloor$. The same $H_n$ is used for real valued functions. We study here the pointwise and uniform convergence of $H_n(f,x)$ to $f(x)$ with rates.

Clearly here $H_n(f) \in C([a,b], \mathbb{C})$.

For convenience, also we call

$$H_n^*(f,x) := \sum_{k=\lceil na \rceil}^{\lfloor nb \rfloor} f\left(\frac{k}{n}\right) M_{q,\lambda}(nx-k), \tag{30}$$

(the same $H_n^*$ can be defined for real valued functions) that is

$$H_n(f,x) := \frac{H_n^*(f,x)}{\sum_{k=\lceil na \rceil}^{\lfloor nb \rfloor} M_{q,\lambda}(nx-k)}. \tag{31}$$

So that

$$H_n(f,x) - f(x) = \frac{H_n^*(f,x)}{\sum_{k=\lceil na \rceil}^{\lfloor nb \rfloor} M_{q,\lambda}(nx-k)} - f(x) = \tag{32}$$

$$\frac{H_n^*(f,x) - f(x)\left(\sum_{k=\lceil na \rceil}^{\lfloor nb \rfloor} M_{q,\lambda}(nx-k)\right)}{\sum_{k=\lceil na \rceil}^{\lfloor nb \rfloor} M_{q,\lambda}(nx-k)}.$$

Consequently, we derive that

$$|H_n(f,x) - f(x)| \le \Delta(q)\left|H_n^*(f,x) - f(x)\left(\sum_{k=\lceil na \rceil}^{\lfloor nb \rfloor} M_{q,\lambda}(nx-k)\right)\right| =$$

$$\Delta(q)\left|\sum_{k=\lceil na \rceil}^{\lfloor nb \rfloor}\left(f\left(\frac{k}{n}\right) - f(x)\right)M_{q,\lambda}(nx-k)\right|, \tag{33}$$

where $\Delta(q)$ as in (26).

We will estimate the right hand side of the last quantity.

For that we need, for $f \in C([a,b], \mathbb{C})$ the first modulus of continuity

$$\omega_1(f,\delta) := \sup_{\substack{x,y \in [a,b] \\ |x-y| \le \delta}} |f(x) - f(y)|, \quad \delta > 0. \tag{34}$$

The fact $f \in C([a,b], \mathbb{C})$ is equivalent to $\lim_{\delta \to 0} \omega_1(f,\delta) = 0$, see [8].

## 7.3  Main Results

We present $\mathbb{C}$-valued neural network high order approximations to a function given with rates. We start with a trigonometric approximation.

***Theorem 7.5***
*Let $f \in C^2\left(\left[a,b\right],\mathbb{C}\right)$, $0 < \alpha < 1$, $n \in \mathbb{N} : n^{1-\alpha} > 2$, $x \in \left[a,b\right]$. Then*
  *1)*

$$\left| H_n\left(f,x\right) - f\left(x\right) \right| \leq \Delta\left(q\right) \left[ \left| f'\left(x\right) \right| \left( \frac{1}{n^{\alpha}} + \left(b-a\right) T e^{-2\lambda n^{(1-\alpha)}} \right) \right. \tag{35}$$

$$+ \frac{\left| f''\left(x\right) \right|}{2} \left( \frac{1}{n^{2\alpha}} + \left(b-a\right)^2 T e^{-2\lambda n^{(1-\alpha)}} \right) +$$

$$\left. \left( \frac{\omega_1\left(f'' + f, \frac{1}{n^{\alpha}}\right)}{2n^{2\alpha}} + \left\| f'' + f \right\|_{\infty} \left(b-a\right)^2 T e^{-2\lambda n^{(1-\alpha)}} \right) \right],$$

  *2) if $f'\left(x\right) = f''\left(x\right) = 0$, we obtain*

$$\left| H_n\left(f,x\right) - f\left(x\right) \right| \leq \Delta\left(q\right) \left[ \frac{\omega_1\left(f'' + f, \frac{1}{n^{\alpha}}\right)}{2n^{2\alpha}} + \left\| f'' + f \right\|_{\infty} \left(b-a\right)^2 T e^{-2\lambda n^{(1-\alpha)}} \right],$$
$$\tag{36}$$

*notice here the high rate of convergence at $n^{-3\alpha}$,*
  *3) furthermore we get*

$$\left\| H_n f - f \right\|_{\infty} \leq \Delta\left(q\right) \left[ \left\| f' \right\|_{\infty} \left( \frac{1}{n^{\alpha}} + \left(b-a\right) T e^{-2\lambda n^{(1-\alpha)}} \right) \right.$$

$$+ \frac{\left\| f'' \right\|_{\infty}}{2} \left( \frac{1}{n^{2\alpha}} + \left(b-a\right)^2 T e^{-2\lambda n^{(1-\alpha)}} \right) +$$

$$\left. \left( \frac{\omega_1\left(f'' + f, \frac{1}{n^{\alpha}}\right)}{2n^{2\alpha}} + \left\| f'' + f \right\|_{\infty} \left(b-a\right)^2 T e^{-2\lambda n^{(1-\alpha)}} \right) \right], \tag{37}$$

*i.e., $\lim\limits_{n \to +\infty} H_n\left(f\right) = f$, pointwise and uniformly,*
  *4) and finally, it holds*

$$\left| H_n\left(f,x\right) - f'\left(x\right) H_n\left(\sin\left(\cdot - x\right), x\right) - 2f''\left(x\right) H_n\left(\sin^2\left(\frac{\cdot - x}{2}\right), x\right) - f\left(x\right) \right| \leq$$
$$\tag{38}$$
$$\Delta\left(q\right) \left[ \frac{\omega_1\left(f'' + f, \frac{1}{n^{\alpha}}\right)}{2n^{2\alpha}} + \left\| f'' + f \right\|_{\infty} \left(b-a\right)^2 T e^{-2\lambda n^{(1-\alpha)}} \right],$$

*again here we achieve high speed of convergence at $n^{-3\alpha}$.*

**Proof 7.1**    Here $f \in C^2\left(\left[a,b\right],\mathbb{C}\right)$, and we apply the trigonometric Taylor's formula for $f \in C^2\left(\left[a,b\right],\mathbb{C}\right)$, see Theorem 6 of [11].
    Let $\frac{k}{n}, x \in \left[a,b\right]$, then

$$f\left(\frac{k}{n}\right) = f\left(x\right) + f'\left(x\right) \sin\left(\frac{k}{n} - x\right) + 2f''\left(x\right) \sin^2\left(\frac{\frac{k}{n} - x}{2}\right) +$$

$$\int_x^{\frac{k}{n}} \left[ \left( f''(t) + f(t) \right) - \left( f''(x) + f(x) \right) \right] \sin \left( \frac{k}{n} - t \right) dt. \tag{39}$$

Hence it holds

$$f\left( \frac{k}{n} \right) M_{q,\lambda}(nx - k) = f(x) M_{q,\lambda}(nx - k) +$$

$$f'(x) \sin \left( \frac{k}{n} - x \right) M_{q,\lambda}(nx - k) + 2f''(x) \sin^2 \left( \frac{\frac{k}{n} - x}{2} \right) M_{q,\lambda}(nx - k) +$$

$$M_{q,\lambda}(nx - k) \left( \int_x^{\frac{k}{n}} \left[ \left( f''(t) + f(t) \right) - \left( f''(x) + f(x) \right) \right] \sin \left( \frac{k}{n} - t \right) dt \right). \tag{40}$$

So that we have

$$\sum_{k=\lceil na \rceil}^{\lfloor nb \rfloor} f\left( \frac{k}{n} \right) M_{q,\lambda}(nx - k) - f(x) \sum_{k=\lceil na \rceil}^{\lfloor nb \rfloor} M_{q,\lambda}(nx - k) =$$

$$f'(x) \sum_{k=\lceil na \rceil}^{\lfloor nb \rfloor} M_{q,\lambda}(nx - k) \sin \left( \frac{k}{n} - x \right) + 2f''(x) \sum_{k=\lceil na \rceil}^{\lfloor nb \rfloor} M_{q,\lambda}(nx - k) \sin^2 \left( \frac{\frac{k}{n} - x}{2} \right) +$$

$$\sum_{k=\lceil na \rceil}^{\lfloor nb \rfloor} M_{q,\lambda}(nx - k) \left( \int_x^{\frac{k}{n}} \left[ \left( f''(t) + f(t) \right) - \left( f''(x) + f(x) \right) \right] \sin \left( \frac{k}{n} - t \right) dt \right). \tag{41}$$

Thus, we obtain

$$H_n^*(f,x) - f(x) \sum_{k=\lceil na \rceil}^{\lfloor nb \rfloor} M_{q,\lambda}(nx - k) =$$

$$f'(x) H_n^* \left( \sin(\cdot - x), x \right) + 2f''(x) H_n^* \left( \sin^2 \left( \frac{\cdot - x}{2} \right), x \right) + \Lambda_n(x), \tag{42}$$

where

$$\Lambda_n(x) := \sum_{k=\lceil na \rceil}^{\lfloor nb \rfloor} M_{q,\lambda}(nx - k) \left( \int_x^{\frac{k}{n}} \left[ \left( f''(t) + f(t) \right) - \left( f''(x) + f(x) \right) \right] \sin \left( \frac{k}{n} - t \right) dt \right).$$

We call

$$R_2(n) := \int_x^{\frac{k}{n}} \left[ \left( f''(t) + f(t) \right) - \left( f''(x) + f(x) \right) \right] \sin \left( \frac{k}{n} - t \right) dt. \tag{43}$$

We assume that $b - a > \frac{1}{n^\alpha}$, which is always the case for large enough $n \in \mathbb{N}$, that is when $n > \left\lceil (b-a)^{-\frac{1}{\alpha}} \right\rceil$.

Thus $\left| \frac{k}{n} - x \right| \le \frac{1}{n^\alpha}$ or $\left| \frac{k}{n} - x \right| > \frac{1}{n^\alpha}$.

In case of $\left|\frac{k}{n} - x\right| \leq \frac{1}{n^\alpha}$, we have the following cases:

i) if $\frac{k}{n} \geq x$, then

$$|R_2(n)| = \left| \int_x^{\frac{k}{n}} \left[ (f''(t) + f(t)) - (f''(x) + f(x)) \right] \sin\left(\frac{k}{n} - t\right) dt \right| \leq$$

$$\int_x^{\frac{k}{n}} \omega_1\left(f'' + f, t - x\right) \left| \sin\left(\frac{k}{n} - t\right) \right| dt \leq \tag{44}$$

(by $|\sin x| \leq |x|, \forall x \in \mathbb{R}$)

$$\int_x^{\frac{k}{n}} \omega_1\left(f'' + f, t - x\right) \left(\frac{k}{n} - t\right) dt \leq \omega_1\left(f'' + f, \frac{k}{n} - x\right) \frac{\left(\frac{k}{n} - x\right)^2}{2}$$

$$\leq \frac{\omega_1\left(f'' + f, \frac{1}{n^\alpha}\right)}{2n^{2\alpha}},$$

that is

$$|R_2(n)| \leq \frac{\omega_1\left(f'' + f, \frac{1}{n^\alpha}\right)}{2n^{2\alpha}}. \tag{45}$$

ii) if $\frac{k}{n} < x$, then

$$|R_2(n)| = \left| \int_x^{\frac{k}{n}} \left[ (f''(t) + f(t)) - (f''(x) + f(x)) \right] \sin\left(\frac{k}{n} - t\right) dt \right| =$$

$$\left| \int_{\frac{k}{n}}^x \left[ (f''(t) + f(t)) - (f''(x) + f(x)) \right] \sin\left(\frac{k}{n} - t\right) dt \right| \leq$$

$$\int_{\frac{k}{n}}^x \left| (f''(t) + f(t)) - (f''(x) + f(x)) \right| \left| \sin\left(\frac{k}{n} - t\right) \right| dt \leq \tag{46}$$

$$\int_{\frac{k}{n}}^x \omega_1\left(f'' + f, x - \frac{k}{n}\right) \left(t - \frac{k}{n}\right) dt \leq \omega_1\left(f'' + f, x - \frac{k}{n}\right) \frac{\left(x - \frac{k}{n}\right)^2}{2}$$

$$\leq \frac{\omega_1\left(f'' + f, \frac{1}{n^\alpha}\right)}{2n^{2\alpha}}.$$

That is

$$|R_2(n)| \leq \frac{\omega_1\left(f'' + f, \frac{1}{n^\alpha}\right)}{2n^{2\alpha}}. \tag{47}$$

So, we have proved when $\left|\frac{k}{n} - x\right| \leq \frac{1}{n^\alpha}$, always it holds

$$|R_2(n)| \leq \frac{\omega_1\left(f'' + f, \frac{1}{n^\alpha}\right)}{2n^{2\alpha}}. \tag{48}$$

Next assume again $\frac{k}{n} \geq x$, then

$$|R_2(n)| = \left| \int_x^{\frac{k}{n}} \left[ (f''(t) + f(t)) - (f''(x) + f(x)) \right] \sin\left( \frac{k}{n} - t \right) dt \right| \leq$$

$$\int_x^{\frac{k}{n}} \left| (f''(t) + f(t)) - (f''(x) + f(x)) \right| \left| \sin\left( \frac{k}{n} - t \right) \right| dt \leq$$

(by $|\sin x| \leq |x|, \, \forall \, x \in \mathbb{R}$)

$$2\|f'' + f\|_\infty \left( \int_x^{\frac{k}{n}} \left| \sin\left( \frac{k}{n} - t \right) \right| dt \right) \leq$$

$$2\|f'' + f\|_\infty \left( \int_x^{\frac{k}{n}} \left( \frac{k}{n} - t \right) dt \right) =$$

$$2\|f'' + f\|_\infty \frac{\left( \frac{k}{n} - x \right)^2}{2} \leq \|f'' + f\|_\infty (b - a)^2 . \tag{49}$$

Hence it is

$$|R_2(n)| \leq \|f'' + f\|_\infty (b - a)^2 . \tag{50}$$

When $\frac{k}{n} < x$, we have

$$|R_2(n)| = \left| \int_x^{\frac{k}{n}} \left[ (f''(t) + f(t)) - (f''(x) + f(x)) \right] \sin\left( \frac{k}{n} - t \right) dt \right| =$$

$$\left| \int_{\frac{k}{n}}^{x} \left[ (f''(x) + f(x)) - (f''(t) + f(t)) \right] \sin\left( \frac{k}{n} - t \right) dt \right| \leq$$

$$\int_{\frac{k}{n}}^{x} \left| (f''(x) + f(x)) - (f''(t) + f(t)) \right| \left| \sin\left( \frac{k}{n} - t \right) \right| dt \leq$$

$$2\|f'' + f\|_\infty \int_{\frac{k}{n}}^{x} \left| \sin\left( \frac{k}{n} - t \right) \right| dt \leq$$

$$2\|f'' + f\|_\infty \int_{\frac{k}{n}}^{x} \left( t - \frac{k}{n} \right) dt = \|f'' + f\|_\infty \left( x - \frac{k}{n} \right)^2 \leq$$

$$\|f'' + f\|_\infty (b - a)^2 . \tag{51}$$

Therefore, it always holds

$$|R_2(n)| \leq \|f'' + f\|_\infty (b - a)^2 . \tag{52}$$

And we have

$$\Lambda_n(x) = \sum_{\substack{k=\lceil na \rceil \\ :\, \left|\frac{k}{n}-x\right| \le \frac{1}{n^\alpha}}}^{\lfloor nb \rfloor} M_{q,\lambda}(nx-k)\, R_2(n) +$$

$$\sum_{\substack{k=\lceil na \rceil \\ :\, \left|\frac{k}{n}-x\right| > \frac{1}{n^\alpha}}}^{\lfloor nb \rfloor} M_{q,\lambda}(nx-k)\, R_2(n). \tag{53}$$

Hence it holds

$$|\Lambda_n(x)| \le \sum_{\substack{k=\lceil na \rceil \\ :\, \left|\frac{k}{n}-x\right| \le \frac{1}{n^\alpha}}}^{\lfloor nb \rfloor} M_{q,\lambda}(nx-k)\, |R_2(n)| +$$

$$\sum_{\substack{k=\lceil na \rceil \\ :\, \left|\frac{k}{n}-x\right| > \frac{1}{n^\alpha}}}^{\lfloor nb \rfloor} M_{q,\lambda}(nx-k)\, |R_2(n)| \le \tag{54}$$

$$\left( \sum_{\substack{k=\lceil na \rceil \\ :\, \left|\frac{k}{n}-x\right| \le \frac{1}{n^\alpha}}}^{\lfloor nb \rfloor} M_{q,\lambda}(nx-k) \right) \frac{\omega_1\left(f''+f,\frac{1}{n^\alpha}\right)}{2n^{2\alpha}} +$$

$$\sum_{\substack{k=\lceil na \rceil \\ :\, \left|\frac{k}{n}-x\right| > \frac{1}{n^\alpha}}}^{\lfloor nb \rfloor} M_{q,\lambda}(nx-k)\, \|f''+f\|_\infty (b-a)^2 \overset{\text{(by (28))}}{\le}$$

$$\frac{\omega_1\left(f''+f,\frac{1}{n^\alpha}\right)}{2n^{2\alpha}} + \|f''+f\|_\infty (b-a)^2$$

$$\left( \sum_{\substack{k=\lceil na \rceil \\ :\, \left|\frac{k}{n}-x\right| > \frac{1}{n^\alpha}}}^{\lfloor nb \rfloor} M_{q,\lambda}(nx-k) \right) \overset{\text{(by Theorem 7.3)}}{\le}$$

$$\frac{\omega_1\left(f''+f,\frac{1}{n^\alpha}\right)}{2n^{2\alpha}} + \|f''+f\|_\infty (b-a)^2\, T e^{-2\lambda n^{(1-\alpha)}}. \tag{55}$$

Consequently, we have derived that

$$|\Lambda_n(x)| \leq \frac{\omega_1\left(f''+f,\frac{1}{n^\alpha}\right)}{2n^{2\alpha}} + \left\|f''+f\right\|_\infty (b-a)^2\, T e^{-2\lambda n^{(1-\alpha)}}. \tag{56}$$

Next we use again $|\sin x| \leq |x|$, $\forall\, x \in \mathbb{R}$.

We have that

$$H_n^*\left(\sin\left(\cdot - x\right), x\right) = \sum_{k=\lceil na\rceil}^{\lfloor nb\rfloor} M_{q,\lambda}\left(nx - k\right)\sin\left(\frac{k}{n} - x\right), \tag{57}$$

and

$$\left|H_n^*\left(\sin\left(\cdot - x\right), x\right)\right| \leq \sum_{k=\lceil na\rceil}^{\lfloor nb\rfloor} M_{q,\lambda}\left(nx - k\right)\left|\sin\left(\frac{k}{n} - x\right)\right| =$$

$$\sum_{\substack{k=\lceil na\rceil \\ :\,\left|\frac{k}{n}-x\right| \leq \frac{1}{n^\alpha}}}^{\lfloor nb\rfloor} M_{q,\lambda}\left(nx - k\right)\left|\sin\left(\frac{k}{n} - x\right)\right| +$$

$$\sum_{\substack{k=\lceil na\rceil \\ :\,\left|\frac{k}{n}-x\right| > \frac{1}{n^\alpha}}}^{\lfloor nb\rfloor} M_{q,\lambda}\left(nx - k\right)\left|\sin\left(\frac{k}{n} - x\right)\right| \leq \tag{58}$$

$$\sum_{\substack{k=\lceil na\rceil \\ :\,\left|\frac{k}{n}-x\right| \leq \frac{1}{n^\alpha}}}^{\lfloor nb\rfloor} M_{q,\lambda}\left(nx - k\right)\left|\frac{k}{n} - x\right| +$$

$$\sum_{\substack{k=\lceil na\rceil \\ :\,\left|\frac{k}{n}-x\right| > \frac{1}{n^\alpha}}}^{\lfloor nb\rfloor} M_{q,\lambda}\left(nx - k\right)\left|\frac{k}{n} - x\right| \leq$$

$$\frac{1}{n^\alpha} + (b-a)\left(\sum_{\substack{k=\lceil na\rceil \\ :\,\left|\frac{k}{n}-x\right| > \frac{1}{n^\alpha}}}^{\lfloor nb\rfloor} M_{q,\lambda}\left(nx - k\right)\right) \overset{\text{(by (25))}}{\leq} \tag{59}$$

$$\frac{1}{n^\alpha} + (b-a)\, T e^{-2\lambda n^{(1-\alpha)}}.$$

We found that

$$\left|H_n^*\left(\sin\left(\cdot - x\right), x\right)\right| \leq \frac{1}{n^\alpha} + (b-a)\, T e^{-2\lambda n^{(1-\alpha)}}. \tag{60}$$

Next we estimate

$$H_n^*\left(\sin^2\left(\frac{\cdot-x}{2}\right),x\right)=\sum_{k=\lceil na\rceil}^{\lfloor nb\rfloor}M_{q,\lambda}\left(nx-k\right)\sin^2\left(\frac{\frac{k}{n}-x}{2}\right),\tag{61}$$

We have that (by $|\sin x|\le|x|,\ \forall\,x\in\mathbb{R}$)

$$H_n^*\left(\sin^2\left(\frac{\cdot-x}{2}\right),x\right)=\sum_{k=\lceil na\rceil}^{\lfloor nb\rfloor}M_{q,\lambda}\left(nx-k\right)\left|\sin\left(\frac{\frac{k}{n}-x}{2}\right)\right|^2\le$$

$$\frac{1}{4}\sum_{k=\lceil na\rceil}^{\lfloor nb\rfloor}M_{q,\lambda}\left(nx-k\right)\left|\frac{k}{n}-x\right|^2=$$

$$\frac{1}{4}\left[\sum_{\substack{k=\lceil na\rceil\\ :\,\left|\frac{k}{n}-x\right|\le\frac{1}{n^\alpha}}}^{\lfloor nb\rfloor}M_{q,\lambda}\left(nx-k\right)\left|\frac{k}{n}-x\right|^2+\right.$$

$$\left.\sum_{\substack{k=\lceil na\rceil\\ :\,\left|\frac{k}{n}-x\right|>\frac{1}{n^\alpha}}}^{\lfloor nb\rfloor}M_{q,\lambda}\left(nx-k\right)\left|\frac{k}{n}-x\right|^2\right]\le\tag{62}$$

$$\frac{1}{4}\left[\frac{1}{n^{2\alpha}}+(b-a)^2\,Te^{-2\lambda n^{(1-\alpha)}}\right].$$

That is

$$H_n^*\left(\sin^2\left(\frac{\cdot-x}{2}\right),x\right)\le\frac{1}{4}\left[\frac{1}{n^{2\alpha}}+(b-a)^2\,Te^{-2\lambda n^{(1-\alpha)}}\right].\tag{63}$$

Consequently we have derived:
1)

$$|H_n\left(f,x\right)-f\left(x\right)|\le\Delta\left(q\right)\left[\left|f'\left(x\right)\right|\left(\frac{1}{n^\alpha}+(b-a)\,Te^{-2\lambda n^{(1-\alpha)}}\right)\right.\tag{64}$$

$$+\frac{\left|f''\left(x\right)\right|}{2}\left(\frac{1}{n^{2\alpha}}+(b-a)^2\,Te^{-2\lambda n^{(1-\alpha)}}\right)+$$

$$\left.\left(\frac{\omega_1\left(f''+f,\frac{1}{n^\alpha}\right)}{2n^{2\alpha}}+\left\|f''+f\right\|_\infty(b-a)^2\,Te^{-2\lambda n^{(1-\alpha)}}\right)\right].$$

2) if $f'(x) = f''(x) = 0$, by (64), we obtain

$$|H_n(f,x) - f(x)| \le \Delta(q) \left[ \frac{\omega_1\left(f''+f, \frac{1}{n^\alpha}\right)}{2n^{2\alpha}} + \|f''+f\|_\infty (b-a)^2 Te^{-2\lambda n^{(1-\alpha)}} \right],$$

(65)

note here the high rate of convergence at $n^{-3\alpha}$.

3) Furthermore, by (64), we get

$$\|H_n f - f\|_\infty \le \Delta(q) \left[ \|f'\|_\infty \left( \frac{1}{n^\alpha} + (b-a) Te^{-2\lambda n^{(1-\alpha)}} \right) + \right.$$

$$\frac{\|f''\|_\infty}{2} \left( \frac{1}{n^{2\alpha}} + (b-a)^2 Te^{-2\lambda n^{(1-\alpha)}} \right) +$$

$$\left. \left( \frac{\omega_1\left(f''+f, \frac{1}{n^\alpha}\right)}{2n^{2\alpha}} + \|f''+f\|_\infty (b-a)^2 Te^{-2\lambda n^{(1-\alpha)}} \right) \right].$$

(66)

We derive that $\lim\limits_{n \to +\infty} H_n(f) = f$, pointwise and uniformly.

We observe that

$$H_n(f,x) - f'(x) H_n(\sin(\cdot - x), x) - 2f''(x) H_n\left(\sin^2\left(\frac{\cdot - x}{2}\right), x\right) - f(x) =$$

$$\frac{H_n^*(f,x)}{\sum_{k=\lceil na \rceil}^{\lfloor nb \rfloor} M_{q,\lambda}(nx-k)} - f'(x) \frac{H_n^*(\sin(\cdot - x), x)}{\sum_{k=\lceil na \rceil}^{\lfloor nb \rfloor} M_{q,\lambda}(nx-k)} -$$

$$2f''(x) \frac{H_n^*\left(\sin^2\left(\frac{\cdot - x}{2}\right), x\right)}{\sum_{k=\lceil na \rceil}^{\lfloor nb \rfloor} M_{q,\lambda}(nx-k)} - f(x) \left( \frac{\sum_{k=\lceil na \rceil}^{\lfloor nb \rfloor} M_{q,\lambda}(nx-k)}{\sum_{k=\lceil na \rceil}^{\lfloor nb \rfloor} M_{q,\lambda}(nx-k)} \right) =$$

(67)

$$\frac{1}{\sum_{k=\lceil na \rceil}^{\lfloor nb \rfloor} M_{q,\lambda}(nx-k)} \left[ H_n^*(f,x) - f'(x) H_n^*(\sin(\cdot - x), x) - \right.$$

$$\left. 2f''(x) H_n^*\left(\sin^2\left(\frac{\cdot - x}{2}\right), x\right) - f(x) \sum_{k=\lceil na \rceil}^{\lfloor nb \rfloor} M_{q,\lambda}(nx-k) \right] =$$

$$\frac{1}{\sum_{k=\lceil na \rceil}^{\lfloor nb \rfloor} M_{q,\lambda}(nx-k)} (\Lambda_n(x)).$$

(68)

Finally, we obtain ($\forall\, x \in [a,b]$, $n \in \mathbb{N}$):

4)

$$\left| H_n(f,x) - f'(x) H_n(\sin(\cdot - x), x) - 2f''(x) H_n\left(\sin^2\left(\frac{\cdot - x}{2}\right), x\right) - f(x) \right| \overset{(26)}{\le}$$

$$\Delta(q) |\Lambda_n(x)| \overset{(56)}{\le}$$

$$\Delta(q)\left[\frac{\omega_1\left(f''+f,\frac{1}{n^\alpha}\right)}{2n^{2\alpha}}+\left\|f''+f\right\|_\infty(b-a)^2\,Te^{-2\lambda n^{(1-\alpha)}}\right]. \tag{69}$$

The theorem is proved.

We continue with a hyperbolic high order neural network approximation.

**Theorem 7.6**
*Let $f \in C^2\left([a,b],\mathbb{C}\right)$, $0 < \alpha < 1$, $n \in \mathbb{N}: n^{1-\alpha} > 2$, $x \in [a,b]$. Then*
*1)*

$$\left|H_n(f,x)-f(x)\right| \le \Delta(q)\cosh(b-a)\left[\left|f'(x)\right|\left(\frac{1}{n^\alpha}+(b-a)\,Te^{-2\lambda n^{(1-\alpha)}}\right)\right.$$

$$\frac{\left|f''(x)\right|}{2}\left(\frac{1}{n^{2\alpha}}+(b-a)^2\,Te^{-2\lambda n^{(1-\alpha)}}\right)+$$

$$\left.\left(\frac{\omega_1\left(f''-f,\frac{1}{n^\alpha}\right)}{2n^{2\alpha}}+\left\|f''-f\right\|_\infty(b-a)^2\,Te^{-2\lambda n^{(1-\alpha)}}\right)\right], \tag{70}$$

*2) if $f'(x)=f''(x)=0$, we obtain*

$$\left|H_n(f,x)-f(x)\right| \le \Delta(q)\cosh(b-a)$$

$$\left[\frac{\omega_1\left(f''-f,\frac{1}{n^\alpha}\right)}{2n^{2\alpha}}+\left\|f''-f\right\|_\infty(b-a)^2\,Te^{-2\lambda n^{(1-\alpha)}}\right], \tag{71}$$

*notice here the high rate of convergence at $n^{-3\alpha}$,*
*3) furthermore, we get*

$$\left\|H_nf-f\right\|_\infty \le \Delta(q)\cosh(b-a)\left[\left\|f'\right\|_\infty\left(\frac{1}{n^\alpha}+(b-a)\,Te^{-2\lambda n^{(1-\alpha)}}\right)\right.$$

$$\frac{\left\|f''\right\|_\infty}{2}\left(\frac{1}{n^{2\alpha}}+(b-a)^2\,Te^{-2\lambda n^{(1-\alpha)}}\right)+$$

$$\left.\left(\frac{\omega_1\left(f''-f,\frac{1}{n^\alpha}\right)}{2n^{2\alpha}}+\left\|f''-f\right\|_\infty(b-a)^2\,Te^{-2\lambda n^{(1-\alpha)}}\right)\right], \tag{72}$$

*it follows that $\lim\limits_{n\to+\infty}H_n(f)=f$, pointwise and uniformly,*
*and*
*4)*

$$\left|H_n(f,x)-f'(x)H_n(\sinh(\cdot-x),x)-2f''(x)H_n\left(\sinh^2\left(\frac{\cdot-x}{2}\right),x\right)-f(x)\right| \le$$

$$\Delta(q)\cosh(b-a)\left[\frac{\omega_1\left(f''-f,\frac{1}{n^\alpha}\right)}{2n^{2\alpha}}+\left\|f''-f\right\|_\infty(b-a)^2\,Te^{-2\lambda n^{(1-\alpha)}}\right], \tag{73}$$

*again here we achieve high speed of convergence at $n^{-3\alpha}$.*

**Proof 7.2**   By the mean value theorem we have that

$$\sinh x = \sinh x - \sinh 0 = (\cosh \xi)(x - 0),$$

for some $\xi$ between $\{0, x\}$, for any $x \in \mathbb{R}$.

Hence

$$|\sinh x| \le \|\cosh\|_{\infty, [-(b-a), b-a]} |x|, \ \forall x \in [-(b-a), b-a]. \tag{74}$$

That is, there exists $M \ge 1$ such that

$$|\sinh x| \le M |x|, \ \forall x \in [-(b-a), b-a], \tag{75}$$

where $M := \|\cosh\|_{\infty, [-(b-a), b-a]} = \cosh(b-a)$.

Here $f \in C^2([a,b], \mathbb{C})$, and we apply the hyperbolic Taylor's formula for $f \in C^2([a,b], \mathbb{C})$, see Theorem 7 of [11].

Let $\frac{k}{n}, x \in [a,b]$, then

$$f\left(\frac{k}{n}\right) = f(x) + f'(x) \sinh\left(\frac{k}{n} - x\right) + 2f''(x) \sinh^2\left(\frac{\frac{k}{n} - x}{2}\right) +$$

$$\int_x^{\frac{k}{n}} \left[(f''(t) - f(t)) - (f''(x) - f(x))\right] \sinh\left(\frac{k}{n} - t\right) dt. \tag{76}$$

Hence it holds

$$f\left(\frac{k}{n}\right) M_{q,\lambda}(nx - k) = f(x) M_{q,\lambda}(nx - k) +$$

$$f'(x) \sinh\left(\frac{k}{n} - x\right) M_{q,\lambda}(nx - k) + 2f''(x) \sinh^2\left(\frac{\frac{k}{n} - x}{2}\right) M_{q,\lambda}(nx - k) +$$

$$M_{q,\lambda}(nx - k)\left(\int_x^{\frac{k}{n}} \left[(f''(t) - f(t)) - (f''(x) - f(x))\right] \sinh\left(\frac{k}{n} - t\right) dt\right). \tag{77}$$

So that we have

$$\sum_{k=\lceil na \rceil}^{\lfloor nb \rfloor} f\left(\frac{k}{n}\right) M_{q,\lambda}(nx - k) - f(x) \sum_{k=\lceil na \rceil}^{\lfloor nb \rfloor} M_{q,\lambda}(nx - k) =$$

$$f'(x) \sum_{k=\lceil na \rceil}^{\lfloor nb \rfloor} M_{q,\lambda}(nx - k) \sinh\left(\frac{k}{n} - x\right) + 2f''(x) \sum_{k=\lceil na \rceil}^{\lfloor nb \rfloor} M_{q,\lambda}(nx - k) \sinh^2\left(\frac{\frac{k}{n} - x}{2}\right) +$$

$$\sum_{k=\lceil na \rceil}^{\lfloor nb \rfloor} M_{q,\lambda}(nx - k)\left(\int_x^{\frac{k}{n}} \left[(f''(t) - f(t)) - (f''(x) - f(x))\right] \sinh\left(\frac{k}{n} - t\right) dt\right). \tag{78}$$

Thus, we obtain

$$H_n^* (f,x) - f(x) \sum_{k=\lceil na \rceil}^{\lfloor nb \rfloor} M_{q,\lambda} (nx - k) =$$

$$f'(x) H_n^* \left( \sinh(\cdot - x), x \right) + 2 f''(x) H_n^* \left( \sinh^2 \left( \frac{\cdot - x}{2} \right), x \right) + \Lambda_n (x), \qquad (79)$$

where

$$\Lambda_n (x) := \sum_{k=\lceil na \rceil}^{\lfloor nb \rfloor} M_{q,\lambda} (nx - k) \left( \int_x^{\frac{k}{n}} \left[ (f''(t) - f(t)) - (f''(x) - f(x)) \right] \sinh \left( \frac{k}{n} - t \right) dt \right).$$

$$(80)$$

We call

$$R_2 (n) := \int_x^{\frac{k}{n}} \left[ (f''(t) - f(t)) - (f''(x) - f(x)) \right] \sinh \left( \frac{k}{n} - t \right) dt. \qquad (81)$$

We assume that $b - a > \frac{1}{n^\alpha}$, which is always the case for large enough $n \in \mathbb{N}$, that is when $n > \left\lceil (b-a)^{-\frac{1}{\alpha}} \right\rceil$.

Thus $\left| \frac{k}{n} - x \right| \le \frac{1}{n^\alpha}$ or $\left| \frac{k}{n} - x \right| > \frac{1}{n^\alpha}$.

In case of $\left| \frac{k}{n} - x \right| \le \frac{1}{n^\alpha}$, we have the following cases:

i) if $\frac{k}{n} \ge x$, then

$$|R_2 (n)| = \left| \int_x^{\frac{k}{n}} \left[ (f''(t) - f(t)) - (f''(x) - f(x)) \right] \sinh \left( \frac{k}{n} - t \right) dt \right| \le$$

$$\int_x^{\frac{k}{n}} \omega_1 (f'' - f, t - x) \left| \sinh \left( \frac{k}{n} - t \right) \right| dt \overset{(75)}{\le}$$

$$\int_x^{\frac{k}{n}} \omega_1 (f'' - f, t - x) M \left( \frac{k}{n} - t \right) dt \le M \omega_1 \left( f'' - f, \frac{k}{n} - x \right) \int_x^{\frac{k}{n}} \left( \frac{k}{n} - t \right) dt =$$

$$(82)$$

$$M \omega_1 \left( f'' - f, \frac{k}{n} - x \right) \frac{\left( \frac{k}{n} - x \right)^2}{2} \le \frac{M \omega_1 \left( f'' - f, \frac{1}{n^\alpha} \right)}{2 n^{2\alpha}},$$

that is

$$|R_2 (n)| \le \frac{M \omega_1 \left( f'' - f, \frac{1}{n^\alpha} \right)}{2 n^{2\alpha}}. \qquad (83)$$

ii) if $\frac{k}{n} < x$, then

$$|R_2 (n)| = \left| \int_x^{\frac{k}{n}} \left[ (f''(t) - f(t)) - (f''(x) - f(x)) \right] \sinh \left( \frac{k}{n} - t \right) dt \right| =$$

$$\left| \int_{\frac{k}{n}}^{x} \left[ \left( f''(t) - f(t) \right) - \left( f''(x) - f(x) \right) \right] \sinh\left( \frac{k}{n} - t \right) dt \right| \le$$

$$\int_{\frac{k}{n}}^{x} \left| \left( f''(t) - f(t) \right) - \left( f''(x) - f(x) \right) \right| \left| \sinh\left( \frac{k}{n} - t \right) \right| dt \le \tag{84}$$

$$M\omega_1 \left( f'' - f, x - \frac{k}{n} \right) \int_{\frac{k}{n}}^{x} \left( t - \frac{k}{n} \right) dt = M\omega_1 \left( f'' - f, x - \frac{k}{n} \right) \frac{\left( x - \frac{k}{n} \right)^2}{2}$$

$$\le \frac{M\omega_1 \left( f'' - f, \frac{1}{n^\alpha} \right)}{2 n^{2\alpha}},$$

that is

$$|R_2(n)| \le \frac{M\omega_1 \left( f'' - f, \frac{1}{n^\alpha} \right)}{2 n^{2\alpha}}. \tag{85}$$

So, we have proved when $\left| \frac{k}{n} - x \right| \le \frac{1}{n^\alpha}$, always it holds

$$|R_2(n)| \le \frac{M\omega_1 \left( f'' - f, \frac{1}{n^\alpha} \right)}{2 n^{2\alpha}}. \tag{86}$$

Next assume again $\frac{k}{n} \ge x$, then

$$|R_2(n)| = \left| \int_{x}^{\frac{k}{n}} \left[ \left( f''(t) - f(t) \right) - \left( f''(x) - f(x) \right) \right] \sinh\left( \frac{k}{n} - t \right) dt \right| \le$$

$$\int_{x}^{\frac{k}{n}} \left| \left( f''(t) - f(t) \right) - \left( f''(x) - f(x) \right) \right| \left| \sinh\left( \frac{k}{n} - t \right) \right| dt \le$$

$$2M \left\| f'' - f \right\|_\infty \int_{x}^{\frac{k}{n}} \left( \frac{k}{n} - t \right) dt = \tag{87}$$

$$2M \left\| f'' - f \right\|_\infty \frac{\left( \frac{k}{n} - x \right)^2}{2} \le M \left\| f'' - f \right\|_\infty (b - a)^2.$$

Hence

$$|R_2(n)| \le M \left\| f'' - f \right\|_\infty (b - a)^2. \tag{88}$$

When $\frac{k}{n} < x$, we have

$$|R_2(n)| = \left| \int_{\frac{k}{n}}^{x} \left[ \left( f''(t) - f(t) \right) - \left( f''(x) - f(x) \right) \right] \sinh\left( \frac{k}{n} - t \right) dt \right| \le$$

$$\int_{\frac{k}{n}}^{x} \left| \left( f''(t) - f(t) \right) - \left( f''(x) - f(x) \right) \right| \left| \sinh\left( \frac{k}{n} - t \right) \right| dt \le$$

$$2M \left\| f'' - f \right\|_\infty \int_{\frac{k}{n}}^{x} \left( t - \frac{k}{n} \right) dt = \tag{89}$$

$$2M\left\|f''-f\right\|_\infty \frac{\left(x-\frac{k}{n}\right)^2}{2} \leq M\left\|f''-f\right\|_\infty (b-a)^2.$$

Therefore, always it holds

$$|R_2(n)| \leq M\left\|f''-f\right\|_\infty (b-a)^2. \tag{90}$$

And we have

$$\Lambda_n(x) = \sum_{\substack{k=\lceil na\rceil \\ :\left|\frac{k}{n}-x\right|\leq\frac{1}{n^\alpha}}}^{\lfloor nb\rfloor} M_{q,\lambda}(nx-k)R_2(n) + \tag{91}$$

$$\sum_{\substack{k=\lceil na\rceil \\ :\left|\frac{k}{n}-x\right|>\frac{1}{n^\alpha}}}^{\lfloor nb\rfloor} M_{q,\lambda}(nx-k)R_2(n).$$

Hence it holds

$$|\Lambda_n(x)| \leq \sum_{\substack{k=\lceil na\rceil \\ :\left|\frac{k}{n}-x\right|\leq\frac{1}{n^\alpha}}}^{\lfloor nb\rfloor} M_{q,\lambda}(nx-k)|R_2(n)| + \tag{92}$$

$$\sum_{\substack{k=\lceil na\rceil \\ :\left|\frac{k}{n}-x\right|>\frac{1}{n^\alpha}}}^{\lfloor nb\rfloor} M_{q,\lambda}(nx-k)|R_2(n)| \leq$$

$$\left(\sum_{\substack{k=\lceil na\rceil \\ :\left|\frac{k}{n}-x\right|\leq\frac{1}{n^\alpha}}}^{\lfloor nb\rfloor} M_{q,\lambda}(nx-k)\right) \frac{\omega_1\left(f''-f,\frac{1}{n^\alpha}\right)M}{2n^{2\alpha}} +$$

$$\left(\sum_{\substack{k=\lceil na\rceil \\ :\left|\frac{k}{n}-x\right|>\frac{1}{n^\alpha}}}^{\lfloor nb\rfloor} M_{q,\lambda}(nx-k)\right) M\left\|f''-f\right\|_\infty (b-a)^2 \overset{((28))}{\leq}$$

$$\frac{M\omega_1\left(f''-f,\frac{1}{n^\alpha}\right)}{2n^{2\alpha}} + M\left\|f''-f\right\|_\infty (b-a)^2$$

$$\left( \sum_{\substack{k=\lceil na \rceil \\ : \left| \frac{k}{n} - x \right| > \frac{1}{n^\alpha}}}^{\lfloor nb \rfloor} M_{q,\lambda} \left( nx - k \right) \right) \overset{\text{(by Theorem 7.3)}}{\leq} \tag{93}$$

$$M \frac{\omega_1 \left( f'' - f, \frac{1}{n^\alpha} \right)}{2n^{2\alpha}} + M \left\| f'' - f \right\|_\infty \left( b - a \right)^2 T e^{-2\lambda n^{(1-\alpha)}}.$$

Consequently, we have derived that

$$\left| \Lambda_n \left( x \right) \right| \leq M \left[ \frac{\omega_1 \left( f'' - f, \frac{1}{n^\alpha} \right)}{2n^{2\alpha}} + \left\| f'' - f \right\|_\infty \left( b - a \right)^2 T e^{-2\lambda n^{(1-\alpha)}} \right]. \tag{94}$$

We have that

$$H_n^* \left( \sinh \left( \cdot - x \right), x \right) = \sum_{k=\lceil na \rceil}^{\lfloor nb \rfloor} M_{q,\lambda} \left( nx - k \right) \sinh \left( \frac{k}{n} - x \right), \tag{95}$$

and

$$\left| H_n^* \left( \sinh \left( \cdot - x \right), x \right) \right| \leq \sum_{k=\lceil na \rceil}^{\lfloor nb \rfloor} M_{q,\lambda} \left( nx - k \right) \left| \sinh \left( \frac{k}{n} - x \right) \right| =$$

$$\sum_{\substack{k=\lceil na \rceil \\ : \left| \frac{k}{n} - x \right| \leq \frac{1}{n^\alpha}}}^{\lfloor nb \rfloor} M_{q,\lambda} \left( nx - k \right) \left| \sinh \left( \frac{k}{n} - x \right) \right| +$$

$$\sum_{\substack{k=\lceil na \rceil \\ : \left| \frac{k}{n} - x \right| > \frac{1}{n^\alpha}}}^{\lfloor nb \rfloor} M_{q,\lambda} \left( nx - k \right) \left| \sinh \left( \frac{k}{n} - x \right) \right| \leq \tag{96}$$

$$M \left[ \sum_{\substack{k=\lceil na \rceil \\ : \left| \frac{k}{n} - x \right| \leq \frac{1}{n^\alpha}}}^{\lfloor nb \rfloor} M_{q,\lambda} \left( nx - k \right) \left| \frac{k}{n} - x \right| + \right.$$

$$\left. \sum_{\substack{k=\lceil na \rceil \\ : \left| \frac{k}{n} - x \right| > \frac{1}{n^\alpha}}}^{\lfloor nb \rfloor} M_{q,\lambda} \left( nx - k \right) \left| \frac{k}{n} - x \right| \right] \leq \tag{97}$$

$$M\left[\frac{1}{n^{\alpha}} + (b-a)\left(\sum_{\substack{k=\lceil na\rceil \\ :\left|\frac{k}{n}-x\right|>\frac{1}{n^{\alpha}}}}^{\lfloor nb\rfloor} M_{q,\lambda}\left(nx-k\right)\right)\right] \overset{(\text{by }(25))}{\leq}$$

$$M\left[\frac{1}{n^{\alpha}} + (b-a)\,T e^{-2\lambda n^{(1-\alpha)}}\right].$$

We found that

$$\left|H_n^*\left(\sinh\left(\cdot - x\right),x\right)\right| \leq M\left[\frac{1}{n^{\alpha}} + (b-a)\,T e^{-2\lambda n^{(1-\alpha)}}\right]. \tag{98}$$

Next we estimate

$$H_n^*\left(\sinh^2\left(\frac{\cdot - x}{2}\right),x\right) = \sum_{k=\lceil na\rceil}^{\lfloor nb\rfloor} M_{q,\lambda}\left(nx-k\right)\sinh^2\left(\frac{\frac{k}{n}-x}{2}\right), \tag{99}$$

We have that

$$H_n^*\left(\sinh^2\left(\frac{\cdot - x}{2}\right),x\right) = \sum_{k=\lceil na\rceil}^{\lfloor nb\rfloor} M_{q,\lambda}\left(nx-k\right)\left(\sinh\left(\frac{\frac{k}{n}-x}{2}\right)\right)^2 \leq$$

$$\frac{M}{4}\sum_{k=\lceil na\rceil}^{\lfloor nb\rfloor} M_{q,\lambda}\left(nx-k\right)\left(\frac{k}{n}-x\right)^2 = \tag{100}$$

$$\frac{M}{4}\left[\sum_{\substack{k=\lceil na\rceil \\ :\left|\frac{k}{n}-x\right|\leq\frac{1}{n^{\alpha}}}}^{\lfloor nb\rfloor} M_{q,\lambda}\left(nx-k\right)\left(\frac{k}{n}-x\right)^2 + \right.$$

$$\left. \sum_{\substack{k=\lceil na\rceil \\ :\left|\frac{k}{n}-x\right|>\frac{1}{n^{\alpha}}}}^{\lfloor nb\rfloor} M_{q,\lambda}\left(nx-k\right)\left(\frac{k}{n}-x\right)^2 \right] \leq$$

$$\frac{M}{4}\left[\frac{1}{n^{2\alpha}} + (b-a)^2\,T e^{-2\lambda n^{(1-\alpha)}}\right].$$

That is

$$H_n^*\left(\sinh^2\left(\frac{\cdot - x}{2}\right),x\right) \leq \frac{M}{4}\left[\frac{1}{n^{2\alpha}} + (b-a)^2\,T e^{-2\lambda n^{(1-\alpha)}}\right]. \tag{101}$$

By (33) and putting together (79), (94), (98) and (101) we derive
1)

$$|H_n(f,x) - f(x)| \le \Delta(q) M \left[ |f'(x)| \left( \frac{1}{n^\alpha} + (b-a) T e^{-2\lambda n^{(1-\alpha)}} \right) + \right.$$

$$\frac{|f''(x)|}{2} \left( \frac{1}{n^{2\alpha}} + (b-a)^2 T e^{-2\lambda n^{(1-\alpha)}} \right) +$$

$$\left. \left( \frac{\omega_1 \left( f'' - f, \frac{1}{n^\alpha} \right)}{2n^{2\alpha}} + \|f'' - f\|_\infty (b-a)^2 T e^{-2\lambda n^{(1-\alpha)}} \right) \right]. \tag{102}$$

2) If $f'(x) = f''(x) = 0$, by (102), we obtain

$$|H_n(f,x) - f(x)| \le$$

$$M\Delta(q) \left[ \frac{\omega_1 \left( f'' - f, \frac{1}{n^\alpha} \right)}{2n^{2\alpha}} + \|f'' - f\|_\infty (b-a)^2 T e^{-2\lambda n^{(1-\alpha)}} \right], \tag{103}$$

notice here the high rate of convergence at $n^{-3\alpha}$.

3) Furthermore, by (102), we get

$$\|H_n f - f\|_\infty \le \Delta(q) M \left[ \|f'\|_\infty \left( \frac{1}{n^\alpha} + (b-a) T e^{-2\lambda n^{(1-\alpha)}} \right) + \right.$$

$$\frac{\|f''\|_\infty}{2} \left( \frac{1}{n^{2\alpha}} + (b-a)^2 T e^{-2\lambda n^{(1-\alpha)}} \right) +$$

$$\left. \left( \frac{\omega_1 \left( f'' - f, \frac{1}{n^\alpha} \right)}{2n^{2\alpha}} + \|f'' - f\|_\infty (b-a)^2 T e^{-2\lambda n^{(1-\alpha)}} \right) \right]. \tag{104}$$

It follows that $\lim_{n \to +\infty} H_n(f) = f$, pointwise and uniformly.

We observe that

$$H_n(f,x) - f'(x) H_n(\sinh(\cdot - x), x) - 2f''(x) H_n \left( \sinh^2 \left( \frac{\cdot - x}{2} \right), x \right) - f(x) =$$

$$\frac{H_n^*(f,x)}{\sum_{k=\lceil na \rceil}^{\lfloor nb \rfloor} M_{q,\lambda}(nx - k)} - f'(x) \frac{H_n^*(\sinh(\cdot - x), x)}{\sum_{k=\lceil na \rceil}^{\lfloor nb \rfloor} M_{q,\lambda}(nx - k)} -$$

$$2f''(x) \frac{H_n^* \left( \sinh^2 \left( \frac{\cdot - x}{2} \right), x \right)}{\sum_{k=\lceil na \rceil}^{\lfloor nb \rfloor} M_{q,\lambda}(nx - k)} - f(x) \left( \frac{\sum_{k=\lceil na \rceil}^{\lfloor nb \rfloor} M_{q,\lambda}(nx - k)}{\sum_{k=\lceil na \rceil}^{\lfloor nb \rfloor} M_{q,\lambda}(nx - k)} \right) = \tag{105}$$

$$\frac{1}{\sum_{k=\lceil na \rceil}^{\lfloor nb \rfloor} M_{q,\lambda}(nx - k)} \left[ H_n^*(f,x) - f'(x) H_n^*(\sinh(\cdot - x), x) - \right.$$

$$2f''(x)H_n^*\left(\sinh^2\left(\frac{\cdot-x}{2}\right),x\right)-f(x)\sum_{k=\lceil na\rceil}^{\lfloor nb\rfloor}M_{q,\lambda}(nx-k)\Bigg] = \tag{106}$$

$$\frac{1}{\sum_{k=\lceil na\rceil}^{\lfloor nb\rfloor}M_{q,\lambda}(nx-k)}\left(\Lambda_n(x)\right).$$

Finally, we obtain ($\forall\, x\in[a,b],\ n\in\mathbb{N}$):

4)

$$\left|H_n(f,x)-f'(x)H_n(\sinh(\cdot-x),x)-2f''(x)H_n\left(\sinh^2\left(\frac{\cdot-x}{2}\right),x\right)-f(x)\right|\overset{(26)}{\le}$$

$$\Delta(q)\,|\Lambda_n(x)|\overset{(94)}{\le}$$

$$\Delta(q)\,M\left[\frac{\omega_1\left(f''-f,\frac{1}{n^\alpha}\right)}{2n^{2\alpha}}+\left\|f''-f\right\|_\infty(b-a)^2\,Te^{-2\lambda n^{(1-\alpha)}}\right]. \tag{107}$$

The theorem is established.

Next follows a mixed hyperbolic-trigonometric high order neural network approximation.

**Theorem 7.7**

*Let $f\in C^4\left([a,b],\mathbb{C}\right)$, $0<\alpha<1$, $n\in\mathbb{N}:n^{1-\alpha}>2$, $x\in[a,b]$. Then*

*1)*

$$\left|H_n(f,x)-f(x)-\frac{f'(x)}{2}H_n\left((\sinh(\cdot-x)+\sin(\cdot-x)),x\right)\right.$$

$$-\frac{f''(x)}{2}H_n\left((\cosh(\cdot-x)-\cos(\cdot-x)),x\right) \tag{108}$$

$$-\frac{f^{(3)}(x)}{2}H_n\left((\sinh(\cdot-x)-\sin(\cdot-x)),x\right)$$

$$\left.-f^{(4)}(x)H_n\left(\left(\sinh^2\left(\frac{\cdot-x}{2}\right)-\sin^2\left(\frac{\cdot-x}{2}\right)\right),x\right)\right|\le$$

$$\frac{\Delta(q)\left(\cosh(b-a)+1\right)}{2}$$

$$\left[\frac{\omega_1\left(f^{(4)}-f,\frac{1}{n^\alpha}\right)}{2n^{2\alpha}}+\left\|f^{(4)}-f\right\|_\infty(b-a)^2\,Te^{-2\lambda n^{(1-\alpha)}}\right], \tag{109}$$

*2) if $f^{(i)}(x)=0$, $i=1,2,3,4$, we get*

$$|H_n(f,x)-f(x)|\le\frac{\Delta(q)\left(\cosh(b-a)+1\right)}{2}$$

$$\left[ \frac{\omega_1\left(f^{(4)} - f, \frac{1}{n^\alpha}\right)}{2n^{2\alpha}} + \left\| f^{(4)} - f \right\|_\infty (b-a)^2 T e^{-2\lambda n^{(1-\alpha)}} \right], \qquad (110)$$

*in the last (110) observe the high speed of convergence at $n^{-3\alpha}$.*

**Proof 7.3**  Here $f \in C^4\left([a,b],\mathbb{C}\right)$, and we apply the hyperbolic-trigonometric Taylor's formula for $f \in C^4\left([a,b],\mathbb{C}\right)$, see Theorem 8 of [11].

Let $\frac{k}{n}, x \in [a,b]$, then

$$f\left(\frac{k}{n}\right) - f(x) - f'(x) \left( \frac{\sinh\left(\frac{k}{n} - x\right) + \sin\left(\frac{k}{n} - x\right)}{2} \right)$$

$$- f''(x) \left( \frac{\cosh\left(\frac{k}{n} - x\right) - \cos\left(\frac{k}{n} - x\right)}{2} \right)$$

$$- f^{(3)}(x) \left( \frac{\sinh\left(\frac{k}{n} - x\right) - \sin\left(\frac{k}{n} - x\right)}{2} \right)$$

$$- f^{(4)}(x) \left( \sinh^2\left(\frac{\frac{k}{n} - x}{2}\right) - \sin^2\left(\frac{\frac{k}{n} - x}{2}\right) \right) = \qquad (111)$$

$$\int_x^{\frac{k}{n}} \left[ \left( f^{(4)}(t) - f(t) \right) - \left( f^{(4)}(x) - f(x) \right) \right] \left( \frac{\sinh\left(\frac{k}{n} - t\right) - \sin\left(\frac{k}{n} - t\right)}{2} \right) dt$$

$$=: R_4\left(\frac{k}{n}, x\right).$$

As in Theorems 7.5, 7.6 we derive

$$H_n^*(f, x) - f(x) \sum_{k=\lceil na \rceil}^{\lfloor nb \rfloor} M_{q,\lambda}(nx - k) -$$

$$\frac{f'(x)}{2} H_n^* \left( (\sinh(\cdot - x) + \sin(\cdot - x)), x \right) -$$

$$\frac{f''(x)}{2} H_n^* \left( (\cosh(\cdot - x) - \cos(\cdot - x)), x \right) -$$

$$\frac{f^{(3)}(x)}{2} H_n^* \left( (\sinh(\cdot - x) - \sin(\cdot - x)), x \right) -$$

$$\frac{f^{(4)}(x)}{2} H_n^* \left( \left( \sinh^2\left(\frac{\cdot - x}{2}\right) - \sin^2\left(\frac{\cdot - x}{2}\right) \right), x \right) = \Phi_n(x), \qquad (112)$$

where

$$\Phi_n(x) := \sum_{k=\lceil na \rceil}^{\lfloor nb \rfloor} M_{q,\lambda}(nx - k) R_4\left(\frac{k}{n}, x\right). \qquad (113)$$

Without loss of generality we can assume that $n > \left\lceil (b-a)^{-\frac{1}{\alpha}} \right\rceil$.

Thus $\left| \frac{k}{n} - x \right| \leq \frac{1}{n^\alpha}$ or $\left| \frac{k}{n} - x \right| > \frac{1}{n^\alpha}$.

In case of $\left| \frac{k}{n} - x \right| \leq \frac{1}{n^\alpha}$, we have the following cases:

i) if $\frac{k}{n} \geq x$, then

$$\left| R_4 \left( \frac{k}{n}, x \right) \right| =$$

$$\left| \frac{1}{2} \int_x^{\frac{k}{n}} \left[ \left( f^{(4)}(t) - f(t) \right) - \left( f^{(4)}(x) - f(x) \right) \right] \left( \sinh \left( \frac{k}{n} - t \right) - \sin \left( \frac{k}{n} - t \right) \right) dt \right| \leq$$

$$\frac{1}{2} \int_x^{\frac{k}{n}} \left| \left( f^{(4)}(t) - f(t) \right) - \left( f^{(4)}(x) - f(x) \right) \right| \left| \sinh \left( \frac{k}{n} - t \right) - \sin \left( \frac{k}{n} - t \right) \right| dt \leq$$

$$\tag{114}$$

$$\frac{1}{2} \int_x^{\frac{k}{n}} \omega_1 \left( f^{(4)} - f, t - x \right) \left( \left| \sinh \left( \frac{k}{n} - t \right) \right| + \left| \sin \left( \frac{k}{n} - t \right) \right| \right) dt \leq$$

$$\frac{\omega_1 \left( f^{(4)} - f, \frac{k}{n} - x \right)}{2} \int_x^{\frac{k}{n}} \left( \cosh (b-a) \left( \frac{k}{n} - t \right) + \left( \frac{k}{n} - t \right) \right) dt =$$

$$\frac{(\cosh(b-a)+1) \, \omega_1 \left( f^{(4)} - f, \frac{k}{n} - x \right)}{2} \int_x^{\frac{k}{n}} \left( \frac{k}{n} - t \right) dt =$$

$$\frac{(\cosh(b-a)+1) \, \omega_1 \left( f^{(4)} - f, \frac{k}{n} - x \right)}{4} \left( \frac{k}{n} - x \right)^2 \leq$$

$$\frac{(\cosh(b-a)+1) \, \omega_1 \left( f^{(4)} - f, \frac{1}{n^\alpha} \right)}{4n^{2\alpha}}. \tag{115}$$

That is , when $\frac{k}{n} \geq x$, then

$$\left| R_4 \left( \frac{k}{n}, x \right) \right| \leq \frac{(\cosh(b-a)+1) \, \omega_1 \left( f^{(4)} - f, \frac{1}{n^\alpha} \right)}{4n^{2\alpha}}. \tag{116}$$

ii) if $\frac{k}{n} < x$, then

$$\left| R_4 \left( \frac{k}{n}, x \right) \right| =$$

$$\left| \frac{1}{2} \int_{\frac{k}{n}}^x \left[ \left( f^{(4)}(t) - f(t) \right) - \left( f^{(4)}(x) - f(x) \right) \right] \left( \sinh \left( \frac{k}{n} - t \right) - \sin \left( \frac{k}{n} - t \right) \right) dt \right| \leq$$

$$\frac{\omega_1 \left( f^{(4)} - f, x - \frac{k}{n} \right)}{2} \int_{\frac{k}{n}}^x \left( \cosh (b-a) \left( t - \frac{k}{n} \right) + \left( t - \frac{k}{n} \right) \right) dt =$$

$$\frac{\left(\cosh\left(b-a\right)+1\right)\omega_1\left(f^{(4)}-f,x-\frac{k}{n}\right)}{2}\int_{\frac{k}{n}}^{x}\left(t-\frac{k}{n}\right)dt=$$

$$\frac{\left(\cosh\left(b-a\right)+1\right)\omega_1\left(f^{(4)}-f,x-\frac{k}{n}\right)}{4}\left(x-\frac{k}{n}\right)^2\leq \qquad (117)$$

$$\leq\frac{\left(\cosh\left(b-a\right)+1\right)\omega_1\left(f^{(4)}-f,\frac{1}{n^\alpha}\right)}{4n^{2\alpha}}.$$

Consequently, when $\left|\frac{k}{n}-x\right|\leq\frac{1}{n^\alpha}$, we always obtain that

$$\left|R_4\left(\frac{k}{n},x\right)\right|\leq\frac{\left(\cosh\left(b-a\right)+1\right)\omega_1\left(f^{(4)}-f,\frac{1}{n^\alpha}\right)}{4n^{2\alpha}}. \qquad (118)$$

Next assume again $\frac{k}{n}\geq x$, then

$$\left|R_4\left(\frac{k}{n},x\right)\right|\leq$$

$$\frac{1}{2}\int_{x}^{\frac{k}{n}}\left|\left(f^{(4)}(t)-f(t)\right)-\left(f^{(4)}(x)-f(x)\right)\right|\left|\sinh\left(\frac{k}{n}-t\right)-\sin\left(\frac{k}{n}-t\right)\right|dt\leq$$

$$\left\|f^{(4)}-f\right\|_\infty\int_{x}^{\frac{k}{n}}\left[\cosh\left(b-a\right)\left(\frac{k}{n}-t\right)+\left(\frac{k}{n}-t\right)\right]dt=$$

$$\left\|f^{(4)}-f\right\|_\infty\left(\cosh\left(b-a\right)+1\right)\int_{x}^{\frac{k}{n}}\left(\frac{k}{n}-t\right)dt=$$

$$\frac{\left\|f^{(4)}-f\right\|_\infty\left(\cosh\left(b-a\right)+1\right)}{2}\left(\frac{k}{n}-x\right)^2\leq \qquad (119)$$

$$\leq\frac{\left\|f^{(4)}-f\right\|_\infty\left(\cosh\left(b-a\right)+1\right)\left(b-a\right)^2}{2}.$$

Hence

$$\left|R_4\left(\frac{k}{n},x\right)\right|\leq\frac{\left\|f^{(4)}-f\right\|_\infty\left(\cosh\left(b-a\right)+1\right)\left(b-a\right)^2}{2}. \qquad (120)$$

When $\frac{k}{n}<x$, we have

$$\left|R_4\left(\frac{k}{n},x\right)\right|\leq$$

$$\frac{1}{2}\int_{\frac{k}{n}}^{x}\left|\left(f^{(4)}(t)-f(t)\right)-\left(f^{(4)}(x)-f(x)\right)\right|\left|\sinh\left(\frac{k}{n}-t\right)-\sin\left(\frac{k}{n}-t\right)\right|dt\leq$$

$$\left\| f^{(4)} - f \right\|_\infty \int_{\frac{k}{n}}^{x} \left[ \cosh\left(b-a\right)\left(t - \frac{k}{n}\right) + \left(t - \frac{k}{n}\right) \right] dt =$$

$$\left\| f^{(4)} - f \right\|_\infty \left(\cosh\left(b-a\right)+1\right) \int_{\frac{k}{n}}^{x} \left(t - \frac{k}{n}\right) dt =$$

$$\frac{\left\| f^{(4)} - f \right\|_\infty \left(\cosh\left(b-a\right)+1\right)}{2} \left(x - \frac{k}{n}\right)^2 \le \tag{121}$$

$$\le \frac{\left\| f^{(4)} - f \right\|_\infty \left(\cosh\left(b-a\right)+1\right)\left(b-a\right)^2}{2}.$$

So, it is always true that

$$\left| R_4\left(\frac{k}{n},x\right) \right| \le \frac{\left\| f^{(4)} - f \right\|_\infty \left(\cosh\left(b-a\right)+1\right)\left(b-a\right)^2}{2}. \tag{122}$$

Thus

$$\left| \Phi_n\left(x\right) \right| \le \sum_{k=\lceil na\rceil}^{\lfloor nb\rfloor} M_{q,\lambda}\left(nx-k\right) \left| R_4\left(\frac{k}{n},x\right) \right| =$$

$$\sum_{\substack{k=\lceil na\rceil \\ :\,\left|\frac{k}{n}-x\right| \le \frac{1}{n^\alpha}}}^{\lfloor nb\rfloor} M_{q,\lambda}\left(nx-k\right) \left| R_4\left(\frac{k}{n},x\right) \right| + \tag{123}$$

$$\sum_{\substack{k=\lceil na\rceil \\ :\,\left|\frac{k}{n}-x\right| > \frac{1}{n^\alpha}}}^{\lfloor nb\rfloor} M_{q,\lambda}\left(nx-k\right) \left| R_4\left(\frac{k}{n},x\right) \right| \le$$

$$\left( \sum_{\substack{k=\lceil na\rceil \\ :\,\left|\frac{k}{n}-x\right| \le \frac{1}{n^\alpha}}}^{\lfloor nb\rfloor} M_{q,\lambda}\left(nx-k\right) \right) \frac{\left(\cosh\left(b-a\right)+1\right)\omega_1\left(f^{(4)}-f,\frac{1}{n^\alpha}\right)}{4n^{2\alpha}} + \tag{124}$$

$$\left( \sum_{\substack{k=\lceil na\rceil \\ :\,\left|\frac{k}{n}-x\right| > \frac{1}{n^\alpha}}}^{\lfloor nb\rfloor} M_{q,\lambda}\left(nx-k\right) \right)$$

$$\left( \frac{\left\| f^{(4)} - f \right\|_\infty \left(\cosh\left(b-a\right)+1\right)\left(b-a\right)^2}{2} \right) \underset{\le}{\text{(by Theorem 7.3)}}$$

$$\frac{\left(\cosh\left(b-a\right)+1\right)\omega_1\left(f^{(4)}-f,\frac{1}{n^\alpha}\right)}{4n^{2\alpha}}+ \tag{125}$$

$$\frac{\left\|f^{(4)}-f\right\|_\infty\left(\cosh\left(b-a\right)+1\right)\left(b-a\right)^2}{2}Te^{-2\lambda n^{(1-\alpha)}}.$$

We have proved that

$$\left|\Phi_n\left(x\right)\right|\leq\frac{\left(\cosh\left(b-a\right)+1\right)}{2}$$

$$\left[\frac{\omega_1\left(f^{(4)}-f,\frac{1}{n^\alpha}\right)}{2n^{2\alpha}}+\left\|f^{(4)}-f\right\|_\infty\left(b-a\right)^2Te^{-2\lambda n^{(1-\alpha)}}\right]. \tag{126}$$

We observe that

$$H_n\left(f,x\right)-f\left(x\right)-\frac{f'\left(x\right)}{2}H_n\left(\left(\sinh\left(\cdot-x\right)+\sin\left(\cdot-x\right)\right),x\right)$$

$$-\frac{f''\left(x\right)}{2}H_n\left(\left(\cosh\left(\cdot-x\right)-\cos\left(\cdot-x\right)\right),x\right) \tag{127}$$

$$-\frac{f^{(3)}\left(x\right)}{2}H_n\left(\left(\sinh\left(\cdot-x\right)-\sin\left(\cdot-x\right)\right),x\right)$$

$$-f^{(4)}\left(x\right)H_n\left(\left(\left(\sinh^2\left(\frac{\cdot-x}{2}\right)-\sin^2\left(\frac{\cdot-x}{2}\right)\right)\right),x\right)=$$

$$\left[H_n^*\left(f,x\right)-f\left(x\right)\sum_{k=\lceil na\rceil}^{\lfloor nb\rfloor}M_{q,\lambda}\left(nx-k\right)-\frac{f'\left(x\right)}{2}H_n^*\left(\left(\sinh\left(\cdot-x\right)+\sin\left(\cdot-x\right)\right),x\right)\right.$$

$$-\frac{f''\left(x\right)}{2}H_n^*\left(\left(\cosh\left(\cdot-x\right)-\cos\left(\cdot-x\right)\right),x\right) \tag{128}$$

$$-\frac{f^{(3)}\left(x\right)}{2}H_n^*\left(\left(\sinh\left(\cdot-x\right)-\sin\left(\cdot-x\right)\right),x\right)$$

$$\left.-f^{(4)}\left(x\right)H_n^*\left(\left(\left(\sinh^2\left(\frac{\cdot-x}{2}\right)-\sin^2\left(\frac{\cdot-x}{2}\right)\right)\right),x\right)\right]\frac{1}{\sum_{k=\lceil na\rceil}^{\lfloor nb\rfloor}M_{q,\lambda}\left(nx-k\right)}$$

$$=\frac{\Phi_n\left(x\right)}{\sum_{k=\lceil na\rceil}^{\lfloor nb\rfloor}M_{q,\lambda}\left(nx-k\right)}.$$

Finally, we obtain ($\forall\,x\in\left[a,b\right],n\in\mathbb{N}$):

$$\left|H_n\left(f,x\right)-f\left(x\right)-\frac{f'\left(x\right)}{2}H_n\left(\left(\sinh\left(\cdot-x\right)+\sin\left(\cdot-x\right)\right),x\right)\right.$$

$$-\frac{f''\left(x\right)}{2}H_n\left(\left(\cosh\left(\cdot-x\right)-\cos\left(\cdot-x\right)\right),x\right) \tag{129}$$

$$-\frac{f^{(3)}(x)}{2}H_n\left(\left(\sinh\left(\cdot-x\right)-\sin\left(\cdot-x\right)\right),x\right)$$

$$-f^{(4)}(x)H_n\left(\left(\sinh^2\left(\frac{\cdot-x}{2}\right)-\sin^2\left(\frac{\cdot-x}{2}\right)\right),x\right)\right|=$$

$$\frac{\left|\Phi_n(x)\right|}{\sum_{k=\lceil na\rceil}^{\lfloor nb\rfloor}M_{q,\lambda}(nx-k)}\leq\Delta(q)\left|\Phi_n(x)\right|\overset{\text{(by (126))}}{\leq}\frac{\Delta(q)\left(\cosh(b-a)+1\right)}{2}$$

$$\left[\frac{\omega_1\left(f^{(4)}-f,\frac{1}{n^\alpha}\right)}{2n^{2\alpha}}+\left\|f^{(4)}-f\right\|_\infty(b-a)^2\,Te^{-2\lambda n^{(1-\alpha)}}\right].\qquad(130)$$

The theorem is proved.

We continue with a general trigonometric result.

**Theorem 7.8**
*Let $f\in C^4\left([a,b],\mathbb{C}\right),\ 0<\alpha<1,\ n\in\mathbb{N}:n^{1-\alpha}>2,\ x\in[a,b]$. Let also $\overline{\alpha},\overline{\beta}\in\mathbb{R}$ with $\overline{\alpha}\overline{\beta}\left(\overline{\alpha}^2-\overline{\beta}^2\right)\neq0$. Then*
*1)*

$$\left|H_n(f,x)-f(x)-\frac{f'(x)}{\overline{\alpha}\overline{\beta}\left(\overline{\beta}^2-\overline{\alpha}^2\right)}H_n\left(\left(\overline{\beta}^3\sin\left(\overline{\alpha}\left(\cdot-x\right)\right)-\overline{\alpha}^3\sin\left(\overline{\beta}\left(\cdot-x\right)\right)\right),x\right)\right.$$

$$-\frac{f''(x)}{\left(\overline{\beta}^2-\overline{\alpha}^2\right)}H_n\left(\left(\cos\left(\overline{\alpha}\left(\cdot-x\right)\right)-\cos\left(\overline{\beta}\left(\cdot-x\right)\right)\right),x\right)$$

$$-\frac{f'''(x)}{\overline{\alpha}\overline{\beta}\left(\overline{\beta}^2-\overline{\alpha}^2\right)}H_n\left(\left(\overline{\beta}\sin\left(\overline{\alpha}\left(\cdot-x\right)\right)-\overline{\alpha}\sin\left(\overline{\beta}\left(\cdot-x\right)\right)\right),x\right)$$

$$-\left(\frac{2f^{(4)}(x)+\left(\overline{\alpha}^2+\overline{\beta}^2\right)f''(x)}{\left(\overline{\alpha}\overline{\beta}\right)^2\left(\overline{\beta}^2-\overline{\alpha}^2\right)}\right)$$

$$H_n\left(\left(\overline{\beta}^2\sin^2\left(\frac{\overline{\alpha}\left(\cdot-x\right)}{2}\right)-\overline{\alpha}^2\sin^2\left(\frac{\overline{\beta}\left(\cdot-x\right)}{2}\right)\right),x\right)\right|\leq\qquad(131)$$

$$\frac{\Delta(q)}{\left|\overline{\beta}^2-\overline{\alpha}^2\right|}\left[\frac{\omega_1\left(\left(f^{(4)}+\left(\overline{\alpha}^2+\overline{\beta}^2\right)f''+\overline{\alpha}^2\overline{\beta}^2f\right),\frac{1}{n^\alpha}\right)}{n^{2\alpha}}+\right.$$

$$2\left\|f^{(4)}+\left(\overline{\alpha}^2+\overline{\beta}^2\right)f''+\overline{\alpha}^2\overline{\beta}^2f\right\|_\infty(b-a)^2\,Te^{-2\lambda n^{(1-\alpha)}}\right],$$

2) if $f^{(i)}(x) = 0$, $i = 1,2,3,4$, *we get*

$$|H_n(f,x) - f(x)| \le \frac{\Delta(q)}{\left|\overline{\beta}^2 - \overline{\alpha}^2\right|}$$

$$\left[\frac{\omega_1\left(\left(f^{(4)} + \left(\overline{\alpha}^2 + \overline{\beta}^2\right)f'' + \overline{\alpha}^2\overline{\beta}^2 f\right), \frac{1}{n^\alpha}\right)}{n^{2\alpha}} + \right. \tag{132}$$

$$\left. 2\left\|f^{(4)} + \left(\overline{\alpha}^2 + \overline{\beta}^2\right)f'' + \overline{\alpha}^2\overline{\beta}^2 f\right\|_\infty (b-a)^2 T e^{-2\lambda n^{(1-\alpha)}}\right].$$

*The high speed of convergence in (1) and (2) is* $n^{-3\alpha}$.

**Proof 7.4**    As similar to Theorem 7.7 is omitted. It is based on Theorem 9 of [11].

We finish with a general hyperbolic result.

***Theorem 7.9***
*Let* $f \in C^4([a,b],\mathbb{C})$, $0 < \alpha < 1$, $n \in \mathbb{N}: n^{1-\alpha} > 2$, $x \in [a,b]$. *Let also* $\overline{\alpha}, \overline{\beta} \in \mathbb{R}$ *with*
$\overline{\alpha}\,\overline{\beta}\left(\overline{\alpha}^2 - \overline{\beta}^2\right) \ne 0$. *Then*
*1)*

$$\left| H_n(f,x) - f(x) - \frac{f'(x)}{\overline{\alpha}\,\overline{\beta}\left(\overline{\beta}^2 - \overline{\alpha}^2\right)} H_n\left(\left(\overline{\beta}^3 \sinh(\overline{\alpha}(\cdot - x)) - \overline{\alpha}^3 \sinh\left(\overline{\beta}(\cdot - x)\right)\right), x\right)\right.$$

$$- \frac{f''(x)}{\overline{\beta}^2 - \overline{\alpha}^2} H_n\left(\left(\cosh\left(\overline{\beta}(\cdot - x)\right) - \cosh(\overline{\alpha}(\cdot - x))\right), x\right)$$

$$- \frac{f'''(x)}{\overline{\alpha}\,\overline{\beta}\left(\overline{\beta}^2 - \overline{\alpha}^2\right)} H_n\left(\left(\overline{\alpha}\sinh\left(\overline{\beta}(\cdot - x)\right) - \overline{\beta}\sinh(\overline{\alpha}(\cdot - x))\right), x\right)$$

$$- \left(\frac{2(f^{(4)}(x) - \left(\overline{\alpha}^2 + \overline{\beta}^2\right)f''(x))}{\left(\overline{\alpha}\,\overline{\beta}\right)^2\left(\overline{\beta}^2 - \overline{\alpha}^2\right)}\right)$$

$$\left. H_n\left(\left(\overline{\alpha}^2 \sinh^2\left(\frac{\overline{\beta}(\cdot - x)}{2}\right) - \overline{\beta}^2 \sinh^2\left(\frac{\overline{\alpha}(\cdot - x)}{2}\right)\right), x\right)\right| \le \tag{133}$$

$$\frac{\Delta(q)\cosh(b-a)}{\left|\overline{\beta}^2 - \overline{\alpha}^2\right|}\left[\frac{\omega_1\left(\left(f^{(4)} - \left(\overline{\alpha}^2 + \overline{\beta}^2\right)f'' + \overline{\alpha}^2\overline{\beta}^2 f\right), \frac{1}{n^\alpha}\right)}{n^{2\alpha}} + \right.$$

$$\left. 2\left\|f^{(4)} - \left(\overline{\alpha}^2 + \overline{\beta}^2\right)f'' + \overline{\alpha}^2\overline{\beta}^2 f\right\|_\infty (b-a)^2 T e^{-2\lambda n^{(1-\alpha)}}\right],$$

*2) if $f^{(i)}(x) = 0$, $i = 1,2,3,4$, we get*

$$|H_n(f,x) - f(x)| \leq \frac{\Delta(q)\cosh(b-a)}{\left|\overline{\beta}^2 - \overline{\alpha}^2\right|} \tag{134}$$

$$\left[ \frac{\omega_1\left(\left(f^{(4)} - \left(\overline{\alpha}^2 + \overline{\beta}^2\right)f'' + \overline{\alpha}^2\overline{\beta}^2 f\right), \frac{1}{n^\alpha}\right)}{n^{2\alpha}} + \right.$$

$$\left. 2\left\|f^{(4)} - \left(\overline{\alpha}^2 + \overline{\beta}^2\right)f'' + \overline{\alpha}^2\overline{\beta}^2 f\right\|_\infty (b-a)^2 \, Te^{-2\lambda n^{(1-\alpha)}} \right].$$

*The high speed of convergence in (1) and (2) is $n^{-3\alpha}$.*

**Proof 7.5**    As similar to Theorem 7.7 is omitted. It is based on Theorem 10 of [11].

# References

[1] Anastassiou, G.A. 1997. Rate of convergence of some neural network operators to the unit-univariate case. J. Math. Anal. Appl. 212: 237–262.

[2] Anastassiou, G.A. 2001. Quantitative Approximations. Chapman & Hall/CRC, Boca Raton, New York.

[3] Anastassiou, G.A. 2011. Univariate hyperbolic tangent neural network approximation. Mathematics and Computer Modelling, 53: 1111–1132.

[4] Anastassiou, G.A. 2011. Multivariate hyperbolic tangent neural network approximation. Computers and Mathematics, 61: 809–821.

[5] Anastassiou, G.A. 2011. Multivariate sigmoidal neural network approximation. Neural Networks, 24: 378–386.

[6] Anastassiou, G.A. 2011. Intelligent Systems: Approximation by Artificial Neural Networks. Intelligent Systems Reference Library, Vol. 19, Springer, Heidelberg.

[7] Anastassiou, G.A. 2012. Univariate sigmoidal neural network approximation. J. of Computational Analysis and Applications 14(4): 659–690.

[8] Anastassiou, G.A. 2017. Vector fractional Korovkin type approximations. Dynamic Systems and Applications, 26: 81–104.

[9] Anastassiou, G.A. 2023. General sigmoid based Banach space valued neural network approximation. J. of Computational Analysis and Applications 31(4): 520–534.

[10] Anastassiou, G.A. 2023. Parametrized, Deformed and General Neural Networks. Accepted for Publication, Springer, Heidelberg, New York.

[11] Anastassiou, G.A. 2023. Opial and Ostrowski type inequalities based on trigonometric and hyperbolic type Taylor formulae. Malaya Journal Mathematik 11S:1–26.

[12] Chen, Z. and Cao, F. 2009. The approximation operators with sigmoidal functions. Computers and Mathematics with Applications, 58: 758–765.

[13] Haykin, S. 1998. Neural Networks: A Comprehensive Foundation. (2 ed.), Prentice Hall, New York.

[14] McCulloch, W. and Pitts, W. 1943. A logical calculus of the ideas immanent in nervous activity. Bulletin of Mathematical Biophysis, 7: 115–133.

[15] Mitchell, T.M. 1997. Machine Learning. WCB-McGraw-Hill, New York.

# Chapter 8

# Nonlinear Exponential Sampling: Approximation Results and Applications

*Danilo Costarelli*

## 8.1 Introduction

The theory of approximating functions by nonlinear operators has been developed since the 1980's, thanks to the crucial contribution of the Polish mathematician J. Musielak (see, e.g., [64, 65, 66, 67, 68]). Further, the theory has been extensively developed in the monograph of Bardaro, Musielak and Vinti (see [20]), in relation to the abstract setting provided by the modular spaces, and with particular emphasis towards singular integrals and sampling-type operators ([47]).

The interest in this topic is due to the fact that nonlinear operators play an important role in Signal Processing. In fact, the above operators are suitable, e.g., in order to describe nonlinear transformations generated by signals that, during their filtering process, produce new frequencies.

Department of Mathematics and Computer Science, University of Perugia, 1, Via Vanvitelli, 06123 Perugia, Italy.
Email: danilo.costarelli@unipg.it

Moreover, the interest in nonlinear tools of mathematical analysis also finds applications, both from the theoretical and numerical point of view, to several well-known problems, such as to the solution of nonlinear integral equations, nonlinear ordinary differential equations, as well as to partial differential equations.

Following the above mentioned field of research, and based on the very recent interest aroused by the study of exponential sampling series, in the present paper we introduce and study the following nonlinear version of the celebrated exponential sampling operators. The proposed definition takes the following form:

$$
(S_w^\chi f)(x) := \sum_{k\in\mathbb{Z}} \chi\left(e^{-t_k} x^w, f\left(e^{t_k/w}\right)\right), \qquad x \in \mathbb{R}^+, \tag{I}
$$

where $f : \mathbb{R}^+ \to \mathbb{R}$ is any function such that the above series is convergent for every $x \in \mathbb{R}^+$, and where the bivariate function $\chi : \mathbb{R}_0^+ \times \mathbb{R} \to \mathbb{R}$ is called a "nonlinear kernel", and satisfies suitable assumptions.

For the introduced operators $S_w^\chi$ we provide a full study concerning their regularization and approximation properties, obtaining a reconstruction theory for signals of the form $f : \mathbb{R}^+ \to \mathbb{R}$.

In order to do this, as we will show later, a crucial role is played by the basic notions of Mellin Analysis, that seems to be the most appropriate topic in order to face the above approximation problems.

The main results here established are two regularization theorems for the operators $S_w^\chi$, in the case of continuous and log-uniformly continuous functions, a pointwise and uniform convergence theorem, and the corresponding quantitative and qualitative estimates for the order of approximation, with respect to the usual uniform norm. The achieved quantitative estimates have been expressed by means of the so-called log-modulus of continuity of the approximated function $f$, that turns out to be, in fact, the Mellin counterparts of the usual modulus of continuity. Similarly, the qualitative order of approximation has been established in case of functions belonging to suitable locally log-Holderian classes. Finally, the last part of the paper is devoted to the analysis of examples of nonlinear kernels, and to numerical examples showing the approximation capabilities of the introduced operators.

Below we give a detailed description of the structure of the paper. In Section 8.2 we provided a historical overview on the sampling theory, and on the main motivations that justified the definitions and the results here established. In Section 8.3 the notion of nonlinear kernel has been given, together with some useful preliminary results, and we finally introduced the definition of the nonlinear exponential sampling operators. In Section 8.4 we established the regularization theory by $S_w^\chi$, while in Section 8.5 we obtained the above mentioned pointwise and uniform convergence theorems. Here also the problem of the so-called "exponential prediction" has been solved. Finally, Section 8.6 has been devoted to quantitative and qualitative theorems regarding the order of approximation. The remaining part of the paper has been related to examples and particular cases. In fact, in Section 8.7 we discussed about examples of product-type kernels, establishing some corollaries that are valid in the considered particular instances. In Section 8.8 we recalled some fundamental results of Mellin Analysis, that have been crucial in order to present the examples given in Section 8.9 and Section 8.10. There, Mellin B-spline and Mellin-Fejer kernels have been presented. Furthermore, numerical examples have been given in Section 8.11, while some final remarks together with some open problems have been discussed in Section 8.12.

## 8.2  Historical Notes on Sampling-type Operators

In order to understand the reasons that brought to the introduction of the sampling-type operators, we have to come back to one of the most important results of the 1900: the celebrated Whittaker-Kotelnikov-Shannon (WKS) sampling theorem (see, e.g., [71, 31, 32, 24]). More precisely, the WKS sampling theorem provides an exact reconstruction (interpolation) formula for band-limited and finite energy signals, starting from a sequence of uniform spaced samples values of the form $f(k/w)$, being $f : \mathbb{R} \to \mathbb{R}$ the signal to be reconstructed and $w > 0$ the parameter defining its band. Such a theorem can be stated as follows.

***Theorem 8.1 WKS sampling theorem***
*Let $f \in L^2(\mathbb{R})$ be a given function, such that its Fourier transform $\widehat{f}$ (in the $L^2$-sense) has:*

$$supp\,\widehat{f} \subset [-\pi w, \pi w], \quad w > 0.$$

*Then, for every $x \in \mathbb{R}$ the following interpolation formula holds:*

$$\sum_{k \in \mathbb{Z}} f(k/w)\,\mathrm{sinc}(wx - k) = f(x), \tag{8.1}$$

*where:*

$$\mathrm{sinc}(x) := \begin{cases} \frac{\sin(\pi x)}{\pi x}, & x \neq 0, \\ 1, & x = 0. \end{cases}$$

It is well-known that, the above theorem, even if it is very elegant from the mathematical point of view, presents some precise disadvantages concerning the possibility to be applied to some concrete problems. Here we list the main issues.

1. Real world signals (functions) have limited duration (i.e., compact support), while the WKS sampling theorem only functions with unbounded support can be reconstructed. This claim is motivated by the well-known fact that, a signal can not be simultaneously duration and band-limited.

2. As a consequence of the Paley-Wiener theorem, the requirement that $f \in L^2(\mathbb{R})$ is band-limited, implies that $f$ must be the restriction to the real

axis of an entire function of the exponential type. This means that $f$ must be a very smooth function. Unfortunately, real world signals generally do not possess such high regularity. Actually, there are several examples of full classes of signals that are discontinuous, such as, for instance, digital images, and several others.

3. The above interpolation formula is characterized by the presence of a series that, of course, in order to be implemented must be truncated. The truncation of the series produces an approximation error, then we get the lost interpolation property.

These issues have brought several authors to consider possible extensions and/or generalizations of the above result. Among the more active authors, we can mention the German mathematician P.L. Butzer. With the help of his students, Butzer had the idea to replace the *sinc* function in the sampling series (8.1) by a, more general, approximate identity.

In practice, considering a suitable function (kernel) $\varphi : \mathbb{R} \to \mathbb{R}$, such that:

$$\sum_{k \in \mathbb{Z}} \varphi(x - k) = 1, \quad x \in \mathbb{R},$$

Butzer introduced the so-called *generalized sampling operators* (see, e.g., [38, 26, 34, 33, 59]), of the form:

$$(G_w f)(x) := \sum_{k \in \mathbb{Z}} f(k/w)\,\varphi(wx - k), \quad x \in \mathbb{R}, \quad w > 0. \tag{8.2}$$

Clearly, the replacement of the *sinc* function by a general kernel function $\varphi$ has transformed an interpolation series into an approximation one. By means of the family of approximation operators $G_w$, Butzer proved that continuous and bounded functions can be reconstructed (instead of $C^\infty$-functions only, i.e., weakening the assumptions on the signal to be approximated), in fact totally and partially solving the issues 1 and 2, respectively.

Moreover, if the kernel function $\varphi$ is assumed to be continuous and with compact support, for the evaluation of the series $G_w f$ at any fixed point $x \in \mathbb{R}$, only a finite number of terms can be considered, thus solving also the issue 3. listed above.

For a detailed overview concerning the above topics, the readers can see, e.g., [29].

Observing that the definition of the generalized sampling operators strongly depends on the pointwise values assumed by the signal $f$, it seems clear that such operators are not the most suitable in order to reconstruct discontinuous signals, such as $L^p$-functions.

For this reason, the operators $G_w f$ have been subsequently generalized by the introduction of their corresponding Kantorovich version ([21]), of the form:

$$(K_w f)(x) := \sum_{k \in \mathbb{Z}} \left[ w \int_{k/w}^{(k+1)/w} f(u)\, du \right] \varphi(wx - k), \quad x \in \mathbb{R}, \quad w > 0, \quad (8.3)$$

for any locally integrable function $f : \mathbb{R} \to \mathbb{R}$.

In practice, in the generalized sampling operators $G_w f$ we replaced the sample values $f(k/w)$ by an average of $f$ on a small interval around $k/w$, namely by $w \int_{k/w}^{(k+1)/w} f(u)\, du$. Practically, more information is usually known around a point rather than precisely at that point; this procedure simultaneously reduces the so-called jitter errors ([21] again).

It has been proved that, the Kantorovich sampling operators $K_w f$ converge to $f$, as $w \to +\infty$, for $f \in L^p(\mathbb{R})$, $1 \le p < +\infty$, with respect to the usual $p$-norm. Then, also the case of the reconstruction of not necessarily continuous signals by sampling-type operators has been completely solved (issue 2). Note that, by the multivariate version of the operators $K_w f$ also real world applications involving digital images (discontinuous signals) have been faced. For some articles on this respect, the readers can see, e.g., [46, 54, 55, 51, 62, 61, 39, 56, 40, 41, 45, 69].

Finally, a further very recent extension of the above mentioned sampling-type operators has been given in [17]. Here, a Durrmeyer version of the above sampling-type series has been introduced as follows:

$$(D_w^{\psi,\varphi} f)(x) := \sum_{k \in \mathbb{Z}} \left[ w \int_{\mathbb{R}} \psi(wu - k) f(u)\, du \right] \varphi(wx - k), \quad x \in \mathbb{R}, \quad w > 0,$$

$$(8.4)$$

where here the new (continuous) kernel function $\psi \in L^1(\mathbb{R})$ and:

$$\int_{\mathbb{R}} \psi(u)\, du = 1.$$

By simple computations, it is not difficult to see that, if we choose $\psi$ equal to the characteristic function of the interval $[0, 1]$, the series $D_w^{\psi,\varphi} f$ reduce to the operators $K_w f$, while, given a distributional interpretation to $D_w^{\psi,\varphi} f$ and choosing as $\psi$ the Dirac delta distribution, we can find again the operators $G_w f$ (see [48, 49, 50]).

Very recently, in Bardaro, Faina and Mantellini introduced a new family of sampling-type operators, the so-called exponential sampling series. The idea behind such a new definition is that, the exponential sampling representation of a Mellin band-limited function, as a series in which the sample values are exponentially spaced, is a power tool in finding solutions of certain inverse problems that have fundamental applications in optical physics phenomena. For instance, in the light scattering, Fraunhofer diffraction and radio astronomy, see, e.g., [22, 42, 70, 23, 60, 57].

Retracing all the above described steps, i.e., starting from WKS sampling theorem since the introduction of the operators $D_w^{\psi;\varphi}$, it is possible to find all the corresponding exponential-type (Mellin-type) versions of the above quoted results. Obviously, in such a study a crucial role is naturally played by the main tools of Mellin Analysis.

For instance, a rigorous version of the exponential sampling theorem for Mellin band-limited functions was firstly given in [35], and subsequently considered in [28, 9, 8]. In practice, the exponential sampling formula corresponding to the classical one given in (8.1) can be formally viewed as its Mellin counterparts, using a suitable change of variables. However, this relation is not only "formal", in the sense that the notion of Mellin band-limited function is concretely different from the corresponding version given in Fourier Analysis ([10]). It is possible to prove that, the two classes of functions are different, i.e., there exist non-trivial functions that can not be simultaneously Mellin and Fourier band-limited. In order to understand that the two theories are different in nature, we can also mention, for instance, that it is possible to provide two different versions of the Paley-Wiener theorem in the two distinct settings. In the Mellin one, such a result involves the study of Riemann surfaces of the logarithm function (see [10] again).

The latter considerations have motivated the study of the Mellin Theory independently from the Fourier one.

The study of the latter topic has been initiated in the following works [63, 27, 28]; the exponential version of the WKS interpolation formula can be written as follows:

$$(E_w f)(x) := \sum_{k \in \mathbb{Z}} f(e^{k/w}) \mathrm{lin}_{c/w}(e^{-k} x^w), \quad w > 0, \quad x \in \mathbb{R}^+, \quad c \in \mathbb{R}, \quad (8.5)$$

where:

$$\mathrm{lin}_c(x) := \begin{cases} x^{-c} \operatorname{sinc}(\log x), & x \neq 1 \\ 1, & x = 1. \end{cases}$$

When $f : \mathbb{R}^+ \to \mathbb{R}$ is Mellin band-limited, and its Mellin-Fourier transform is supported in $c + i[-\pi w, \pi w]$, then the following interpolation formula holds:

$$(E_w f)(x) = f(x), \quad x \in \mathbb{R}^+.$$

Retracing the history of the sampling-type series, the following exponential sampling type operators have been recently introduced and studied from the point of view of Approximation Theory. More precisely, in the literature we can find the following:

$$(EG_w f)(x) := \sum_{k \in \mathbb{Z}} f(e^{k/w}) \varphi(e^{-k} x^w), \quad x \in \mathbb{R}, \quad w > 0, \quad (8.6)$$

where $EG_w$ are known with the name of generalized exponential sampling series (see [11]),

$$(EK_w f)(x) := \sum_{k \in \mathbb{Z}} \left[ w \int_{k/w}^{(k+1)/w} f(e^u)\,du \right] \varphi(e^{-k}x^w), \quad x \in \mathbb{R}, \quad w > 0, \quad (8.7)$$

where $EK_w$ are known with the name of exponential Kantorovich sampling series (see [3, 1, 2]), and finally:

$$(ED_w^{\psi,\varphi} f)(x) := \sum_{k \in \mathbb{Z}} \left[ w \int_0^{+\infty} \psi(e^{-k}u^w)f(u)\frac{du}{u} \right] \varphi(e^{-k}x^w), \quad x \in \mathbb{R}, \quad w > 0,$$

$$(8.8)$$

where $ED_w^{\psi,\varphi}$ are the exponential Durrmeyer sampling operators (see [15, 6]). Note that, the integrals

$$w \int_0^{+\infty} \psi(e^{-k}u^w)f(u)\frac{du}{u}$$

(see, e.g., [13]) represent the Mellin analogous of the classical singular (convolution) integrals.

For general useful references on sampling-type operators, one can also consult [44, 7, 4, 43, 5, 19, 25].


## 8.3  Preliminary Definitions and Results

In what follows, we denote by $C^0(\mathbb{R}^+)$ the set of all bounded and continuous functions $f : \mathbb{R}^+ \to \mathbb{R}$ endowed with the usual sup-norm $\|\cdot\|_\infty$.

We will say that a function $f \in C^0(\mathbb{R}^+)$ is *log-uniformly continuous*, if for any $\varepsilon > 0$ there exists $\gamma > 0$ such that $|f(x) - f(y)| < \varepsilon$ for every $x, y \in \mathbb{R}^+$, with $|\ln x - \ln y| \le \gamma$.

With the symbol $C(\mathbb{R}^+)$ we denote the subspace of $C^0(\mathbb{R}^+)$ of log-uniformly continuous functions.

Note that, it is not difficult to see that the notion of log-uniformly continuous functions is in general different from the usual definition of uniformly continuous functions. It is also clear that the two concepts are equivalent on compact intervals of $\mathbb{R}^+$.

Let now $\Pi = (t_k)_{k \in \mathbb{Z}}$ be a sequence of (not necessarily equally spaced) real numbers such that $-\infty < t_k < t_{k+1} < +\infty$, $\lim_{k \to \pm\infty} t_k = \pm\infty$ and such that $\delta \le \Delta_k := t_{k+1} - t_k \le \Delta$, for every $k \in \mathbb{Z}$ and for some fixed $\delta, \Delta > 0$.

In what follows, a function $\chi : \mathbb{R}^+ \times \mathbb{R} \to \mathbb{R}$ will be called a *nonlinear kernel* if it satisfies the following conditions:

$(\chi 1)$  $k \mapsto \chi(e^{-t_k}x^w, u) \in \ell^1(\mathbb{Z})$, for every $(x, u) \in \mathbb{R}^+ \times \mathbb{R}$ and $w > 0$;

$(\chi 2)$  $\chi(x, 0) = 0$, for every $x \in \mathbb{R}^+$;

$(\chi 3)$    $\chi$ is a $(L, \psi)$-Lipschitz kernel, i.e., there exists a measurable function $L : \mathbb{R}^+ \to \mathbb{R}_0^+$ and a continuous function $\psi : \mathbb{R}_0^+ \to \mathbb{R}_0^+$, with:

$(\psi 1)$    $\psi(0) = 0$, and $\psi(u) > 0$ for $u > 0$;

$(\psi 2)$    $\displaystyle\lim_{u \to +\infty} \psi(u) = +\infty$;

and such that

$$|\chi(x, u) - \chi(x, v)| \leq L(x)\, \psi(|u - v|),$$

for every $x \in \mathbb{R}^+$ and $u,\, v \in \mathbb{R}$;

$(\chi 4)$    there exists a parameter $\theta > 0$ such that:

$$\mathscr{T}_w(x, u) := \left| \sum_{k \in \mathbb{Z}} \chi(e^{-t_k} x^w, u) - u \right| = \mathcal{O}(w^{-\theta})$$

as $w \to +\infty$, uniformly with respect to $x \in \mathbb{R}^+$ and $u \in \mathbb{R}$.

Moreover, we will assume that the function $L$ of condition $(\chi 3)$ satisfies the following properties:

$(L1)$    $L$ is continuous and bounded;

$(L2)$    there exists $\beta > 0$ such that

$$M_{\beta,\Pi}(L) := \sup_{x \in \mathbb{R}^+} \sum_{k \in \mathbb{Z}} L(e^{-t_k} x) \left| \ln\left(e^{-t_k} x\right) \right|^\beta$$

$$= \sup_{x \in \mathbb{R}^+} \sum_{k \in \mathbb{Z}} L(e^{-t_k} x) \left| \ln x - t_k \right|^\beta < +\infty.$$

The constant $M_{\beta,\Pi}(L)$ can be called the *log-discrete absolute moment of order* $\beta$.

**Remark 8.1**    Note that, when we deal with the study of approximation results in the nonlinear frame, the requirement of a generalized Lipschitz condition (assumption $(\chi 3)$) and of a generalized singularity (assumption $(\chi 4)$) is very tipycal. For more details regarding the introduction of the $(L, \psi)$-condition $(\chi 3)$ the readers can see [20]. Concerning condition $(\chi 4)$, in the literature there are available several versions of such property (see, e.g., [20, 16, 72, 73]). Here, we considered a new quantitative version of the assumption first introduced in [12], p. 396. ∎

Now, we can prove the following lemma.

*Lemma 8.1*

*Under the above assumptions on L, we have:*

*(i)*   $M_{0,\Pi}(L) := \sup_{x \in \mathbb{R}^+} \sum_{k \in \mathbb{Z}} L(e^{-t_k}x) < +\infty;$

*(ii)*   *for every $\gamma > 0$ it turns out that*

$$\sum_{|t_k - w\ln x| > \gamma w} L(e^{-t_k}x^w) \leq \frac{1}{\gamma^\beta w^\beta} M_{\beta,\Pi}(L)$$

*for every $x \in \mathbb{R}^+$, from which we get:*

$$\lim_{w \to +\infty} \sum_{|t_k - w\ln x| > \gamma w} L(e^{-t_k}x^w) = 0,$$

*uniformly with respect to $x \in \mathbb{R}^+$.*

**Proof 8.1**   We begin proving (i). For every fixed $x \in \mathbb{R}^+$, we can write what follows:

$$\sum_{k \in \mathbb{Z}} L(e^{-t_k}x) \leq \sum_{|\ln(e^{-t_k}x)| \leq M} L(e^{-t_k}x) + \sum_{|\ln(e^{-t_k}x)| > M} L(e^{-t_k}x),$$

where $M > 1$ is any fixed sufficiently large parameter. Now, to consider $|\ln(e^{-t_k}x)| \leq M$ means to take into account all $k$ for which $\ln x - M \leq t_k \leq \ln x + M$, and in view of the assumptions of $\Pi$, such indexes are of a number less or equal to $(2M\delta^{-1} + 1)$. Hence, using $(L2)$ we can finally get:

$$\sum_{k \in \mathbb{Z}} L(e^{-t_k}x) \leq \|L\|_\infty (2M\delta^{-1} + 1) + \sum_{|\ln(e^{-t_k}x)| > M} L(e^{-t_k}x)|\ln(e^{-t_k}x)|^\beta$$

$$= \|L\|_\infty (2M\delta^{-1} + 1) + \sum_{|\ln(e^{-t_k}x)| > M} L(e^{-t_k}x)|\ln x - t_k|^\beta$$

$$\leq \|L\|_\infty (2M\delta^{-1} + 1) + M_{\beta,\Pi}(L) < +\infty.$$

Now, passing to the supremum with respect to $x \in \mathbb{R}^+$ in the above inequality we immediately obtain the thesis. Concerning (ii), for any fixed $\gamma > 0$, and proceeding as above, we have:

$$\sum_{|t_k - w\ln x| > \gamma w} L(e^{-t_k}x^w) = \sum_{|\ln(e^{-t_k}x^w)| > \gamma w} L(e^{-t_k}x^w) = \sum_{|\ln(e^{-t_k}x^w)| > \gamma w} L(e^{-t_k}x^w) \frac{|\ln(e^{-t_k}x^w)|^\beta}{|\ln(e^{-t_k}x^w)|^\beta}$$

$$\leq \frac{1}{\gamma^\beta w^\beta} \sum_{|\ln(e^{-t_k}x^w)| > \gamma w} L(e^{-t_k}x^w)|\ln x - t_k|^\beta \leq \frac{1}{\gamma^\beta w^\beta} M_{\beta,\Pi}(L),$$

for every sufficiently large $w > 0$. This completes the proof.

Note that, using Lemma 8.1 we can also prove the following result.

***Lemma 8.2***

*Under the above assumptions on L, we have that:*

$$M_{v,\Pi}(L) < +\infty,$$

*for every $0 \leq v \leq \beta$.*

**Proof 8.2**    The proof follows immediately observing that:

$$\sum_{k \in \mathbb{Z}} L(e^{-t_k}x)\,|\ln x - t_k|^v \leq \|L\|_\infty(2\delta^{-1}+1) + \sum_{|\ln(e^{-t_k}x)|>1} L(e^{-t_k}x)\,|\ln(e^{-t_k}x)|^v$$

$$\leq \|L\|_\infty(2\delta^{-1}+1) + \sum_{|\ln(e^{-t_k}x)|>1} L(e^{-t_k}x)\,|\ln x - t_k|^\beta$$

$$\leq \|L\|_\infty(2\delta^{-1}+1) + M_{\beta,\Pi}(L) < +\infty.$$

Now, we are able to introduce the following definition.

**Definition 8.1**    We define the nonlinear exponential sampling operators for a given nonlinear kernel $\chi$, as follows

$$(S_w^\chi f)(x) := \sum_{k \in \mathbb{Z}} \chi\left(e^{-t_k}x^w,\, f\left(e^{t_k/w}\right)\right), \qquad x \in \mathbb{R}^+,$$

where $f : \mathbb{R}^+ \to \mathbb{R}$ is any function such that the above series is convergent for every $x \in \mathbb{R}^+$.

Note that, for any bounded function $f : \mathbb{R}^+ \to \mathbb{R}$, the nonlinear exponential sampling operators turn out to be well-defined. In fact, using assumption $(\chi 3)$ and Lemma 8.1 (i), we obtain:

$$|(S_w^\chi f)(x)| \leq \sum_{k \in \mathbb{Z}} L\left(e^{-t_k}x^w\right)\,\psi\left(\left|f\left(e^{t_k/w}\right)\right|\right) \leq \psi(\|f\|_\infty)M_{0,\Pi}(L) < +\infty.$$

$$(8.9)$$

For references concerning nonlinear-type operators, see, e.g., [20, 52, 53, 72, 73].

## 8.4    Regularization Theorems

In this section we want to investigate the regularity properties of the nonlinear exponential sampling operators.

We can prove the following.

***Theorem 8.2***

*Let $\chi$ be a continuous nonlinear kernel, and $f \in C^0(\mathbb{R}^+)$ be fixed. Then $S_w^\chi f \in C^0(\mathbb{R}^+)$, for every fixed $w > 0$.*

**Proof 8.3**  For every $m \in \mathbb{N}^+$ we consider the sequence:

$$h_m^w(x) := \sum_{|k| \le m} \chi\left(e^{-t_k} x^w, f\left(e^{t_k/w}\right)\right), \qquad x \in \mathbb{R}^+.$$

Then, we can write the following inequality:

$$\left|(S_w^\chi f)(x) - h_m^w(x)\right| \le \sum_{|k| > m} \left|\chi\left(e^{-t_k} x^w, f\left(e^{t_k/w}\right)\right)\right|, \qquad x \in \mathbb{R}^+.$$

Let now $x \in \mathbb{R}^+$ be fixed. Using condition $(\chi 3)$ and the boundedness of $f$, we can write what follows:

$$\left|(S_w^\chi f)(x) - h_m^w(x)\right| \le \sum_{|k| > m} L\left(e^{-t_k} x^w\right) \psi\left(\left|f\left(e^{t_k/w}\right)\right|\right)$$

$$\le \psi(\|f\|_\infty) \sum_{|k| > m} L\left(e^{-t_k} x^w\right). \tag{8.10}$$

Now, recalling Lemma 8.1 (i), we can immediately observe that the series in the last term of the above inequality is the remainder of an absolutely convergent series on $\mathbb{R}^+$. Hence, getting $m \to +\infty$ in the above inequality we provide the uniform convergence of the sequence $(h_m^w)_{m \in \mathbb{N}^+}$ to $S_w^\chi f$ on the whole $\mathbb{R}^+$. Now, observing that any $h_m^w$ is continuous on $\mathbb{R}^+$ as a finite sum of continuous functions, we finally get that also $S_w^\chi f$ is continuous on $\mathbb{R}^+$. Finally, the boundedness of the operators as a function of $x$ follows immediately by (8.9).

### Theorem 8.3

*Let $\chi$ be a continuous nonlinear kernel, which is log-uniformly continuous with respect to the first variable. Further, let $f \in C^0(\mathbb{R}^+)$ be fixed. Then $S_w^\chi f \in C(\mathbb{R}^+)$, for every fixed $w > 0$.*

**Proof 8.4**  If $\chi$ is log-uniformly continuous with respect to the first variable, for any fixed $w > 0$, $\varepsilon > 0$ and $x, y \in \mathbb{R}^+$, using (8.10) and Lemma 8.1 (i), we can write:

$$\left|(S_w^\chi f)(x) - (S_w^\chi f)(y)\right| \le \left|(S_w^\chi f)(x) - h_m^w(x)\right| + \left|h_m^w(x) - h_w^w(y)\right|$$

$$+ \left|h_m^w(y) - (S_w^\chi f)(y)\right| \le \frac{2}{3}\varepsilon + \left|h_m^w(x) - h_w^w(y)\right|,$$

for a fixed sufficiently large $m \in \mathbb{N}^+$, where $h_m^w$ is defined as in the proof of Theorem 8.2. Now, we have that $h_m^w$ are log-uniformly continuous on $\mathbb{R}^+$ since they are defined as a finite sum of log-uniformly continuous functions, thus, if we choose the parameter $\gamma > 0$ of the uniform continuity of $h_m^w$ corresponding to $\varepsilon/3$, and $|\ln x - \ln y| < \gamma$, we finally get: $\left|(S_w^\chi f)(x) - (S_w^\chi f)(y)\right| \le \varepsilon$, i.e., $S_w^\chi f \in C(\mathbb{R}^+)$. Finally, the boundedness of $S_w^\chi f$ follows as in the proof of Theorem 8.2.

## 8.5   Convergence Theorems

We can prove the following convergence theorem.

***Theorem 8.4***
*Let $\chi$ be a given nonlinear kernel and $f : \mathbb{R}^+ \to \mathbb{R}$ a bounded function. Then:*

$$\lim_{w \to +\infty} (S_w^{\chi} f)(x) = f(x),$$

*at any point $x \in \mathbb{R}^+$ of continuity of $f$.*

**Proof 8.5**   Let $x \in \mathbb{R}^+$ be a point of continuity of $f$. For every fixed $\varepsilon > 0$, we denote by $\gamma > 0$ the corresponding parameter for which $|f(x) - f(e^y)| = |f(e^{\ln x}) - f(e^y)| < \varepsilon$, if $|\ln x - y| \leq \gamma$. Now, we can write what follows:

$$|(S_w^{\chi} f)(x) - f(x)| \leq \left| (S_w^{\chi} f)(x) - \sum_{k \in \mathbb{Z}} \chi \left( e^{-t_k} x^w, f(x) \right) \right|$$

$$+ \left| \sum_{k \in \mathbb{Z}} \chi \left( e^{-t_k} x^w, f(x) \right) - f(x) \right| =: I_1 + I_2.$$

Concerning $I_1$, using $(\chi 3)$, we have:

$$I_1 \leq \sum_{k \in \mathbb{Z}} L \left( e^{-t_k} x^w \right) \psi \left( \left| f(e^{t_k/w}) - f(x) \right| \right)$$

$$= \sum_{|t_k - w \ln x| \leq \gamma w} L \left( e^{-t_k} x^w \right) \psi \left( \left| f(e^{t_k/w}) - f(x) \right| \right)$$

$$+ \sum_{|t_k - w \ln x| > \gamma w} L \left( e^{-t_k} x^w \right) \psi \left( \left| f(e^{t_k/w}) - f(x) \right| \right) =: I_{1,1} + I_{1,2}.$$

Noting that, if $|t_k - w \ln x| \leq \gamma w$, we also have $|t_k/w - \ln x| \leq \gamma$, hence recalling Lemma 8.1 (i) and the properties of $\psi$, for $I_{1,1}$ we obtain:

$$I_{1,1} \leq \psi(\varepsilon) \sum_{|t_k - w \ln x| \leq \gamma w} L \left( e^{-t_k} x^w \right) \leq \psi(\varepsilon) M_{0,\Pi}(L) < +\infty.$$

Moreover, using the boundedness of $f$ and Lemma 8.1 (ii), we immediately obtain:

$$I_{1,2} \leq \varepsilon \, \psi(2\|f\|_\infty),$$

for $w > 0$ sufficiently large. Finally, in order to estimate $I_2$, we can simply consider the case $f(x) \neq 0$ since if $f(x) = 0$ we immediately have $I_2 = 0$. Then, using $(\chi 4)$ we get:

$$I_2 \leq \mathscr{T}_w(x, f(x)) < \varepsilon, \tag{8.11}$$

as $w \to +\infty$. This completes the proof.

Now, we can prove the following uniform convergence theorem.

**Theorem 8.5**
*Let $\chi$ be a given nonlinear kernel and $f \in C(\mathbb{R}^+)$. Then:*

$$\lim_{w \to +\infty} \|S_w^\chi f - f\|_\infty = 0.$$

**Proof 8.6**  The proof follows by the same arguments of Theorem 8.4, taking into account that if $f \in C(\mathbb{R}^+)$ we can choose $\gamma > 0$ as the parameter of the log-uniform continuity of $f$ corresponding to $\varepsilon$, that is:

$$|f(x) - f(e^y)| = |f(e^{\ln x}) - f(e^y)| < \varepsilon,$$

for every $x, y \in \mathbb{R}^+$ such that $|\ln x - y| < \gamma$. In fact, in this way it is immediate to see that both the estimates of the terms $I_{1,1}$ and $I_{1,2}$ defined in the proof of Theorem 8.4 holds uniformly with respect to $x \in \mathbb{R}^+$. Finally, recalling again $(\chi 4)$, we can note also that the estimate for $I_2$ holds uniformly with respect to $x \in \mathbb{R}^+$.

The nonlinear exponential series can be now viewed as a reconstruction method for both bounded continuous and log-uniformly continuous signals.

Note that, if the nonlinear kernel $\chi$ has the following property:

$(\chi 5)$  for every $u \in \mathbb{R}$:

$$\chi(x, u) = 0, \quad \text{for} \quad x \in (0, 1],$$

we can prove what follows.

**Theorem 8.6**
*Let $\chi$ be a nonlinear kernel satisfying $(\chi 5)$. Then, for any bounded function $f$ : $\mathbb{R}^+ \to \mathbb{R}$ the nonlinear exponential series reduces to the following:*

$$(S_w^\chi f)(x) = \sum_{t_k/w < \ln x} \chi\left(e^{-t_k} x^w, f(e^{t_k/w})\right), \quad x \in \mathbb{R}^+,$$

$w > 0$.

**Proof 8.7**  The proof follows immediately from the fact that, the series $S_w^\chi f$ is non-null only if $e^{-t_k} x^w > 1$, i.e., if $t_k/w < \ln x$.

Note that, what established in Theorem 8.6 can be seen as the phenomenon that in signal theory is known with the name of *prediction by samples from the past*.

More precisely, in [30], for the classical generalized sampling series, it has been proved that any continuous and bounded signal can be reconstructed by sample values taken only from the past with respect to a given time $x$; in that case we always refers to *linear prediction from the past*. Clearly, what happens in Theorem 8.6 can be viewed as an *exponential prediction by samples from the past*.

## 8.6 Quantitative and Qualitative Estimates for the Aliasing Errors

In this section, based on the convergence results obtained in Section 8.5, we will consider the problem of obtaining quantitative, as well as qualitative, estimations for the aliasing errors (order of approximation) in case of the approximation of continuous functions by the nonlinear exponential sampling series.

In order to do this, we preliminarily recall the well-known notion of the log-modulus of continuity of any given function $f : \mathbb{R}^+ \to \mathbb{R}$. We denote by:

$$\widetilde{\omega}(f,\delta) := \sup\left\{ |f(x) - f(y)| : \ |\ln x - \ln y| \leq \delta, \ x, y \in \mathbb{R}^+ \right\}, \qquad (8.12)$$

for any positive parameter $\delta > 0$. The notion of log-modulus of continuity has been firstly introduced in [14]; it is easy to see that $\widetilde{\omega}$ satisfies the classical properties of the moduli of continuity, such as the following:

$$\widetilde{\omega}(f,\delta) \ \leq \ \widetilde{\omega}(f,\delta'), \quad \text{with } \delta \leq \delta', \qquad (8.13)$$

and

$$\widetilde{\omega}(f,\lambda\delta) \ \leq \ (1+\lambda)\,\widetilde{\omega}(f,\delta), \qquad (8.14)$$

for any $\lambda$, $\delta > 0$. Obviously, as in the classical case (see [58]), it is possible to prove that, if $f \in C(\mathbb{R}^+)$ it turns out that:

$$\lim_{\delta \to 0^+} \widetilde{\omega}(f,\delta) \ = \ 0. \qquad (8.15)$$

Through this section, in order to reach quantitative estimates, a crucial role is played by the function $\psi$ arising in $(\chi 3)$ and by $(\chi 4)$. First, for the sake of convenience, we write explicitly all the constants involved in the statement of these conditions on $\chi$.

We say that condition $(\chi 4)$ is satisfied if there exists a parameter $\theta > 0$ and constants $C > 0$, $N > 0$, such that:

$$\mathscr{T}_w(x,u) := \left| \sum_{k\in\mathbb{Z}} \chi(e^{-t_k}x^w,u) - u \right| \leq Cw^{-\theta}, \qquad (8.16)$$

with $x \in \mathbb{R}^+$, $u \in \mathbb{R}$, for every $w > N$.

Moreover, we also introduce the following additional condition on $\psi$ of $(\chi 3)$:

$(\psi 3)$ we suppose that $\psi$ is concave on $\mathbb{R}_0^+$.

Now, we can prove the following.

### Theorem 8.7

*Let $\chi$ be a given nonlinear kernel satisfying $(\chi 3)$ with a function $\psi$ for which condition $(\psi 3)$ holds, and with a function $L$ that satisfies $(L2)$ with $\beta \geq 1$. Hence, for any $f \in C(\mathbb{R}^+)$ we have:*

$$\|S_w^\chi f - f\|_\infty \leq (M_{0,\Pi}(L) + M_{1,\Pi}(L))\, \psi\left(\widetilde{\omega}\left(f, \frac{1}{w}\right)\right) + Cw^{-\theta},$$

*for every $w > 0$ sufficiently large, where $C > 0$ and $\theta > 0$ are the parameters of condition $(\chi 4)$ explicitly introduced in (8.16).*

**Proof 8.8**  For any fixed $x \in \mathbb{R}^+$ we can consider the term $I_1$ and $I_2$ defined in the proof of Theorem 8.4. Concerning the term $I_2$, recalling (8.16) and inequality (8.11) we immediately have that:

$$I_2 \leq Cw^{-\theta},$$

for every $w > 0$ sufficiently large, with $C > 0$. Furthermore, for $I_1$, using $(\chi 3)$ we can write what follows:

$$
\begin{aligned}
I_1 &\leq \sum_{k \in \mathbb{Z}} L\left(e^{-t_k}x^w\right)\, \psi\left(\left|f(e^{t_k/w}) - f(x)\right|\right) \\
&= \sum_{k \in \mathbb{Z}} L\left(e^{-t_k}x^w\right)\, \psi\left(\left|f(e^{t_k/w}) - f(e^{\ln x})\right|\right) \\
&\leq \sum_{k \in \mathbb{Z}} L\left(e^{-t_k}x^w\right)\, \psi\left(\widetilde{\omega}\left(f, \left|\ln x - \frac{t_k}{w}\right|\right)\right) \qquad (8.17) \\
&\leq \sum_{k \in \mathbb{Z}} L\left(e^{-t_k}x^w\right)\, \psi\left(\left[w\left|\ln x - \frac{t_k}{w}\right| + 1\right]\widetilde{\omega}\left(f, \frac{1}{w}\right)\right).
\end{aligned}
$$

Now, recalling that the function $\psi$ satisfies $(\psi 1)$ and $(\psi 3)$, it also satisfies the following general inequality:

$$x\,\psi(y) = x\,\psi\left(\frac{xy}{x}\right) \geq x\left[\frac{1}{x}\psi(xy)\right] = \psi(xy), \quad x \geq 1, \quad y \geq 0, \qquad (8.18)$$

with $1/x \in [0,1]$, and $xy \geq 0$. Using the above properties of $\psi$ for any fixed $k$, $w$ and $x$, we have:

$$
\begin{aligned}
I_1 &\leq \sum_{k \in \mathbb{Z}} L\left(e^{-t_k}x^w\right)\left[w\left|\ln x - \frac{t_k}{w}\right| + 1\right]\psi\left(\widetilde{\omega}\left(f, \frac{1}{w}\right)\right) \\
&\leq \sum_{k \in \mathbb{Z}} L\left(e^{-t_k}x^w\right)|t_k - w\ln x|\,\psi\left(\widetilde{\omega}\left(f, \frac{1}{w}\right)\right) + M_{0,\Pi}(L)\,\psi\left(\widetilde{\omega}\left(f, \frac{1}{w}\right)\right)
\end{aligned}
$$

$$=: I_{1,A} + I_{1,B}.$$

In order to conclude the proof we have to estimate $I_{1,A}$, and we immediately obtain:

$$I_{1,A} \le \psi\left(\widetilde{\omega}\left(f, \frac{1}{w}\right)\right) M_{1,\Pi}(L) < +\infty.$$

The proof can now be completed observing that both $M_{0,\Pi}(L)$ and $M_{1,\Pi}(L)$ are finite in view of Lemma 8.2, since here $\beta \ge 1$.

Note that, the quantitative estimate given in Theorem 8.7 holds only in case of nonlinear kernel for which the corresponding function $L$ arising from assumption $(\chi 2)$ satisfies condition $(L2)$ with $\beta \ge 1$. Hence, it is quite natural to ask what happens in the case $0 < \beta < 1$. The answer can be given by the following theorem.

### Theorem 8.8

*Let $\chi$ be a given nonlinear kernel satisfying $(\chi 3)$ with a function $\psi$ for which condition $(\psi 3)$ holds, and with a function $L$ that satisfies $(L2)$ with $0 < \beta < 1$. Hence, for any $f \in C(\mathbb{R}^+)$ we have:*

$$\|S_w^\chi f - f\|_\infty \le \left(M_{0,\Pi}(L) + M_{\beta,\Pi}(L)\right) \psi\left(\widetilde{\omega}\left(f, \frac{1}{w^\beta}\right)\right)$$

$$+ \psi(2\|f\|_\infty) M_{\beta,\Pi}(L) w^{-\beta} + C w^{-\theta},$$

*for every $w > 0$ sufficiently large, where $C > 0$ and $\theta > 0$ are the parameters of condition $(\chi 4)$ explicitly introduced in (8.16).*

**Proof 8.9**   Let $x \in \mathbb{R}^+$ be fixed, and then consider $I_1$ and $I_2$ as made in the proof of Theorem 8.7. Obviously, for what concerns $I_2$ we consider the same estimate achieved in the proof of Theorem 8.7. Hence we have to obtain only an upper bound for $I_1$. Recalling the estimates given in (8.17), we can write what follows:

$$I_1 \le \sum_{k\in\mathbb{Z}} L\left(e^{-t_k} x^w\right) \psi\left(\left|f(e^{t_k/w}) - f(e^{\ln x})\right|\right)$$

$$= \left\{ \sum_{\left|\ln x - \frac{t_k}{w}\right| \le 1} + \sum_{\left|\ln x - \frac{t_k}{w}\right| > 1} \right\} L\left(e^{-t_k} x^w\right) \psi\left(\left|f(e^{t_k/w}) - f(e^{\ln x})\right|\right)$$

$$\le \sum_{\left|\ln x - \frac{t_k}{w}\right| \le 1} L\left(e^{-t_k} x^w\right) \psi\left(\widetilde{\omega}\left(f, \left|\ln x - \frac{t_k}{w}\right|\right)\right)$$

$$+ \sum_{\left|\ln x - \frac{t_k}{w}\right| > 1} L\left(e^{-t_k} x^w\right) \psi\left(\left|f(e^{t_k/w}) - f(e^{\ln x})\right|\right) =: \mathscr{I}_{1,1} + \mathscr{I}_{1,2}.$$

By the property (8.13), and since $0 < \beta < 1$, if $\left|\ln x - \frac{t_k}{w}\right| \leq 1$ we have:

$$\psi\left(\widetilde{\omega}\left(f, \left|\ln x - \frac{t_k}{w}\right|\right)\right) \leq \psi\left(\widetilde{\omega}\left(f, \left|\ln x - \frac{t_k}{w}\right|^{\beta}\right)\right),$$

and reasoning as in the proof of Theorem 8.7, together with the application of (8.14) and (8.18), we get:

$$\mathscr{I}_{1,1} \leq \sum_{\left|\ln x - \frac{t_k}{w}\right| \leq 1} L\left(e^{-t_k}x^w\right) \psi\left(\widetilde{\omega}\left(f, \left|\ln x - \frac{t_k}{w}\right|^{\beta}\right)\right)$$

$$\leq \sum_{\left|\ln x - \frac{t_k}{w}\right| \leq 1} L\left(e^{-t_k}x^w\right) \psi\left(\left[w^{\beta}\left|\ln x - \frac{t_k}{w}\right|^{\beta} + 1\right]\widetilde{\omega}\left(f, \frac{1}{w^{\beta}}\right)\right)$$

$$\leq \sum_{\left|\ln x - \frac{t_k}{w}\right| \leq 1} L\left(e^{-t_k}x^w\right) \psi\left(\widetilde{\omega}\left(f, \frac{1}{w^{\beta}}\right)\right)\left[w^{\beta}\left|\ln x - \frac{t_k}{w}\right|^{\beta} + 1\right]$$

$$= \psi\left(\widetilde{\omega}\left(f, \frac{1}{w^{\beta}}\right)\right)\sum_{\left|\ln x - \frac{t_k}{w}\right| \leq 1} L\left(e^{-t_k}x^w\right)\left|w\ln x - t_k\right|^{\beta}$$

$$+ \psi\left(\widetilde{\omega}\left(f, \frac{1}{w^{\beta}}\right)\right)\sum_{\left|\ln x - \frac{t_k}{w}\right| \leq 1} L\left(e^{-t_k}x^w\right) \leq \psi\left(\widetilde{\omega}\left(f, \frac{1}{w^{\beta}}\right)\right)\left[M_{\beta,\Pi}(L) + M_{0,\Pi}(L)\right].$$

Finally, using Lemma 8.1 (with $\gamma = 1$) we obtain:

$$\mathscr{I}_2 \leq \psi\left(2\|f\|_{\infty}\right)\sum_{|w\ln x - t_k| > w} L\left(e^{-t_k}x^w\right) \leq \psi\left(2\|f\|_{\infty}\right)M_{\beta,\Pi}(L)\,w^{-\beta},$$

for every $w > 0$ sufficiently large. This completes the proof.

Now, recalling that a function $f : \mathbb{R}^+ \to \mathbb{R}$ is called *locally log-Holderian of order* $0 < \nu \leq 1$, if:

$$\widetilde{\omega}(f, \delta) = \mathscr{O}(\delta^{\nu}), \quad as \quad \delta \to 0^+, \tag{8.19}$$

we can immediately prove the following.

### Theorem 8.9

*Let $\chi$ be a given nonlinear kernel satisfying $(\chi3)$ with a function $\psi$ for which condition $(\psi3)$ holds, and let $f : \mathbb{R}^+ \to \mathbb{R}$ be a fixed locally log-Holderian function of order $0 < \nu \leq 1$. Then:*

1. *If the function L of condition ($\chi 2$) satisfies assumption (L2) with $0 < \beta < 1$, we have:*

$$\|S_w^{\chi} f - f\|_{\infty} = \mathcal{O}\left(\psi(w^{-\beta \nu}) + w^{-\eta}\right), \quad w \to +\infty,$$

*with $\eta := \min\{\beta, \theta\}$.*

2. *If the function L of condition ($\chi 2$) satisfies assumption (L2) with $\beta \geq 1$, we have:*

$$\|S_w^{\chi} f - f\|_{\infty} = \mathcal{O}\left(\psi(w^{-\nu}) + w^{-\theta}\right), \quad w \to +\infty.$$

**Proof 8.10**   The proof immediately follows the definition of locally log-Holderian function, Theorem 8.7 and Theorem 8.8.

## 8.7   Examples of Nonlinear Kernels

In this section we discuss examples of suitable nonlinear kernels for the nonlinear exponential sampling series. More precisely, we propose the following natural procedure to construct kernels satisfying the assumptions given in Section 8.3.

We first consider a net of functions of the following product-type form:

$$\chi_w(x, u) := L(x) g_w(u), \quad x \in \mathbb{R}^+, \quad u \in \mathbb{R}, \tag{8.20}$$

$w > 0$, where $g_w : \mathbb{R} \to \mathbb{R}$ are functions with the property that:

$$g_w(u) \to u, \quad as \quad w \to +\infty, \tag{8.21}$$

uniformly with respect to $u \in \mathbb{R}$, and such that there exists a function $\psi$ (satisfying the properties listed in condition ($\chi 3$)) for which:

$$|g_w(u) - g_w(v)| \leq \psi(|u - v|), \quad u, v \in \mathbb{R}, w > 0. \tag{8.22}$$

Hence, in the case of nonlinear kernel functions generated using the formula in (8.20), the assumptions ($\chi_i$), $i = 1, 2, 3$, and ($L_j$), $j = 1, 2$, can be reformulated as follows:

($\mathscr{L}1$)   the map $k \mapsto L(e^{-t_k} x^w) \in \ell^1(\mathbb{Z})$ for every $x \in \mathbb{R}^+$ and $w > 0$, $L$ is bounded in $\mathbb{R}^+$ and there exists $\beta > 0$ such that

$$M_{\beta, \Pi}(L) := \sup_{x \in \mathbb{R}^+} \sum_{k \in \mathbb{Z}} L(e^{-t_k} x)|t_k - \ln x|^{\beta} < +\infty;$$

($\mathscr{L}2$)   $g_w(0) = 0$, for every $w > 0$;

($\mathscr{L}3$)   there exists $\theta > 0$ such that:

$$\mathscr{T}_w(x, u) := \left| g_w(u) \sum_{k \in \mathbb{Z}} L(e^{-t_k} x^w) - u \right| = \mathcal{O}(w^{-\theta}),$$

as $w \to +\infty$, uniformly with respect to $x \in \mathbb{R}^+, u \in \mathbb{R}$.

Now, in order to concretely construct nonlinear kernels we begin providing an example of net of functions $(g_w)_{w>0}$ satisfying the previous conditions.

We define $g_w : \mathbb{R} \to \mathbb{R}$:

$$g_w(u) = \begin{cases} u^{1-1/w}, & u \in (0,1), \\ u, & \text{otherwise,} \end{cases} \qquad (8.23)$$

$w > 0$. Clearly, it is easy to see that $g_w(u) \to u$ uniformly on $\mathbb{R}$, as $w \to +\infty$.

In this case, condition $(\mathscr{L}3)$ holds for $\theta = 1$. In fact, the function $g_w(u) - u$ on $(0,1)$ achieves its maximum at $u_0 := (\frac{w-1}{w})^w$, for sufficiently large $w > 0$ ($g_w(u) - u = 0$ otherwise), then we have, for every $u \in \mathbb{R}$

$$|g_w(u) - u| \leq |g_w(u_0) - u_0| = \left(\frac{w-1}{w}\right)^w \left(\frac{1}{w-1}\right) \leq \frac{\tilde{C}}{w-1} = \mathcal{O}\left(w^{-1}\right),$$

as $w \to +\infty$, and for some constant $\tilde{C} > 0$.

Moreover, it easy to see that the functions $g_w$, $w > 0$, defined in (8.25), satisfy condition (8.22) for sufficiently large $w > 0$. In fact, if we consider, for instance, the function $\psi : \mathbb{R}_0^+ \to \mathbb{R}$:

$$\psi(u) := g_2|_{\mathbb{R}_0^+}(u) = \begin{cases} \sqrt{u}, & u \in (0,1), \\ u, & \text{otherwise.} \end{cases} \qquad (8.24)$$

we can prove that inequality (8.22) holds. In details, for every $w \geq 2$, if we consider $u, v \geq 1$, or $u, v \leq 0$, with $|u - v| \geq 1$ we have:

$$|g_w(u) - g_w(v)| = |u - v| = \psi(|u - v|);$$

if we consider $u, v \geq 1$, or $u, v \leq 0$, with $|u - v| \in (0,1)$ we obtain:

$$|g_w(u) - g_w(v)| = |u - v| \leq \sqrt{|u - v|} = \psi(|u - v|);$$

if we consider $u, v \in (0,1)$ with $|u - v| \in (0,1)$, and since $g_w$ are concave on $(0,1)$, we can write:

$$|g_w(u) - g_w(v)| = |u^{1-1/w} - v^{1-1/w}| \leq |u - v|^{1-1/w} \leq \sqrt{|u - v|} = \psi(|u - v|);$$

if we take $u \in (0,1)$, $v \geq 1$ (or conversely) with $|u - v| \geq 1$, we get:

$$|g_w(v) - g_w(u)| = v - u^{1-1/w} \leq v - u = \psi(|u - v|);$$

finally, if we assume $u \in (0,1)$, $v \geq 1$ (or conversely) with $|u - v| \in (0,1)$, we obtain:

$$|g_w(v) - g_w(u)| = v - u^{1-1/w} \leq v - u \leq (v - u)^{1-1/w} \leq \sqrt{|u - v|} = \psi(|u - v|).$$

Obviously, it is also immediate to see that the above $\psi$ does not satisfy condition $(\psi 3)$ of Section 8.6 since it is only piecewise concave and not globally concave.

In order to provide an example of sequence $g_w$ satisfying $(\psi 3)$ we can define the following $g_w : \mathbb{R} \to \mathbb{R}$:

$$g_w(u) = \begin{cases} u^{1+1/w}, & u \in (0, 1), \\ u, & \text{otherwise,} \end{cases} \tag{8.25}$$

$w > 0$. It is easy to see that, the functions defined in (8.25) converge uniformly on $\mathbb{R}$ to $u$, as $w \to +\infty$.

Also in this case, condition $(\mathscr{L}3)$ holds with $\theta = 1$. In fact, the function $u - g_w(u)$ on $(0, 1)$ achieves its maximum at $u_0 := (\frac{w}{w+1})^w$, for sufficiently large $w > 0$ (we have again $g_w(u) - u = 0$ on the other points $u$), then we have, for every $u \in \mathbb{R}$:

$$|g_w(u) - u| \leq u_0 - g_w(u_0) = \left(\frac{w}{w+1}\right)^w \left(\frac{1}{w+1}\right) \leq \frac{1}{w} = \mathscr{O}\left(w^{-1}\right),$$

as $w \to +\infty$.

Moreover, setting: $\psi(u) := \frac{3}{2}u$, $u \in \mathbb{R}_0^+$, we immediately obtain:

$$|g_w(u) - g_w(v)| = |u - v| \leq \psi(|u - v|),$$

if $|u|, |v| \geq 1$; furthermore, if $u, v \in (0, 1)$, using the Langrange theorem we get:

$$|g_w(u) - g_w(v)| = |u^{1+1/w} - v^{1+1/w}| \leq \frac{3}{2}|u - v| = \psi(|u - v|),$$

for $w \geq 2$; finally, for $u \in (0, 1)$ and $|v| \geq 1$ (or conversely), using again the Lagrange theorem, we obtain:

$$|g_w(u) - g_w(v)| = v - u^{1+1/w} = (v - 1) + (1 - u^{1+1/w})$$

$$\leq (v - 1) + \frac{3}{2}(1 - u) \leq \frac{3}{2}(v - 1 + 1 - u) = \psi(|v - u|),$$

if $v \geq 1$, while if $v \leq -1$:

$$|g_w(u) - g_w(v)| = u^{1+1/w} - v \leq \frac{3}{2}u - v = \frac{3}{2}u - v - \frac{1}{2}v + \frac{1}{2}v$$

$$= \frac{3}{2}u - \frac{3}{2}v + \frac{1}{2}v \leq \frac{3}{2}u - \frac{3}{2}v = \psi(|u - v|).$$

Note that, since the above function $\psi$ is linear, hence assumption $(\psi 3)$ holds.

Hence, the nonlinear exponential sampling series, takes now the following form:

$$(S_w^\chi f)(x) = \sum_{k \in \mathbb{Z}} L(e^{-t_k}x^w) g_w\left(f(e^{t_k/w})\right), \quad x \in \mathbb{R}^+, \quad w > 0, \tag{8.26}$$

with $g_w$, defined as, e.g., in (8.25) (we have a similar expression in case of $g_w$ defined in (8.23)). For the nonlinear exponential sampling series in (8.26) we can formulate the following corollary, as a consequence of Theorem 8.4 and Theorem 8.5.

### Corollary 8.1

*Let $f : \mathbb{R}^+ \to \mathbb{R}$ a bounded function and let $x \in \mathbb{R}^+$ a point of continuity. Then:*

$$\lim_{w \to +\infty} \sum_{k \in \mathbb{Z}} L(e^{-t_k} x^w) g_w \left( f(e^{t_k/w}) \right) = f(x).$$

*In particular, if $f \in C(\mathbb{R}^+)$ we have:*

$$\lim_{w \to +\infty} \| \sum_{k \in \mathbb{Z}} L(e^{-t_k}(\cdot)^w) g_w \left( f(e^{t_k/w}) \right) - f(\cdot) \|_\infty = 0.$$

Furthermore, the following quantitative estimates can be deduced for (8.26).

### Corollary 8.2

*Let $\chi$ defined in (8.20) with $L$ which satisfies condition $(\mathcal{L}1)$ with $\beta \geq 1$ and $g_w$ defined in (8.25). Hence, for any $f \in C(\mathbb{R}^+)$ we have:*

$$\| S_w^\chi f - f \|_\infty \leq \frac{3}{2} \left( M_{0,\Pi}(L) + M_{1,\Pi}(L) \right) \widetilde{\omega} \left( f, \frac{1}{w} \right) + C w^{-1},$$

*for every $w > 0$ sufficiently large, where $C > 0$ is a suitable constant independent from $w$ arising from $(\mathcal{L}3)$.*

*While, if $L$ satisfying condition $(\mathcal{L}1)$ with $0 < \beta < 1$, and $f \in C(\mathbb{R}^+)$ we have:*

$$\| S_w^\chi f - f \|_\infty \leq \frac{3}{2} \left( M_{0,\Pi}(L) + M_{\beta,\Pi}(L) \right) \widetilde{\omega} \left( f, \frac{1}{w^\beta} \right)$$

$$+ 3 \| f \|_\infty M_{\beta,\Pi}(L) w^{-\beta} + C w^{-1},$$

*for every $w > 0$ sufficiently large, with $C > 0$ arising (again) from $(\mathcal{L}3)$.*

Obviously, if we assume that the nonlinear product kernel introduced in (8.20) is of the following form:

$$\chi(x,u) := L(x)u, \quad x \in \mathbb{R}^+, \quad u \in \mathbb{R}, \tag{8.27}$$

all the previous conditions $(\mathcal{L}i)$, $i = 1,...,4$ turns out to be satisfied, and the corresponding nonlinear exponential sampling series reduces to the linear ones already considered in [11, 18] and also recalled in (8.6) with $L = \varphi$, of the form:

$$(S_w^\chi f)(x) = \sum_{k \in \mathbb{Z}} L(e^{-t_k} x^w) f(e^{t_k/w}), \quad x \in \mathbb{R}^+, \quad w > 0. \tag{8.28}$$

Note that, in the linear case (8.28) condition $(\mathscr{L}3)$ becomes:

$$\mathscr{T}_w(x,u) = |u| \cdot \left| \sum_{k\in\mathbb{Z}} L(e^{-t_k}x^w) - 1 \right| = \mathscr{O}(w^{-\theta}), \quad as\ w \to +\infty, \qquad (8.29)$$

uniformly with respect to $x \in \mathbb{R}^+$, $u \in \mathbb{R}$. Moreover, (8.22) holds with $\psi(u) = u$ (strong Lipschitz condition) and $(\psi3)$ becomes trivial.

**Remark 8.2** Note that, for the relation written in (8.29), in order to get the uniformity with respect to the variable $u$, it is clear that we should have:

$$\sum_{k\in\mathbb{Z}} L(e^{-t_k}x^w) = 1, \qquad (8.30)$$

for every $x \in \mathbb{R}^+$, $w > 0$, i.e., we need a classical (strong) singularity assumption for the function $L$. This equality will be proved for some examples of functions $L$ in the next sections. ■

Then, for the (linear) exponential sampling operators in (8.28), other than the convergence results (as made in Corollary 8.1 in case of (8.26)) we can also formulate the following corollary concerning the order of approximation, as a consequence of Theorem 8.7 and Theorem 8.8, and tacking into account that condition (8.30) must be satisfied.

*Corollary 8.3*
*Let $\chi$ defined in (8.27) with $L$ satisfying condition $(\mathscr{L}1)$ with $\beta \geq 1$. Hence, for any $f \in C(\mathbb{R}^+)$ we have:*

$$\|S_w^{\chi}f - f\|_{\infty} \leq (M_{0,\Pi}(L) + M_{1,\Pi}(L))\, \widetilde{\omega}\left(f, \frac{1}{w}\right),$$

*for every $w > 0$ sufficiently large.*
*While, if $L$ satisfying condition $(\mathscr{L}1)$ with $0 < \beta < 1$, and $f \in C(\mathbb{R}^+)$ we have:*

$$\|S_w^{\chi}f - f\|_{\infty} \leq (M_{0,\Pi}(L) + M_{\beta,\Pi}(L))\, \widetilde{\omega}\left(f, \frac{1}{w^{\beta}}\right) + 2\|f\|_{\infty} M_{\beta,\Pi}(L)w^{-\beta},$$

*for every $w > 0$ sufficiently large.*

**Remark 8.3** Note that, the first estimate of Corollary 8.3 is similar to that one achieved in Corollary 3.4.1 of [11], but here it has been proved under slightly different assumptions. Furthermore, the second estimate of Corollary 8.3 is completely new and allows us to provide a quantitative estimate for the order of approximation also in case of kernels of the form (8.27) for which $M_{1,\Pi}(L) = +\infty$. The latter case has not been considered in [11]. ■

In conclusion of this section, we have to provide some examples of functions $L$ for which the above results hold. In order to do this, we first have to recall some basic notions of Mellin Analysis, which turn out to be crucial for proving that $L$ satisfies all the required assumptions.

## 8.8   Some Fundamental Results in Mellin Analysis

For $c \in \mathbb{R}$ we define the space:

$$X_c := \left\{ f : \mathbb{R}^+ \to \mathbb{C} : f(\cdot)(\cdot)^{c-1} \in L^1(\mathbb{R}^+) \right\}, \tag{8.31}$$

endowed with the following norm:

$$\|f\|_{X_c} := \|f(\cdot)(\cdot)^{c-1}\|_1 = \int_0^{+\infty} |f(u)| |u|^{c-1} du, \tag{8.32}$$

where $\|\cdot\|_1$ denotes the usual $L^1$-norm. The space $X_c$ can be also equivalently expressed as the space of all functions $f$, such that $f(\cdot)(\cdot)^c \in L^1_\mu(\mathbb{R}^+)$, where $L^1_\mu(\mathbb{R}^+)$ is the Lebesgue space with respect to the invariant measure:

$$\mu(A) = \int_A \frac{dt}{t}, \tag{8.33}$$

for any measurable set $A \subset \mathbb{R}^+$.

The Mellin transform of a function $f \in X_c$ is defined by (see, e.g., [63, 27]):

$$\widehat{[f]}_M(s) := \int_0^{+\infty} u^{s-1} f(u) \, du, \quad s = c + it, \, t \in \mathbb{R},$$

Below, we recall some basic properties of the Mellin transform:

$$[\widehat{af(\cdot) + bg(\cdot)}]_M(s) := a\widehat{[f]}_M(s) + b\widehat{[g]}_M(s), \quad f, g \in X_c, \quad a, b \in \mathbb{R},$$

and

$$|\widehat{[f]}_M(s)| \leq \|f\|_{X_c}, \quad s = c + it.$$

The inverse Mellin transform $M_c^{-1}[g]$ of a function $g \in L^1(\{c\} \times i\mathbb{R})$, is defined by:

$$M_c^{-1}[g](x) = M_c^{-1}[g(c+it)](x) := \frac{x^{-c}}{2\pi} \int_{-\infty}^{+\infty} g(c+it)x^{-it} \, dt, \quad x \in \mathbb{R}^+,$$

where by the symbol $L^p(\{c\} \times i\mathbb{R})$, $p \geq 1$, we denote the space of functions of the form $h : \{c\} \times i\mathbb{R} \to \mathbb{C}$, with $h(c + i\cdot) \in L^p(\mathbb{R})$.

The Mellin translation operator $\tau_h^c$, with $h \in \mathbb{R}^+$, $c \in \mathbb{R}$, and $f : \mathbb{R}^+ \to \mathbb{C}$ can be defined as follows:

$$(\tau_h^c)(x) := h^c f(hx), \quad x \in \mathbb{R}^+.$$

Obviously, it is easy to see that:

$$(\tau_h^c f)(x) = h^c(\tau_h^0 f)(x), \quad x \in \mathbb{R}^+,$$

and that the following relation holds:

$$\|\tau_h^c f\|_{X_c} = \|f\|_{X_c}.$$

Finally, we recall the notion of Mellin convolution product of two functions $f, g : \mathbb{R}^+ \to \mathbb{C}$:

$$(f * g)(x) := \int_0^{+\infty} g\left(\frac{x}{u}\right) f(u) \frac{du}{u} = \int_0^{+\infty} g(u)\, u^c (\tau_{1/u}^c f)(x) \frac{du}{u},$$

if the above integral exists. For more details about basic tools of Mellin Analysis, the readers can see, e.g., [63, 11].

Now, we are able to present some well-known examples of Mellin-type kernel functions.

## 8.9　Mellin Spline Kernels

Remarkable examples of kernels are given by the well-known Mellin B-splines of order $n \in \mathbb{N}$, that are the Mellin analogous of the famous central B-splines of order $n$. Such kernel functions are defined as follows:

$$M_n(x) := \frac{1}{(n-1)!} \sum_{i=0}^n (-1)^i \binom{n}{i} \left(\frac{n}{2} + \ln x - i\right)_+^{n-1}, \quad x \in \mathbb{R}^+,$$

where the function $(x)_+ := \max\{x, 0\}$ denotes the positive part of $x \in \mathbb{R}$ (see [37]).

More generally, for any parameter $c \in \mathbb{R}$ one can consider the functions

$$M_{c,n}(x) := x^{-c} M_n(x), \quad x \in \mathbb{R}^+.$$

Since the functions $M_{c,n}$ are continuous and have compact support in $\mathbb{R}^+$, it turns out that $M_{c,n} \in X_c$, for any $c \in \mathbb{R}$. Hence, we can get:

$$\widehat{[M_{c,n}]}_M(s) = \int_0^{+\infty} M_{c,n}(t)\, t^{c+iv-1}\, dt = \int_0^{+\infty} M_{c,n}(t)\, e^{(c+iv)\ln t} \frac{dt}{t}, \quad s = c+iv,$$

$v \in \mathbb{R}$, according to the definition recalled in Section 8.8. Now, by the change of variable $z = \ln t$, and setting:

$$\widetilde{M}_n(z) := M_n(e^z), \quad z \in \mathbb{R},$$

we immediately obtain:

$$\widehat{[M_{c,n}]}_M(s) = \widehat{\widetilde{M}_n}(-v), \quad s = c+iv,$$

where the symbol $\widehat{f}$ denotes the usual Fourier transform of an $L^1(\mathbb{R})$ function, that is:

$$\widehat{f}(v) := \int_{\mathbb{R}} f(x)e^{-ivx}\,dx, \quad v \in \mathbb{R}.$$

In practice, observing that $\widetilde{M}_n$ coincides with the classical central B-splines of order $n$, i.e.,

$$\widetilde{M}_n(t) = \frac{1}{(n-1)!}\sum_{i=0}^{n}(-1)^i \binom{n}{i}\left(\frac{n}{2}+t-i\right)_+^{n-1}, \quad t \in \mathbb{R},$$

we can deduce from all the above relations that:

$$\widehat{[M_{c,n}]}_M(c+iv) = \left(\frac{\sin(v/2)}{v/2}\right)^n, \quad v \in \mathbb{R}. \tag{8.34}$$

The expression (8.34) can be used in order to show that assumption $(\mathscr{L}3)$ is satisfied. This will be proved in the special case of the uniform spaced sequence $\Pi = (t_k)_{k\in\mathbb{Z}}$, $t_k = k$. We can proceed as follows.

Consider $c = 0$ and we set, for the sake of simplicity $M_{0,n} := M_n$. We claim that:

$$\sum_{k\in\mathbb{Z}} M_n(e^k x) = 1, \quad x \in \mathbb{R}^+. \tag{8.35}$$

In order to prove (8.35), we will use the Mellin-Poisson summation formula (see, e.g., [36, 35], that is the Mellin equivalent of the celebrated Poisson summation formula of Fourier Analysis). Setting:

$$g(x) := \sum_{k\in\mathbb{Z}} M_n(e^k x),$$

this function has the property that $g(x) = g(ex)$ ($g$ is said to be *recurrent*) we can write its Mellin-Fourier series:

$$q \approx \sum_{k\in\mathbb{Z}} g_{k,M}\, x^{-2\pi ki},$$

where the Mellin-Fourier coefficients are:

$$g_{k,M} := \int_{1/e}^{e} g(x)x^{2\pi ki-1}\,dx, \quad k \in \mathbb{Z}.$$

Then, it results that:

$$g_{k,M} = \widehat{[M_n]}_M(2k\pi i), \quad k \in \mathbb{Z},$$

from which we obtain the following expression for the Mellin-Poisson summation formula:

$$\sum_{k\in\mathbb{Z}} M_n(e^k x) = \sum_{k\in\mathbb{Z}} \widehat{[M_n]}_M(2k\pi i)\, x^{-2\pi ki},$$

that together with (8.34) immediately implies (8.35). Similar results can be achieved in the case of Mellin splines $M_{c,n}$ with $c \neq 0$; for more details see [11].

From (8.35), it is clear that condition $(\mathscr{L}3)$ reduces to:

$$\mathscr{T}_w(x,u) = |g_w(u) - u| = \mathcal{O}(w^{-\theta}), \quad as\ w \to +\infty,$$

uniformly with respect to $x \in \mathbb{R}^+$ and $u \in \mathbb{R}$, for a suitable $\theta > 0$, and in the linear case $g_w(u) = u$, it becomes trivially satisfied for every $\theta > 0$, since any $\mathscr{T}_w(x,u) = 0$.

Moreover, observing that the functions $M_n$ have compact support, we have that the discrete absolute moments $M_{\beta,\Pi}(M_n)$ of any order $\beta$ are finite, i.e., assumption $(\mathscr{L}1)$ is fullfilled.

In conclusion of this section we explicitly provide examples of Mellin B-spline. For instance, we can consider:

$$M_2(x) := \begin{cases} 1 - \ln x, & 1 < x < e, \\ 1 + \ln x, & e^{-1} < x < 1, \\ 0, & \text{otherwise.} \end{cases} \tag{8.36}$$

The plot of $M_2$ is given in Fig. 8.1, together with the plot of $\widetilde{M_2}$, in order to see the main differences between the two associated instances of kernels. Finally, the third order Mellin B-spline is:

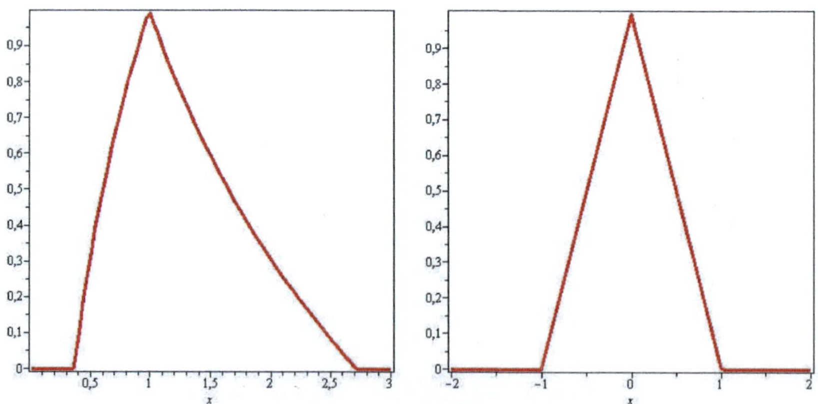

**Figure 8.1:** On the left, we have the plot of the Mellin B-spline $M_2$ defined in (8.36), while on right we have the plot of the classical central B-spline $\widetilde{M}_2$.

$$M_3(x) := \begin{cases} \frac{3}{4} - \ln^2 x, & e^{-1/2} \leq x \leq e^{1/2}, \\[2ex] \frac{1}{2}\left(\frac{3}{2} - |\ln x|\right)^2, & e^{1/2} < x \leq e^{3/2} \text{ or } e^{-3/2} \leq x < e^{-1/2}, \\[2ex] 0, & \text{otherwise.} \end{cases}$$

$$(8.37)$$

The plot of $M_3$ is given in Fig. 8.2, together with the plot of $\widetilde{M}_3$, as above.

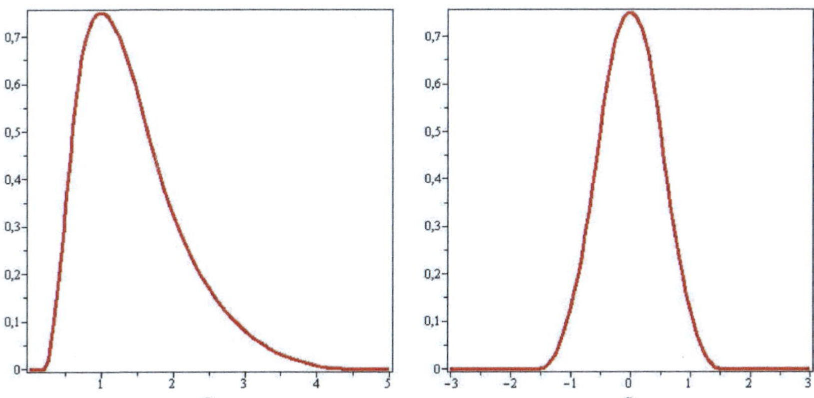

**Figure 8.2:** On the left, we have the plot of the Mellin B-spline $M_3$ defined in (8.37), while on the right we have the plot of the classical central B-spline $\widetilde{M}_3$.

Based on the above considerations, we can easy deduce the following corollary which is valid for the nonlinear exponential sampling series, in the case of product-type kernels (8.20) generated with $L = M_n$ and $g_w$ defined in (8.25), or generated with (8.27) always with $L = M_n$.

***Corollary 8.4***

*Let $\chi^n$ denotes the nonlinear kernel defined as in (8.20) with $L = M_n$ and $g_w$ defined as in (8.25). Then:*

$$\lim_{w \to +\infty} (S_w^{\chi^n} f)(x) = f(x),$$

*at any point of continuity $x \in \mathbb{R}^+$ of a given bounded function $f : \mathbb{R}^+ \to \mathbb{R}$. Moreover, if $f \in C(\mathbb{R}^+)$ we have:*

$$\lim_{w \to +\infty} \|S_w^{\chi^n} f - f\|_\infty = 0,$$

*and the following estimate holds:*

$$\|S_w^{\chi_n} f - f\|_\infty \;\leq\; \frac{3}{2}\,[M_{0,\Pi}(M_n) + M_{1,\Pi}(M_n)]\,\widetilde{\omega}\left(f,\frac{1}{w}\right) \;+\; Cw^{-1},$$

*for $w > 0$ sufficiently large, where $C > 0$ is a suitable constant.*

*Finally, if $\widetilde{\chi}_n$ denotes the (linear) kernel defined as in (8.27) with $L = M_n$, other than the pointwise and uniform convergence (as for the case of $\chi^n$) we can also provide the following quantitative estimate:*

$$\|S_w^{\widetilde{\chi}_n} f - f\|_\infty \;\leq\; [M_{0,\Pi}(M_n) + M_{1,\Pi}(M_n)]\,\widetilde{\omega}\left(f,\frac{1}{w}\right),$$

*for $w > 0$.*

Finally, since the examples of functions $L$ provided in this section have compact support, it is not difficult to see that if we consider suitable shift of these functions we are able to obtain other examples of kernels, for which $supp\,L \subset (1,+\infty)$. Hence, for kernels of the product-form (8.20) generated by $L = M_n(\cdot - \ell)$, with suitable $\ell \in \mathbb{N}^+$, condition $(\chi 5)$ of Section 8.5 turns out to be satisfied. For instance, we consider $L(x) = M_2(x - 2)$, $x \in \mathbb{R}^+$ (see Fig. 8.3). The main advantage of this fact resides in the possibility to apply Theorem 8.6, i.e., to find examples of nonlinear kernels for which it is possible to achieve the so-called exponential prediction by samples from the past.

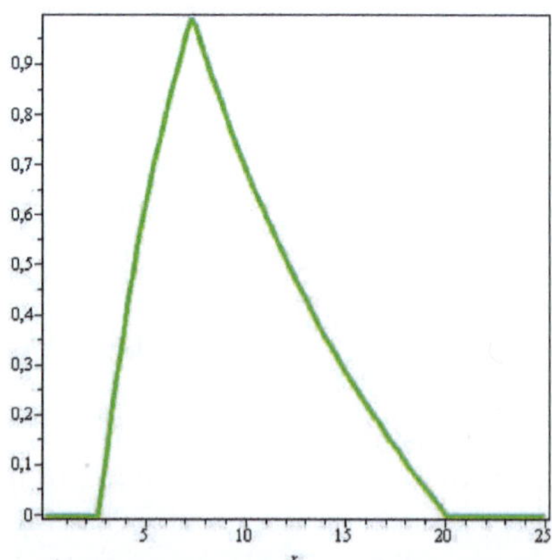

**Figure 8.3:** The plot of the shifted Mellin B-spline $M_2(\cdot - 2)$ for which it is possible to have the phenomenon of the exponential prediction by samples from the past.

Obviously, for the shifted version of the Mellin B-splines all the assumptions $(\chi 1), ..., (\chi 4)$ remain satisfied.

## 8.10  Mellin-Fejér Kernels

As a second class of functions that can be used as $L$ in order to construct product type nonlinear kernels, is given by the families of Mellin-Fejér kernels, defined by:

$$F_\rho^c(x) := \begin{cases} \frac{x^{-c}}{2\pi} \rho \, \mathrm{sinc}^2\left(\frac{\rho}{\pi} \ln \sqrt{x}\right), & x \neq 1, \\ \frac{\rho}{2\pi}, & x = 1, \end{cases} \tag{8.38}$$

$c \in \mathbb{R}$, $\rho > 0$, $x \in \mathbb{R}^+$, where the *sinc*-function is defined as follows:

$$\mathrm{sinc}(x) := \begin{cases} \frac{\sin(\pi x)}{\pi x}, & x \neq 0, \\ 1, & x = 0. \end{cases} \tag{8.39}$$

The Mellin-Fejér type kernels are not with compact support, but it can be proved that $F_\rho^c \in X_c$. Hence, their Mellin transforms can be computed as follows:

$$\widehat{[F_\rho^c]}_M(c + iv) = \begin{cases} 1 - \frac{|v|}{\rho}, & |v| \leq \rho, \\ 0, & |v| > \rho. \end{cases}$$

Now, using the above expressions of the Mellin transforms and following a procedure similar to that one used in Section 8.9 based on the computation of the Mellin-Fourier series and on the application of the Mellin-Poisson summation formula, we can get what follows. Setting:

$$g_c^\rho(x) := \sum_{k \in \mathbb{Z}} F_\rho^c(e^k x), \quad x \in \mathbb{R}^+,$$

and considering for simplicity the case $c = 0$ and $\rho = 1$, we have that $g_0^1(ex) = g_0^1(x)$, and:

$$\sum_{k \in \mathbb{Z}} F_1^0(e^k x) = \sum_{k \in \mathbb{Z}} \widehat{[F_1^0]}_M(2k\pi i) x^{-2k\pi i} = \widehat{[F_1^0]}_M(0) = 1,$$

obtaining that $(\mathscr{L}3)$ holds, with considerations similar to those ones given in Section 8.9.

Concerning the discrete absolute moments, it is not difficult to see that $M_{\beta,\Pi}(F_1^0) < +\infty$, with $0 < \beta < 1$ and $M_{1,\Pi}(F_1^0) = +\infty$. The plot of $F_1^0$ is given in Fig. 8.4.

Based on the above considerations, we can easy deduce the following corollary which is valid for the exponential sampling series, in the case of product-type kernels (8.20) generated with $L = F_1^0$ and $g_w$ defined in (8.25), or generated with (8.27) always with $L = F_1^0$.

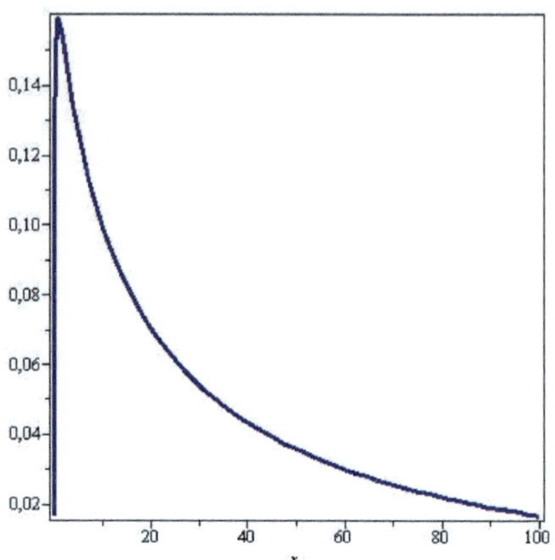

**Figure 8.4:** The plot of the Mellin-Fejér kernel $F_1^0$.

***Corollary 8.5***

*Let $\chi_1^0$ denotes the nonlinear kernel defined as in (8.20) with $L = F_1^0$ and $g_w$ defined as in (8.25). Then:*

$$\lim_{w \to +\infty} (S_w^{\chi_1^0} f)(x) \,=\, f(x),$$

*at any point of continuity $x \in \mathbb{R}^+$ of a given bounded function $f : \mathbb{R}^+ \to \mathbb{R}$. Moreover, if $f \in C(\mathbb{R}^+)$ we have:*

$$\lim_{w \to +\infty} \|S_w^{\chi_1^0} f - f\|_\infty \,=\, 0,$$

*and, for every fixed $0 < \beta < 1$ the following estimate holds:*

$$\|S_w^{\chi_1^0} f - f\|_\infty \,\leq\, \frac{3}{2} \left( M_{0,\Pi}(F_1^0) + M_{\beta,\Pi}(F_1^0) \right) \widetilde{\omega}\left( f, \frac{1}{w^\beta} \right)$$

$$+ 3\|f\|_\infty M_{\beta,\Pi}(F_1^0)\, w^{-\beta} + Cw^{-1},$$

*for $w > 0$ sufficiently large, and where $C > 0$ is a suitable constant.*

*Finally, if $\widetilde{\chi_1^0}$ denotes the nonlinear kernel defined as in (8.27) with $L = F_1^0$ we can also provide the following estimate:*

$$\|S_w^{\widetilde{\chi_1^0}} f - f\|_\infty \,\leq\, \left( M_{0,\Pi}(F_1^0) + M_{\beta,\Pi}(F_1^0) \right) \widetilde{\omega}\left( f, \frac{1}{w^\beta} \right) + 2\|f\|_\infty M_{\beta,\Pi}(F_1^0)\, w^{-\beta},$$

*for every $w > 0$, with a fixed $0 < \beta < 1$.*

## 8.11   Numerical Examples

In this section, we provide some numerical examples in order to show the reconstruction performances of the proposed nonlinear exponential sampling operators.

Here we consider the nonlinear exponential sampling series defined in (8.20) generated with $g_w$ defined in (8.25) and choosing as function $L$ the Mellin-splines $M_2$ and $M_3$ recalled in Section 8.9 and the Mellin-Fejer kernel $F_1^0$ recalled in Section 8.10.

The above approximation operators will be tested on the following function on $\mathbb{R}^+$. For instance, we consider:

$$f_1(x) := \sqrt{x} + \sin x, \qquad x \in \mathbb{R}^+. \tag{8.40}$$

In Figs. 8.5, 8.6 and 8.7, we plotted the approximations of $f_1$ achieved by the nonlinear exponential sampling operators generated by the nonlinear kernel $\chi_w(x,u) = L(x)\,g_w(u)$, with $g_w$ defined in (8.25) and $L$ respectively $M_2$, $M_3$ and $F_1^0$.

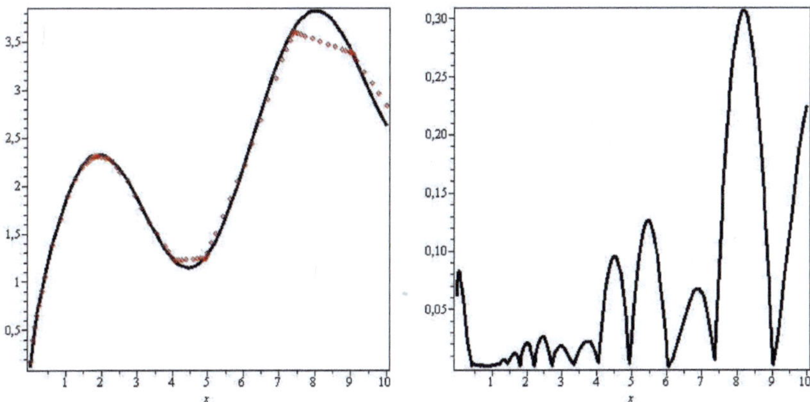

**Figure 8.5:** On the left: the approximation of $f_1$ (black line) achieved by the nonlinear exponential sampling operator with $\chi_w(x,u) = M_2(x)g_w(u)$, with $w = 5$ (red dots). On the right: the corresponding approximation errors.

Note that, from the analysis of the plots given on the right in Fig.s 8.5, 8.6 and 8.7, it seems clear that the numerical results are coherent with the theoretical ones, in particular those concerning the order of approximation, in which we proved that in the case of the Mellin-Fejer kernel the rate of convergence is in general slower than the cases of Mellin B-splines.

## 8.12   Final Remarks and Open Problems

In this paper, motivated by the wide current interest on exponential type operators (see, e.g., [11]) we considered a new family of nonlinear sampling series of the exponential type. The proposed definition generalizes to the nonlinear frame that was given by Bardaro, Faina and Mantellini in 2017.

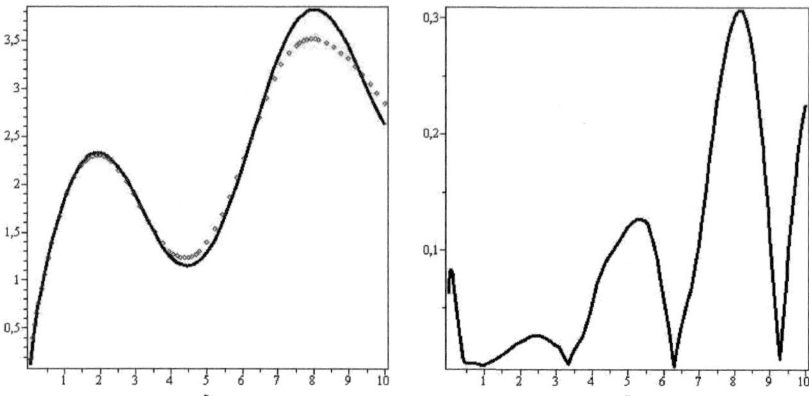

**Figure 8.6:** On the left: the approximation of $f_1$ (black line) achieved by the nonlinear exponential sampling operator with $\chi_w(x,u) = M_3(x)g_w(u)$, with $w = 5$ (red dots). On the right: the corresponding approximation errors.

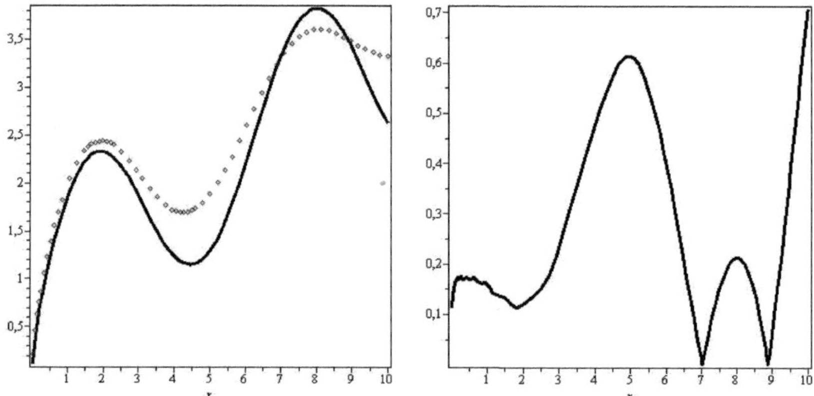

**Figure 8.7:** On the left: the approximation of $f_1$ (black line) achieved by the nonlinear exponential sampling operator with $\chi_w(x,u) = F_1^0(x)g_w(u)$, with $w = 15$ (red dots). On the right: the corresponding approximation errors.

As a future work, a Kantorovich version (see, e.g., [52, 53]) of the above operators could be studied according to the linear version recalled in (8.7). In particular, as deeply discussed in Section 8.2, such operators could be suitable in order to prove approximation results in $L^p$-spaces, or more in general, in Orlicz or modular spaces, with the aim to obtain approximation results in the case of not-necessarily continuous signals.

The latter operators could also be extended to the multivariate setting; this can be useful in order to face applications to the reconstruction of multivariate signals, such as digital images. The latter topic received a lot of interest in the last years, in relation to the study of real world problems (see, e.g., [39, 69]).

Finally, in order to provide a complete overview concerning the nonlinear sampling operators of the exponential type, also a Durrmeyer version could be introduced and studied, according to the linear version recalled in (8.8).

## Acknowledgments

The author is member of the Gruppo Nazionale per l'Analisi Matematica, la Probabilità e le loro Applicazioni (GNAMPA) of the Istituto Nazionale di Alta Matematica (INdAM), of the network RITA (Research ITalian network on Approximation), and of the UMI (Unione Matematica Italiana) group T.A.A. (Teoria dell'Approssimazione e Applicazioni).

## Funding

The author has been partially supported within the (1) 2022 GNAMPA-INdAM Project "Enhancement e segmentazione di immagini mediante operatori di tipo campionamento e metodi variazionali per lo studio di applicazioni biomediche", (2) 2023 GNAMPA-INdAM Project "Approssimazione costruttiva e astratta mediante operatori di tipo sampling e loro applicazioni", (3) "Metodiche di Imaging non invasivo mediante angiografia OCT sequenziale per lo studio delle Retinopatie degenerative dell'Anziano (M.I.R.A.)", funded by the Fondazione Cassa di Risparmio di Perugia (FCRP), 2019, and (4) "National Innovation Ecosystem grant ECS00000041 - VITALITY", funded by the European Union - NextGenerationEU under the Italian Ministry of University and Research (MUR).

## Conflict of Interest/Competing Interests

The author declares that he has no conflict of interest and competing interest.

## Availability of Data and Material and Code Availability

Not applicable.

# References


[1] Acar, T. and Kursun, S. 2023. Pointwise convergence of generalized Kantorovich exponential sampling series. Dolomites Research Notes on Approximation 16(1).

[2] Acar, T., Kursun, S. and Turgay, M. 2023. Multidimensional Kantorovich modifications of exponential sampling series. Quaestiones Mathematicae 46(1): 57–72.

[3] Aral, A., Acar, T. and Kursun, S. 2022. Generalized Kantorovich forms of exponential sampling series. Analysis and Mathematical Physics 12(2): 50.

[4] Bajpeyi, S. 2023. Order of approximation for exponential sampling type neural network operators. Results in Mathematics 78(3): 99.

[5] Bajpeyi, S., Kumar, A. and Devaraj, P. 2022. An inverse approximation and saturation order for Kantorovich exponential sampling series. arXiv preprint arXiv:2212.06006.

[6] Bajpeyi, S., Kumar, A. and Mantellini, I. 2022. Approximation by durrmeyer type exponential sampling operators. Numerical Functional Analysis and Optimization 43(1): 16–34.

[7] Balsamo, S. and Mantellini, I. 2019. On linear combinations of general exponential sampling series. Results in Mathematics 74(4): 180.

[8] Bardaro, C., Butzer, P. and Mantellini, I. 2016. The Mellin–Parseval formula and its interconnections with the exponential sampling theorem of optical physics. Integral Transforms and Special Functions, 27(1): 17–29.

[9] Bardaro, C., Butzer, P. and Mantellini, I. 2014. The exponential sampling theorem of signal analysis and the reproducing kernel formula in the mellin

transform setting. Sampling Theory in Signal and Image Processing, 13: 35–66.

[10] Bardaro, C., Butzer, P., Mantellini, I. and Schmeisser, G. 2016. On the Paley–Wiener theorem in the mellin transform setting. Journal of Approximation Theory 207: 60–75.

[11] Bardaro, C., Faina, L., and Mantellini, I. 2017. A generalization of the exponential sampling series and its approximation properties. Mathematica Slovaca 67(6): 1481–1496.

[12] Bardaro, C. and Mantellini, I. 2006. Approximation properties in abstract modular spaces for a class of general sampling-type operators. Applicable Analysis, 85(4): 383–413.

[13] Bardaro, C. and Mantellini, I. 2011. A note on the Voronovskaja theorem for mellin–fejer convolution operators. Applied Mathematics Letters, 24(12): 2064–2067.

[14] Bardaro, C. and Mantellini, I. 2014. On Mellin convolution operators: A direct approach to the asymptotic formulae. Integral Transforms and Special Functions 25(3): 182–195.

[15] Bardaro, C. and Mantellini, I. 2021. On a Durrmeyer-type modification of the exponential sampling series. Rendiconti del Circolo Matematico di Palermo Series 2, 70: 1289–1304.

[16] Bardaro, C. and Mantellini, I. 2002. On a singularity concept for kernels of nonlinear integral operators. International Mathematical Journal 1(3): 239–254.

[17] Bardaro, C. and Mantellini, I. 2014. Asymptotic expansion of generalized Durrmeyer sampling type series. Jaen Journal on Approximation, 6(2): 143–165.

[18] Bardaro, C., Mantellini, I. and Schmeisser, G. 2019. Exponential sampling series: Convergence in Mellin–Lebesgue spaces. Results in Mathematics 74: 1–20.

[19] Bardaro, C., Mantellini, I. and Schmeisser, G. 2023. Convergence of semi-discrete exponential sampling operators in Mellin–Lebesgue spaces. Revista de la Real Academia de Ciencias Exactas, Físicas y Naturales. Serie A. Matemáticas 117(1): 30.

[20] Bardaro, C., Musielak, J. and Vinti, G. 2003. Nonlinear Integral Operators and Applications. Walter de Gruyter.

[21] Bardaro, C., Butzer, P.L., Stens, R.L. and Vinti, G. 2007. Kantorovich-type generalized sampling series in the setting of Orlicz spaces. Sampling Theory in Signal and Image Processing 6: 29–52.

[22] Barna, J.C.J., Laue, E.D., Mayger, R., Skilling, J. and Worrall, S.J.P. 1987. Exponential sampling, an alternative method for sampling in two-dimensional NMR experiments. Journal of Magnetic Resonance (1969), 73(1): 69–77.

[23] Bertero, M. and Pike, E.R. 1991. Exponential-sampling method for laplace and other dilationally invariant transforms: II. Examples in photon correlation spectroscopy and fraunhofer diffraction. Inverse Problems 7(1): 21.

[24] Bezuglaya, L. and Katsnelson, V. 1993. The sampling theorem for functions with limited multi-band spectrum I. Zeitschrift für Analysis und ihre Anwendungen 12(3): 511–534.

[25] Boccali, L., Costarelli, D. and Vinti, G. 2023. Convergence results in Orlicz spaces for sequences of max-product Kantorovich sampling operators. Journal of Computational and Applied Mathematics 449(2024): 115957.

[26] Butzer, P 1988. The sampling theorem and linear prediction in signal analysis. Jahresbericht der Deutschen Mathematiker-Vereinigung, 90: 1–70.

[27] Butzer, P. and Jansche, S. 1997. A direct approach to the Mellin transform. Journal of Fourier Analysis and Applications 3: 325–376.

[28] Butzer, P. and Jansche, S. 1999. A self-contained approach to Mellin transform analysis for square integrable functions; applications. Integral Transforms and Special Functions 8(3-4): 175–198.

[29] Butzer, P. and Stens, R.L. 1992. Sampling theory for not necessarily band-limited functions: A historical overview. SIAM Review 34(1): 40–5.

[30] Butzer, P. and Stens, R.L. 1993. Linear prediction by samples from the past. Advanced Topics in Shannon Sampling and Interpolation Theory, (Editor R.J. Marks II).

[31] Butzer, P. 1983. A survey of the Whittaker- Shannon sampling theorem and some of its extensions. In Journal of Mathematical Research and Exposition, volume 3.

[32] Butzer, P., Engels, W., Ries, S. and Stens, R.L. 1986. The Shannon sampling series and the reconstruction of signals in terms of linear, quadratic and cubic splines. SIAM Journal on Applied Mathematics, 46(2): 299–323.

[33] Butzer, P., Fischer, A. and Stens, R.L. 1990. Generalized sampling approximation of multivariate signals; Theory and some applications. Note di Matematica, 10(suppl. 1): 173–191.

[34] Butzer, P. and Hinsen, G. 1989. Reconstruction of bounded signals from pseudo-periodic, irregularly spaced samples. Signal Processing, 17(1): 1–17.

[35] Butzer, P. and Jansche, S. 1998. The exponential sampling theorem of signal analysis. Atti Sem. Mat. Fis. Univ. Moden. 46: 99–122.

[36] Butzer, P. and Jansche, S. 1998. Mellin transform, the Mellin-Poisson summation formula and the exponential sampling theorem. Atti Sem. Mat. Fis. Univ. Modena, Suppl. 46: 99–122.

[37] Butzer, P. and Nessel, R.J. 1971. Fourier analysis and approximations, vol. I. Birkhuser.

[38] Butzer, P., Ries, S. and Stens, R.L. 1987. Approximation of continuous and discontinuous functions by generalized sampling series. Journal of Approximation Theory 50(1): 25–39.

[39] Cagini, C., Costarelli, D., Gujar, R., Lupidi, M., Lutty, G.A., Seracini, M. and Vinti, G. 2022. Improvement of retinal OCT angiograms by sampling Kantorovich algorithm in the assessment of retinal and choroidal perfusion. Applied Mathematics and Computation 427: 127152.

[40] Cantarini, M., Costarelli, D. and Vinti, G. 2022. Approximation of differentiable and not differentiable signals by the first derivative of sampling Kantorovich operators. Journal of Mathematical Analysis and Applications 509(1): 125913.

[41] Cantarini, M., Costarelli, D. and Vinti, G. 2022. Convergence of a class of generalized sampling Kantorovich operators perturbed by multiplicative noise. In Recent Advances in Mathematical Analysis: Celebrating the 70th Anniversary of Francesco Altomare, pp. 249–267. Springer, 2022.

[42] Casasent, D. 1978. Optical Data Processing, volume 6. Springer Berlin.

[43] Çetin, N., Costarelli, D., Natale, M. and Vinti, G. 2022. Nonlinear multivariate sampling Kantorovich operators: Quantitative estimates in functional spaces. Dolomites Research Notes on Approximation, 15(3).

[44] Coroianu, L. and Gal, S.G. 2010. Approximation by nonlinear generalized sampling operators of max-product kind. Sampling Theory in Signal and Image Processing 9: 59–75.

[45] Costarelli, D., De Angelis, E. and Vinti, G. 2023. Convergence of perturbed sampling Kantorovich operators in modular spaces. Results in Mathematics 78(2023): Art. Numb. 239.

[46] Costarelli, D., Minotti, A.M. and Vinti, G. 2017. Approximation of discontinuous signals by sampling Kantorovich series. Journal of Mathematical Analysis and Applications 450(2): 1083–1103.

[47] Costarelli, D., Natale, M. and Vinti, G. 2023. Convergence results for nonlinear sampling Kantorovich operators in modular spaces. Numerical Functional Analysis and Optimization 44(12): 1276–1299.

[48] Costarelli, D., Piconi, M. and Vinti, G. 2023. The multivariate Durrmeyer-sampling type operators in functional spaces. Dolomites Research Notes on Approximation 15(5).

[49] Costarelli, D., Piconi, M. and Vinti, G. 2023. On the convergence properties of sampling Durrmeyer-type operators in Orlicz spaces. Mathematische Nachrichten 296(2): 588–609.

[50] Costarelli, D., Piconi, M. and Vinti, G. 2023. Quantitative estimates for Durrmeyer-sampling series in Orlicz spaces. Sampling Theory, Signal Processing, and Data Analysis 21(1): 3.

[51] Costarelli, D., Seracini, M. and Vinti, G. 2020. A comparison between the sampling Kantorovich algorithm for digital image processing with some interpolation and quasi-interpolation methods. Applied Mathematics and Computation, 374: 125046.

[52] Costarelli, D. and Vinti, G. 2013. Approximation by nonlinear multivariate sampling Kantorovich type operators and applications to image processing. Numerical Functional Analysis and Optimization 34(8): 819–844.

[53] Costarelli, D. and Vinti, G. 2015. Degree of approximation for nonlinear multivariate sampling Kantorovich operators on some functions spaces. Numerical Functional Analysis and Optimization 36(8): 964–990.

[54] Costarelli, D. and Vinti, G. 2019. Inverse results of approximation and the saturation order for the sampling Kantorovich series. Journal of Approximation Theory 242: 64–82.

[55] Costarelli, D. and Vinti, G. 2019. Saturation by the Fourier transform method for the sampling Kantorovich series based on bandlimited kernels. Analysis and Mathematical Physics 9(4): 2263–2280.

[56] Costarelli, D. and Vinti, G. 2022. Approximation properties of the sampling Kantorovich operators: Regularization, saturation, inverse results and Favard classes in Lp-spaces. Journal of Fourier Analysis and Applications 28(3): 49.

[57] David, L. and Wool, A. 2022. Rank estimation with bounded error via exponential sampling. Journal of Cryptographic Engineering, 12(2): 151–168.

[58] DeVore, R.A. and Lorentz, G.G. 1993. Constructive Approximation, volume 303. Springer Science & Business Media.

[59] Draganov, B. 2022. A fast converging sampling operator. Constructive Mathematical Analysis 5(4): 190–201.

[60] Gori, F. 1993. Sampling in optics. In Advanced Topics in Shannon Sampling and Interpolation Theory, pp. 37–83. Springer.

[61] Kolomoitsev, Yu and Skopina, M. 2021. Quasi-projection operators in weighted Lp spaces. Applied and Computational Harmonic Analysis 52: 165–197.

[62] Kolomoitsev, Y. and Skopina, M. 2020. Approximation by sampling-type operators in Lp-spaces. Mathematical Methods in the Applied Sciences 43(16): 9358–9374.

[63] Mamedov, R.G. 1991. The Mellin transform and approximation theory (russian) 'elm'.

[64] Musielak, J. 1981. On some approximation problems in modular spaces. Constructive Function Theory 81: 455–461.

[65] Musielak, J. 1991. Approximation by nonlinear singular integral operators in generalized Orlicz spaces. Comment. Math. Prace Mat. 31: 79–88.

[66] Musielak, J. 1993. Nonlinear approximation in some modular function spaces I. Mathematica Japonica 38(1): 83–90.

[67] Musielak, J. 1993. On the approximation by nonlinear integral operators with generalized Lipschitz kernel over a compact abelian group. Comment. Math. Prace. Mat, 33: 99–104.

[68] Musielak, J. 1999. On nonlinear integral operators. Atti del Seminario Matematico e Fisico Universita di Modena 47: 183–190.

[69] Osowska-Kurczab, A., Les, T., Markiewicz, T., Dziekiewicz, M., Lorent, M., Cierniak, S., Costarelli, D., Seracini, M. and Vinti, G. 2023. Improvement of renal image recognition through resolution enhancement. Expert Systems with Applications 213: 118836.

[70] Ostrowsky, N., Sornette, D., Parker, P. and Pike, E.R. 1981. Exponential sampling method for light scattering polydispersity analysis. Optica Acta: International Journal of Optics 28(8): 1059–1070.

[71] Shannon, C.E. 1949. Communication in the presence of noise. Proceedings of the IRE 37(1): 10–21.

[72] Vinti, G. and Zampogni, L. 2009. Approximation by means of nonlinear Kantorovich sampling type operators in Orlicz spaces. Journal of Approximation Theory 161(2): 511–528.

[73] Vinti, G. and Zampogni, L. 2018. A general approximation approach for the simultaneous treatment of integral and discrete operators. Advanced Nonlinear Studies 18(4): 705–724.

# Chapter 9

# Refinements and Reverses of Some Inequalities for the Normalized Determinants of Sequences of Positive Operators in Hilbert Spaces

*Silvestru Sever Dragomir*

## 9.1 Introduction

Let $Q(H)$ be the space of all bounded linear operators on a Hilbert space $H$, and $I$ stands for the identity operator on $H$. An operator $P$ in $Q(H)$ is said to be positive (in symbol: $P \geq 0$) if $\langle Pu, u \rangle \geq 0$ for all $u \in H$. In particular, $P > 0$ means that $P$ is positive and invertible. For a pair $P$, $Q$ of selfadjoint operators the order relation $P \geq Q$ means as usual that $P - Q$ is positive.

Victoria University, Melbourne, Australia and University of the Witwatersrand, Johannesburg, South Africa.

In 1998, Fujii et al. [3], [4], introduced the *normalized determinant* $\Delta_u(P)$ for positive invertible operators $P$ on a Hilbert space $H$ and a fixed unit vector $u \in H$, namely $\|u\| = 1$, defined by $\Delta_u(P) := \exp \langle \ln Pu, u \rangle$ and discussed it as a continuous geometric mean and observed some inequalities around the determinant from this point of view.

Some of the fundamental properties of normalized determinant are as follows, [3].

For each unit vector $u \in H$, see also [5], we have:

(i) *continuity*: the map $P \to \Delta_u(P)$ is norm continuous;

(ii) *bounds*: $\langle P^{-1}u, u \rangle^{-1} \leq \Delta_u(P) \leq \langle Pu, u \rangle$;

(iii) *continuous mean*: $\langle P^p u, u \rangle^{1/p} \downarrow \Delta_u(P)$ for $p \downarrow 0$ and $\langle P^p u, u \rangle^{1/p} \uparrow \Delta_u(P)$ for $p \uparrow 0$;

(iv) *power equality:* $\Delta_u(P^t) = \Delta_u(P)^t$ for all $t > 0$;

(v) *homogeneity*: $\Delta_u(tP) = t\Delta_u(P)$ and $\Delta_u(tI) = t$ for all $t > 0$;

(vi) *monotonicity*: $0 < P \leq Q$ implies $\Delta_u(P) \leq \Delta_u(Q)$;

(vii) *multiplicativity*: $\Delta_u(PQ) = \Delta_u(P)\Delta_u(Q)$ for commuting $P$ and $Q$;

(viii) *Ky Fan type inequality*: $\Delta_u((1-\alpha)P + \alpha Q) \geq \Delta_u(P)^{1-\alpha}\Delta_u(Q)^\alpha$ for $0 < \alpha < 1$.

We define the logarithmic mean of two positive numbers $a, b$ by

$$
L(a,b) := \begin{cases} \dfrac{b-a}{\ln b - \ln a} & \text{if } b \neq a, \\[2mm] a & \text{if } b = a. \end{cases}
$$

In [3] the authors obtained the following additive reverse inequality for the operator $P$ which satisfies the condition $0 < \gamma I \leq P \leq \Gamma I$, where $\gamma, \Gamma$ are positive numbers,

$$
0 \leq \langle Pu, u \rangle - \Delta_u(P) \leq L(\gamma, \Gamma)\left[\ln L(\gamma, \Gamma) + \frac{\Gamma \ln \gamma - \gamma \ln \Gamma}{\Gamma - \gamma} - 1\right] \tag{9.1}
$$

for all $u \in H$, $\|u\| = 1$.

We recall that *Specht's ratio* is defined by [6]

$$
S(h) := \begin{cases} \dfrac{h^{\frac{1}{h-1}}}{e \ln\left(h^{\frac{1}{h-1}}\right)} & \text{if } h \in (0,1) \cup (1,\infty), \\[4mm] 1 & \text{if } h = 1. \end{cases} \tag{9.2}
$$

It is well known that $\lim_{h\to 1} S(h) = 1$, $S(h) = S\left(\frac{1}{h}\right) > 1$ for $h > 0$, $h \neq 1$. The function is decreasing on $(0,1)$ and increasing on $(1,\infty)$.

In [4], the authors obtained the following multiplicative reverse inequality as well

$$1 \leq \frac{\langle Pu, u \rangle}{\Delta_u(P)} \leq S\left(\frac{\Gamma}{\gamma}\right) \tag{9.3}$$

for $0 < \gamma I \leq P \leq \Gamma I$ and $u \in H$, $\|u\| = 1$.

In this chapter we obtain several refinements and reverses for the normalized determinant of a sequence of operators that have the spectra in a positive interval $[\gamma, \Gamma]$. For this purpose we used some Jensen's type inequalities for twice differentiable functions obtained by the author in [1].

## 9.2   Inequalities for $p \in (-\infty, 0) \cup (1, \infty)$

Assume that $P > 0$. For a vector $v \neq 0$ we can extend the normalized determinant as $\tilde{\Delta}_v(P) := \exp \langle \ln Pv, v \rangle$. We observe that

$$\tilde{\Delta}_v(P) := \exp \langle \ln Pv, v \rangle = \exp\left(\|v\|^2 \left\langle \ln P\frac{v}{\|v\|}, \frac{v}{\|v\|} \right\rangle\right)$$

$$= \left[\exp\left(\left\langle \ln P\frac{v}{\|v\|}, \frac{v}{\|v\|} \right\rangle\right)\right]^{\|v\|^2} = \left[\Delta_{\frac{v}{\|v\|}}(P)\right]^{\|v\|^2}$$

for any $v \neq 0$.

**Theorem 9.1**
*Assume that $P_j$ are operators such that $0 < \gamma \leq P_j \leq \Gamma$, $j \in \{1, ..., n\}$. Define*

$$\gamma_p(\gamma, \Gamma) := \begin{cases} \dfrac{\Gamma^{-p}}{p(p-1)} \text{ for } p \in (1,\infty), \\[2mm] \dfrac{\gamma^{-p}}{p(p-1)} \text{ for } p \in (-\infty, 0) \end{cases}$$

*and*

$$\Gamma_p(\gamma, \Gamma) := \begin{cases} \dfrac{\gamma^{-p}}{p(p-1)} \text{ for } p \in (1,\infty), \\[2mm] \dfrac{\Gamma^{-p}}{p(p-1)} \text{ for } p \in (-\infty, 0). \end{cases}$$

*Then*

$$1 \leq \exp\left( \gamma_p(\gamma,\Gamma) \left[ \sum_{j=1}^{n} \left\langle P_j^p u_j, u_j \right\rangle - \left( \sum_{j=1}^{n} \langle P_j u_j, u_j \rangle \right)^p \right] \right) \tag{9.4}$$

$$\leq \frac{\frac{\sum_{j=1}^{n} \langle P_j u_j, u_j \rangle}{n}}{\prod_{j=1}^{n} \tilde{\Delta}_{u_j}(P_j)}$$

$$\leq \exp\left( \Gamma_p(\gamma,\Gamma) \left[ \sum_{j=1}^{n} \left\langle P_j^p u_j, u_j \right\rangle - \left( \sum_{j=1}^{n} \langle P_j u_j, u_j \rangle \right)^p \right] \right),$$

*for each $u_j \in H$, $j \in \{1,...,n\}$ with $\sum_{j=1}^{n} \|u_j\|^2 = 1$.*

**Proof 9.1**  Let $P_j$ be positive definite operators with $\mathrm{Sp}(P_j) \subseteq [\gamma,\Gamma] \subset (0,\infty)$ $j \in \{1,...,n\}$. If $f$ is a twice differentiable function on $(\gamma,\Gamma)$ and for $p \in (-\infty,0) \cup (1,\infty)$ we have for some $\gamma < \Gamma$ that

$$\gamma \leq g(t) := \frac{t^{2-p}}{p(p-1)} f''(t) \leq \Gamma \text{ for any } t \in (\gamma,\Gamma), \tag{9.5}$$

then, see [1],

$$\gamma \left[ \sum_{j=1}^{n} \left\langle P_j^p u_j, u_j \right\rangle - \left( \sum_{j=1}^{n} \langle P_j u_j, u_j \rangle \right)^p \right] \tag{9.6}$$

$$\leq \sum_{j=1}^{n} \langle f(P_j) u_j, u_j \rangle - f\left( \sum_{j=1}^{n} \langle P_j u_j, u_j \rangle \right)$$

$$\leq \Gamma \left[ \sum_{j=1}^{n} \left\langle P_j^p u_j, u_j \right\rangle - \left( \sum_{j=1}^{n} \langle P_j u_j, u_j \rangle \right)^p \right]$$

for each $u_j \in H$, $j \in \{1,...,n\}$ with $\sum_{j=1}^{n} \|u_j\|^2 = 1$.

We consider the convex function $f(t) = -\ln t$, $t \in [\gamma,\Gamma] \subset (0,\infty)$. Then

$$g(t) = \frac{t^{2-p}}{p(p-1)} \frac{1}{t^2} = \frac{1}{p(p-1)t^p}.$$

For $p \in (1,\infty)$, we have

$$\sup_{t \in [\gamma,\Gamma]} g(t) = \frac{\gamma^{-p}}{p(p-1)} \text{ and } \inf_{t \in [\gamma,\Gamma]} g(t) = \frac{\Gamma^{-p}}{p(p-1)}$$

and for $p \in (-\infty, 0)$

$$\sup_{t \in [\gamma, \Gamma]} g(t) = \sup_{t \in [\gamma, \Gamma]} \frac{t^{-p}}{p(p-1)} = \frac{\Gamma^{-p}}{p(p-1)}$$

and

$$\inf_{t \in [\gamma, \Gamma]} g(t) = \inf_{t \in [\gamma, \Gamma]} \frac{t^{-p}}{p(p-1)} = \frac{\gamma^{-p}}{p(p-1)}.$$

Therefore by (9.6) we get

$$0 \le \gamma_p(\gamma, \Gamma) \left[ \sum_{j=1}^{n} \left\langle P_j^p u_j, u_j \right\rangle - \left( \sum_{j=1}^{n} \langle P_j u_j, u_j \rangle \right)^p \right] \tag{9.7}$$

$$\le \ln \left( \sum_{j=1}^{n} \langle P_j u_j, u_j \rangle \right) - \sum_{j=1}^{n} \langle \ln P_j u_j, u_j \rangle$$

$$\le \Gamma_p(\gamma, \Gamma) \left[ \sum_{j=1}^{n} \left\langle P_j^p u_j, u_j \right\rangle - \left( \sum_{j=1}^{n} \langle P_j u_j, u_j \rangle \right)^p \right],$$

where $\gamma_p(\gamma, \Gamma)$ and $\Gamma_p(\gamma, \Gamma)$ are given above.

If we take the exponential in (9.7), then we get

$$1 \le \exp \left( \gamma_p(\gamma, \Gamma) \left[ \sum_{j=1}^{n} \left\langle P_j^p u_j, u_j \right\rangle - \left( \sum_{j=1}^{n} \langle P_j u_j, u_j \rangle \right)^p \right] \right) \tag{9.8}$$

$$\le \frac{\exp \ln \left( \sum_{j=1}^{n} \langle P_j u_j, u_j \rangle \right)}{\exp \left( \sum_{j=1}^{n} \langle \ln P_j u_j, u_j \rangle \right)}$$

$$\le \exp \left( \Gamma_p(\gamma, \Gamma) \left[ \sum_{j=1}^{n} \left\langle P_j^p u_j, u_j \right\rangle - \left( \sum_{j=1}^{n} \langle P_j u_j, u_j \rangle \right)^p \right] \right),$$

for each $u_j \in H$, $j \in \{1, ..., n\}$ with $\sum_{j=1}^{n} \|u_j\|^2 = 1$.

Since

$$\exp \left( \sum_{j=1}^{n} \langle \ln P_j u_j, u_j \rangle \right) = \prod_{j=1}^{n} \exp \langle \ln P_j u_j, u_j \rangle = \prod_{j=1}^{n} \Delta_{u_j}(P_j),$$

hence by (9.8) we derive (9.4).

***Corollary 9.1***
*With the assumptions of Theorem 9.1, define*

$$\tilde{\gamma}_p(\gamma,\Gamma) := \begin{cases} \dfrac{\gamma^p}{p(p-1)} & \text{for } p \in (1,\infty), \\[2ex] \dfrac{\Gamma^p}{p(p-1)} & \text{for } p \in (-\infty,0) \end{cases}$$

*and*

$$\tilde{\Gamma}_p(\gamma,\Gamma) := \begin{cases} \dfrac{\Gamma^p}{p(p-1)} & \text{for } p \in (1,\infty), \\[2ex] \dfrac{\gamma^p}{p(p-1)} & \text{for } p \in (-\infty,0). \end{cases}$$

*Then*

$$1 \le \exp\left(\tilde{\gamma}_p(\gamma,\Gamma)\left[\sum_{j=1}^{n}\left\langle P_j^{-p}u_j,u_j\right\rangle - \left(\sum_{j=1}^{n}\left\langle P_j^{-1}u_j,u_j\right\rangle\right)^p\right]\right) \tag{9.9}$$

$$\le \frac{\displaystyle\prod_{j=1}^{n}\tilde{\Delta}_{u_j}(P_j)}{\left(\sum_{j=1}^{n}\left\langle P_j^{-1}u_j,u_j\right\rangle\right)^{-1}}$$

$$\le \exp\left(\tilde{\Gamma}_p(\gamma,\Gamma)\left[\sum_{j=1}^{n}\left\langle P_j^{-p}u_j,u_j\right\rangle - \left(\sum_{j=1}^{n}\left\langle P_j^{-1}u_j,u_j\right\rangle\right)^p\right]\right),$$

*for each $u_j \in H$, $j \in \{1,...,n\}$ with $\sum_{j=1}^{n}\|u_j\|^2 = 1$.*

The proof follows from Theorem 9.1 by writing for $\Gamma^{-1} \le P_j^{-1} \le \gamma^{-1}$, $j \in \{1,...,n\}$.

**Remark 9.1**  Assume that $P_j$ are operators such that $0 < \gamma \le P_j \le \Gamma$, $j \in \{1,...,n\}$. If we take $p = 2$ in (9.4), then we get

$$1 \le \exp\left(\frac{1}{2\Gamma^2}\left[\sum_{j=1}^{n}\langle P_j^2 u_j,u_j\rangle - \left(\sum_{j=1}^{n}\langle P_j u_j,u_j\rangle\right)^2\right]\right) \tag{9.10}$$

$$\le \frac{\dfrac{\sum_{j=1}^{n}\langle P_j u_j,u_j\rangle}{n}}{\displaystyle\prod_{j=1}^{n}\tilde{\Delta}_{u_j}(P_j)}$$

$$\le \exp\left(\frac{1}{2\gamma^2}\left[\sum_{j=1}^{n}\langle P_j^2 u_j,u_j\rangle - \left(\sum_{j=1}^{n}\langle P_j u_j,u_j\rangle\right)^2\right]\right)$$

and from (9.9),

$$1 \le \exp\left(\frac{\gamma^2}{2}\left[\sum_{j=1}^{n}\left\langle P_j^{-2}u_j,u_j\right\rangle - \left(\sum_{j=1}^{n}\left\langle P_j^{-1}u_j,u_j\right\rangle\right)^2\right]\right) \tag{9.11}$$

$$\le \frac{\prod_{j=1}^{n}\tilde{\Delta}_{u_j}(P_j)}{\left(\sum_{j=1}^{n}\left\langle P_j^{-1}u_j,u_j\right\rangle\right)^{-1}}$$

$$\le \exp\left(\frac{\Gamma^2}{2}\left[\sum_{j=1}^{n}\left\langle P_j^{-2}u_j,u_j\right\rangle - \left(\sum_{j=1}^{n}\left\langle P_j^{-1}u_j,u_j\right\rangle\right)^2\right]\right),$$

for each $u_j \in H$, $j \in \{1,...,n\}$ with $\sum_{j=1}^{n}\|u_j\|^2 = 1$.

If we take $p = -1$ in (9.4), then we get

$$1 \le \exp\left(\frac{\gamma}{2}\left[\sum_{j=1}^{n}\left\langle P_j^{-1}u_j,u_j\right\rangle - \left(\sum_{j=1}^{n}\left\langle P_ju_j,u_j\right\rangle\right)^{-1}\right]\right) \tag{9.12}$$

$$\le \frac{\sum_{j=1}^{n}\left\langle P_ju_j,u_j\right\rangle}{\prod_{j=1}^{n}\tilde{\Delta}_{u_j}(P_j)}$$

$$\le \exp\left(\frac{\Gamma}{2}\left[\sum_{j=1}^{n}\left\langle P_j^{-1}u_j,u_j\right\rangle - \left(\sum_{j=1}^{n}\left\langle P_ju_j,u_j\right\rangle\right)^{-1}\right]\right)$$

and from (9.9)

$$1 \le \exp\left(\frac{1}{2\Gamma}\left[\sum_{j=1}^{n}\left\langle P_ju_j,u_j\right\rangle - \left(\sum_{j=1}^{n}\left\langle P_j^{-1}u_j,u_j\right\rangle\right)^{-1}\right]\right) \tag{9.13}$$

$$\le \frac{\prod_{j=1}^{n}\tilde{\Delta}_{u_j}(P_j)}{\left(\sum_{j=1}^{n}\left\langle P_j^{-1}u_j,u_j\right\rangle\right)^{-1}}$$

$$\le \exp\left(\frac{1}{2\gamma}\left[\sum_{j=1}^{n}\left\langle P_ju_j,u_j\right\rangle - \left(\sum_{j=1}^{n}\left\langle P_j^{-1}u_j,u_j\right\rangle\right)^{-1}\right]\right),$$

for each $u_j \in H$, $j \in \{1,...,n\}$ with $\sum_{j=1}^{n}\|u_j\|^2 = 1$. ∎

The case of normalized determinant is as follows:

**Corollary 9.2**
*Assume that $p_j \geq 0$ with $\sum_{j=1}^{n} p_j = 1$, then*

$$1 \leq \exp\left(\gamma_p(\gamma,\Gamma)\left[\sum_{j=1}^{n} p_j \left\langle P_j^p u, u\right\rangle - \left(\sum_{j=1}^{n} p_j \left\langle P_j u, u\right\rangle\right)^p\right]\right) \tag{9.14}$$

$$\leq \frac{\sum_{j=1}^{n} p_j \left\langle P_j u, u\right\rangle}{\prod_{j=1}^{n} [\Delta_u(P_j)]^{p_j}}$$

$$\leq \exp\left(\Gamma_p(\gamma,\Gamma)\left[\sum_{j=1}^{n} p_j \left\langle P_j^p u, u\right\rangle - \left(\sum_{j=1}^{n} p_j \left\langle P_j u, u\right\rangle\right)^p\right]\right),$$

*and*

$$1 \leq \exp\left(\tilde{\gamma}_p(\gamma,\Gamma)\left[\sum_{j=1}^{n} p_j \left\langle P_j^{-p} u, u\right\rangle - \left(\sum_{j=1}^{n} p_j \left\langle P_j^{-1} u, u\right\rangle\right)^p\right]\right) \tag{9.15}$$

$$\leq \frac{\prod_{j=1}^{n} [\Delta_u(P_j)]^{p_j}}{\left(\sum_{j=1}^{n} p_j \left\langle P_j^{-1} u, u\right\rangle\right)^{-1}}$$

$$\leq \exp\left(\tilde{\Gamma}_p(\gamma,\Gamma)\left[\sum_{j=1}^{n} p_j \left\langle P_j^{-p} u, u\right\rangle - \left(\sum_{j=1}^{n} p_j \left\langle P_j^{-1} u, u\right\rangle\right)^p\right]\right)$$

*for $u \in H$, $\|u\| = 1$.*

The proof follows by (9.4) and (9.9) by taking $u_j = \sqrt{p_j}u$, $u \in H$, $\|u\| = 1$ and observing that

$$\prod_{j=1}^{n} \tilde{\Delta}_{u_j}(P_j) = \prod_{j=1}^{n} \exp\left\langle \ln P_j \sqrt{p_j}u, \sqrt{p_j}u\right\rangle = \prod_{j=1}^{n} \exp\left[p_j \left\langle \ln P_j u, u\right\rangle\right]$$

$$= \prod_{j=1}^{n} \exp\left[\left\langle \ln P_j u, u\right\rangle\right]^{p_j} = \prod_{j=1}^{n} [\Delta_u(P_j)]^{p_j}.$$

If we take $p = 2$ in (9.14), then we get

$$1 \leq \exp\left(\frac{1}{2\Gamma^2}\left[\sum_{j=1}^{n} p_j \langle P_j^2 u, u \rangle - \left(\sum_{j=1}^{n} \langle P_j u, u \rangle\right)^2\right]\right) \qquad (9.16)$$

$$\leq \frac{\sum_{j=1}^{n} p_j \langle P_j u, u \rangle}{\prod_{j=1}^{n} [\Delta_u(P_j)]^{p_j}}$$

$$\leq \exp\left(\frac{1}{2\gamma^2}\left[\sum_{j=1}^{n} p_j \langle P_j^2 u, u \rangle - \left(\sum_{j=1}^{n} \langle P_j u, u \rangle\right)^2\right]\right)$$

and from (9.15),

$$1 \leq \exp\left(\frac{\gamma^2}{2}\left[\sum_{j=1}^{n} \langle P_j^{-2} u_j, u_j \rangle - \left(\sum_{j=1}^{n} \langle P_j^{-1} u_j, u_j \rangle\right)^2\right]\right) \qquad (9.17)$$

$$\leq \frac{\prod_{j=1}^{n} [\Delta_u(P_j)]^{p_j}}{\left(\sum_{j=1}^{n} p_j \langle P_j^{-1} u, u \rangle\right)^{-1}}$$

$$\leq \exp\left(\frac{\Gamma^2}{2}\left[\sum_{j=1}^{n} \langle P_j^{-2} u_j, u_j \rangle - \left(\sum_{j=1}^{n} \langle P_j^{-1} u_j, u_j \rangle\right)^2\right]\right),$$

for each $u \in H$, $\|u\| = 1$.

If we take $p = -1$ in (9.14), then we get

$$1 \leq \exp\left(\frac{\gamma}{2}\left[\sum_{j=1}^{n} p_j \langle P_j^{-1} u, u \rangle - \left(\sum_{j=1}^{n} p_j \langle P_j u, u \rangle\right)^{-1}\right]\right) \qquad (9.18)$$

$$\leq \frac{\sum_{j=1}^{n} p_j \langle P_j u, u \rangle}{\prod_{j=1}^{n} [\Delta_u(P_j)]^{p_j}}$$

$$\leq \exp\left(\frac{\Gamma}{2}\left[\sum_{j=1}^{n} p_j \langle P_j^{-1} u, u \rangle - \left(\sum_{j=1}^{n} p_j \langle P_j u, u \rangle\right)^{-1}\right]\right)$$

and from (9.15)

$$1 \leq \exp\left(\frac{1}{2\Gamma}\left[\sum_{j=1}^{n} p_j \langle P_j u, u\rangle - \left(\sum_{j=1}^{n} p_j \langle P_j^{-1} u, u\rangle\right)^{-1}\right]\right) \tag{9.19}$$

$$\leq \frac{\prod_{j=1}^{n} [\Delta_u(P_j)]^{p_j}}{\left(\sum_{j=1}^{n} p_j \langle P_j^{-1} u, u\rangle\right)^{-1}}$$

$$\leq \exp\left(\frac{1}{2\gamma}\left[\sum_{j=1}^{n} p_j \langle P_j u, u\rangle - \left(\sum_{j=1}^{n} p_j \langle P_j^{-1} u, u\rangle\right)^{-1}\right]\right),$$

for $u \in H$, $\|u\| = 1$.

The case of two operators is as follows. Assume that $0 < \gamma \leq P, Q \leq \Gamma$, and $t \in [0,1]$. Then

$$1 \leq \exp\left\{\gamma_p(\gamma, \Gamma) \tag{9.20}\right.$$
$$\left. \times \left[\langle[(1-t)P^p + tQ^p]u, u\rangle - (\langle[(1-t)P + tQ]u, u\rangle)^p\right]\right\}$$
$$\leq \frac{\langle[(1-t)P + tQ]u, u\rangle}{[\Delta_u(P)]^{(1-t)}[\Delta_u(Q)]^t}$$
$$\leq \exp\left\{\Gamma_p(\gamma, \Gamma)\right.$$
$$\left. \times \left[\langle[(1-t)P^p + tQ^p]u, u\rangle - (\langle[(1-t)P + tQ]u, u\rangle)^p\right]\right\}$$

and

$$1 \leq \exp\left\{\tilde{\gamma}_p(\gamma, \Gamma) \tag{9.21}\right.$$
$$\left. \times \left[\langle[(1-t)P^{-p} + tQ^{-p}]u, u\rangle - (\langle[(1-t)P^{-1} + tQ^{-1}]u, u\rangle)^p\right]\right\}$$
$$\leq \frac{[\Delta_u(P)]^{(1-t)}[\Delta_u(Q)]^t}{(\langle[(1-t)P^{-1} + tQ^{-1}]u, u\rangle)^{-1}}$$
$$\leq \exp\left\{\tilde{\Gamma}_p(\gamma, \Gamma)\right.$$
$$\left. \times \left[\langle[(1-t)P^{-p} + tQ^{-p}]u, u\rangle - (\langle[(1-t)P^{-1} + tQ^{-1}]u, u\rangle)^p\right]\right\}$$

If $Q = P$, then we get

$$1 \leq \exp\left\{\left(\gamma_p(\gamma, \Gamma)\left[\langle P^p u, u\rangle - \langle P u, u\rangle^p\right]\right)\right\} \tag{9.22}$$

$$\leq \frac{\langle P u, u\rangle}{\Delta_u(P)} \leq \exp\left\{\Gamma_p(\gamma, \Gamma)\left[\langle P^p u, u\rangle - \langle P u, u\rangle^p\right]\right\}$$

and

$$1 \le \exp\left\{ \tilde{\gamma}_p\left(\gamma,\Gamma\right) \left[ \langle P^{-p}u,u \rangle - \langle P^{-1}u,u \rangle^p \right] \right\} \tag{9.23}$$

$$\le \frac{\Delta_u(P)}{\left(\langle P^{-1}u,u \rangle\right)^{-1}} \le \exp\left\{ \tilde{\Gamma}_p\left(\gamma,\Gamma\right) \left[ \langle P^{-p}u,u \rangle - \langle P^{-1}u,u \rangle^p \right] \right\}$$

for $u \in H$, $\|u\| = 1$.

For $p = 2$ we have

$$1 \le \exp\left\{ \left( \frac{1}{2\Gamma^2} \left[ \langle P^2 u,u \rangle - \langle Pu,u \rangle^2 \right] \right) \right\} \tag{9.24}$$

$$\le \frac{\langle Pu,u \rangle}{\Delta_u(P)} \le \exp\left\{ \frac{1}{2\gamma^2} \left[ \langle P^2 u,u \rangle - \langle Pu,u \rangle^2 \right] \right\}$$

and

$$1 \le \exp\left\{ \frac{\gamma^2}{2} \left[ \langle P^{-2}u,u \rangle - \langle P^{-1}u,u \rangle^2 \right] \right\} \tag{9.25}$$

$$\le \frac{\Delta_u(P)}{\left(\langle P^{-1}u,u \rangle\right)^{-1}} \le \exp\left\{ \frac{\Gamma^2}{2} \left[ \langle P^{-2}u,u \rangle - \langle P^{-1}u,u \rangle^2 \right] \right\}$$

for $u \in H$, $\|u\| = 1$.

For $p = -1$ we derive

$$1 \le \exp\left\{ \left( \frac{\gamma}{2} \left[ \langle P^{-1}u,u \rangle - \langle Pu,u \rangle^{-1} \right] \right) \right\} \tag{9.26}$$

$$\le \frac{\langle Pu,u \rangle}{\Delta_u(P)} \le \exp\left\{ \frac{\Gamma}{2} \left[ \langle P^{-1}u,u \rangle - \langle Pu,u \rangle^{-1} \right] \right\}$$

and

$$1 \le \exp\left\{ \frac{1}{2\Gamma} \left[ \langle Pu,u \rangle - \langle P^{-1}u,u \rangle^{-1} \right] \right\} \tag{9.27}$$

$$\le \frac{\Delta_u(P)}{\left(\langle P^{-1}u,u \rangle\right)^{-1}} \le \exp\left\{ \frac{1}{2\gamma} \left[ \langle Pu,u \rangle - \langle P^{-1}u,u \rangle^{-1} \right] \right\}$$

for $u \in H$, $\|u\| = 1$.

We observe that the above inequalities (9.22)-(9.27) provide refinements and reverse of the fundamental bounds for the normalized determinant incorporated in (ii) from the introduction.

It is well known that, see for instance [2, p. 28],

$$\langle P^2 u,u \rangle - \langle Pu,u \rangle^2 \le \frac{1}{4}\left(\Gamma - \gamma\right)^2 \tag{9.28}$$

for $u \in H$, $\|u\| = 1$.

Then by (9.24) we get

$$\frac{\langle Pu,u\rangle}{\Delta_u(P)} \leq \exp\left\{\frac{1}{2\gamma^2}\left[\langle P^2u,u\rangle - \langle Pu,u\rangle^2\right]\right\} \leq \exp\left\{\frac{1}{8}\left(\frac{\Gamma}{\gamma}-1\right)^2\right\} \qquad (9.29)$$

for $u \in H$, $\|u\| = 1$.

Since $0 < \Gamma^{-1} \leq P^{-1} \leq \gamma^{-1}$,

$$\langle P^{-2}u,u\rangle - \langle P^{-1}u,u\rangle^2 \leq \frac{1}{4}\left(\frac{\Gamma-\gamma}{\gamma\Gamma}\right)^2$$

holds and by (9.25) we get

$$\frac{\Delta_u(P)}{\left(\langle P^{-1}u,u\rangle\right)^{-1}} \leq \exp\left\{\frac{\Gamma^2}{2}\left[\langle P^{-2}u,u\rangle - \langle P^{-1}u,u\rangle^2\right]\right\} \qquad (9.30)$$

$$\leq \exp\left\{\frac{1}{8}\left(\frac{\Gamma}{\gamma}-1\right)^2\right\}$$

for $u \in H$, $\|u\| = 1$.

We also use the well known inequality, see for instance [2, p. 28],

$$\langle P^{-1}u,u\rangle - \langle Pu,u\rangle^{-1} \leq \frac{\left(\sqrt{\Gamma}-\sqrt{\gamma}\right)^2}{\gamma\Gamma} \qquad (9.31)$$

for $u \in H$, $\|u\| = 1$.

Then by (9.26) we obtain

$$\frac{\langle Pu,u\rangle}{\Delta_u(P)} \leq \exp\left\{\frac{\Gamma}{2}\left[\langle P^{-1}u,u\rangle - \langle Pu,u\rangle^{-1}\right]\right\} \qquad (9.32)$$

$$\leq \exp\left\{\frac{1}{2}\left(\sqrt{\frac{\Gamma}{\gamma}}-1\right)^2\right\}$$

for $u \in H$, $\|u\| = 1$.

Also, by (9.27) we derive

$$\frac{\Delta_u(P)}{\left(\langle P^{-1}u,u\rangle\right)^{-1}} \leq \exp\left\{\frac{1}{2\gamma}\left[\langle Pu,u\rangle - \langle P^{-1}u,u\rangle^{-1}\right]\right\}$$

$$\leq \exp\left\{\frac{1}{2}\left(\sqrt{\frac{\Gamma}{\gamma}}-1\right)^2\right\}.$$

These inequalities provide simple upper bounds related to the fundamental inequalities incorporated in (ii).

## 9.3   Related Results

We also have the following reverse of Ky Fan's inequality (viii):

***Theorem 9.2***
*Assume that $P_j > 0$ and $p_j \geq 0$, $j \in \{1,...,n\}$ with $\sum_{j=1}^{n} p_j = 1$, then we have the following reverse of Ky Fan's inequality*

$$1 \leq \frac{\Delta_u \left( \sum_{j=1}^{n} p_j P_j \right)}{\prod_{j=1}^{n} [\Delta_u (P_j)]^{p_j}} \leq \frac{\sum_{j=1}^{n} p_j \langle P_j u, u \rangle}{\prod_{j=1}^{n} [\Delta_u (P_j)]^{p_j}} \tag{9.33}$$

$$\leq \exp \left( \Gamma_p (\gamma, \Gamma) \left[ \sum_{j=1}^{n} p_j \left\langle P_j^p u, u \right\rangle - \left( \sum_{j=1}^{n} p_j \langle P_j u, u \rangle \right)^p \right] \right)$$

*for $u \in H$, $\|u\| = 1$.*
  *In particular, we have*

$$1 \leq \frac{\Delta_u \left( \sum_{j=1}^{n} p_j P_j \right)}{\prod_{j=1}^{n} [\Delta_u (P_j)]^{p_j}} \leq \frac{\sum_{j=1}^{n} p_j \langle P_j u, u \rangle}{\prod_{j=1}^{n} [\Delta_u (P_j)]^{p_j}} \tag{9.34}$$

$$\leq \exp \left( \frac{1}{2\gamma^2} \left[ \sum_{j=1}^{n} p_j \left\langle P_j^2 u, u \right\rangle - \left( \sum_{j=1}^{n} p_j \langle P_j u, u \rangle \right)^2 \right] \right)$$

*and*

$$1 \leq \frac{\Delta_u \left( \sum_{j=1}^{n} p_j P_j \right)}{\prod_{j=1}^{n} [\Delta_u (P_j)]^{p_j}} \leq \frac{\sum_{j=1}^{n} p_j \langle P_j u, u \rangle}{\prod_{j=1}^{n} [\Delta_u (P_j)]^{p_j}} \tag{9.35}$$

$$\leq \exp \left( \frac{\Gamma}{2} \left[ \sum_{j=1}^{n} p_j \left\langle P_j^{-1} u, u \right\rangle - \left( \sum_{j=1}^{n} p_j \langle P_j u, u \rangle \right)^{-1} \right] \right)$$

*for $u \in H$, $\|u\| = 1$.*

**Proof 9.2**　Observe that, by Jensen's inequality for the concave function ln

$$\sum_{j=1}^{n} p_j \langle P_j u, u \rangle = \left\langle \left( \sum_{j=1}^{n} p_j P_j \right) u, u \right\rangle = \exp\left( \ln \left\langle \left( \sum_{j=1}^{n} p_j P_j \right) u, u \right\rangle \right)$$

$$\geq \exp\left\langle \ln \left( \sum_{j=1}^{n} p_j P_j \right) u, u \right\rangle = \Delta_u \left( \sum_{j=1}^{n} p_j P_j \right)$$

for $u \in H$, $\|u\| = 1$.

By the second inequality in (ii) from introduction and (9.14) we then get

$$\frac{\Delta_u \left( \sum_{j=1}^{n} p_j P_j \right)}{\prod_{j=1}^{n} [\Delta_u (P_j)]^{p_j}} \leq \frac{\sum_{j=1}^{n} p_j \langle P_j u, u \rangle}{\prod_{j=1}^{n} [\Delta_u (P_j)]^{p_j}}$$

$$\leq \exp\left( \Gamma_p (\gamma, \Gamma) \left[ \sum_{j=1}^{n} p_j \left\langle P_j^p u, u \right\rangle - \left( \sum_{j=1}^{n} p_j \langle P_j u, u \rangle \right)^p \right] \right)$$

for $u \in H$, $\|u\| = 1$.

If we use Ky Fan's type inequality (viii) and a standard induction argument we also have

$$1 \leq \frac{\Delta_u \left( \sum_{j=1}^{n} p_j P_j \right)}{\prod_{j=1}^{n} [\Delta_u (P_j)]^{p_j}}$$

for $u \in H$, $\|u\| = 1$.

These prove the desired result (9.33).

Assume that $0 < \gamma \leq P_j \leq \Gamma$, $j \in \{1,...,n\}$ and $u_j \in H$, $j \in \{1,...,n\}$ with $\sum_{j=1}^{n} \|u_j\|^2 = 1$. As in [2, p. 6], if we put

$$\widetilde{P} := \begin{pmatrix} P_1 & . & . & . & 0 \\ & . & & & \\ & & . & & \\ & & & . & \\ 0 & . & . & . & P_n \end{pmatrix} \quad \text{and} \quad \widetilde{u} = \begin{pmatrix} u_1 \\ . \\ . \\ . \\ u_n \end{pmatrix}$$

then we have $Sp\left( \widetilde{P} \right) \subseteq [\gamma, \Gamma]$, $\|\widetilde{u}\|_2 = 1$ and

$$\left\langle f\left( \widetilde{P} \right) \widetilde{u}, \widetilde{u} \right\rangle = \sum_{j=1}^{n} \langle f(P_j) u_j, u_j \rangle, \left\langle \widetilde{P} \widetilde{u}, \widetilde{u} \right\rangle = \sum_{j=1}^{n} \langle P_j u_j, u_j \rangle$$

for any continuous function $f$ on $[\gamma, \Gamma]$.

Therefore, by (9.28) and (9.31) we derive

$$\sum_{j=1}^{n} \langle P_j^2 u_j, u_j \rangle - \left( \sum_{j=1}^{n} \langle P_j u_j, u_j \rangle \right)^2 \leq \frac{1}{4} (\Gamma - \gamma)^2 \tag{9.36}$$

and

$$\sum_{j=1}^{n} \langle P_j u_j, u_j \rangle - \left( \sum_{j=1}^{n} \langle P_j^{-1} u_j, u_j \rangle \right)^{-1} \leq \frac{\left( \sqrt{\Gamma} - \sqrt{\gamma} \right)^2}{\gamma \Gamma} \tag{9.37}$$

provided that $P_j$ are operators such that $0 < \gamma \leq P_j \leq \Gamma$, $j \in \{1, ..., n\}$ and $u_j \in H$, $j \in \{1, ..., n\}$ with $\sum_{j=1}^{n} \|u_j\|^2 = 1$.

Using (9.36) and (9.37) we get

$$\sum_{j=1}^{n} p_j \langle P_j^2 u, u \rangle - \left( \sum_{j=1}^{n} p_j \langle P_j u, u \rangle \right)^2 \leq \frac{1}{4} (\Gamma - \gamma)^2 \tag{9.38}$$

and

$$\sum_{j=1}^{n} p_j \langle P_j u, u \rangle - \left( \sum_{j=1}^{n} p_j \langle P_j^{-1} u, u \rangle \right)^{-1} \leq \frac{\left( \sqrt{\Gamma} - \sqrt{\gamma} \right)^2}{\gamma \Gamma} \tag{9.39}$$

provided that $P_j$ are operators such that $0 < \gamma \leq P_j \leq \Gamma$, $j \in \{1, ..., n\}$, $p_j \geq 0$, $j \in \{1, ..., n\}$ with $\sum_{j=1}^{n} p_j = 1$ and $u \in H$, $\|u\| = 1$.

### Corollary 9.3
*With the assumptions of Theorem 9.2 we have the following reverses of Ky Fan's inequality*

$$1 \leq \frac{\Delta_u \left( \sum_{j=1}^{n} p_j P_j \right)}{\prod_{j=1}^{n} [\Delta_u (P_j)]^{p_j}} \leq \frac{\sum_{j=1}^{n} p_j \langle P_j u, u \rangle}{\prod_{j=1}^{n} [\Delta_u (P_j)]^{p_j}} \tag{9.40}$$

$$\leq \exp \left( \frac{1}{2\gamma^2} \left[ \sum_{j=1}^{n} p_j \langle P_j^2 u, u \rangle - \left( \sum_{j=1}^{n} p_j \langle P_j u, u \rangle \right)^2 \right] \right)$$

$$\leq \exp \left\{ \frac{1}{8} \left( \frac{\Gamma}{\gamma} - 1 \right)^2 \right\}$$

*and*

$$1 \leq \frac{\Delta_u \left( \sum_{j=1}^n p_j P_j \right)}{\prod\limits_{j=1}^n \left[ \Delta_u \left( P_j \right) \right]^{p_j}} \leq \frac{\sum_{j=1}^n p_j \left\langle P_j u, u \right\rangle}{\prod\limits_{j=1}^n \left[ \Delta_u \left( P_j \right) \right]^{p_j}} \tag{9.41}$$

$$\leq \exp \left( \frac{\Gamma}{2} \left[ \sum_{j=1}^n p_j \left\langle P_j^{-1} u, u \right\rangle - \left( \sum_{j=1}^n p_j \left\langle P_j u, u \right\rangle \right)^{-1} \right] \right)$$

$$\leq \exp \left\{ \frac{1}{2} \left( \sqrt{\frac{\Gamma}{\gamma}} - 1 \right)^2 \right\}$$

*for all* $u \in H$, $\|u\| = 1$.

## 9.4  Inequalities for $p \in (0,1)$

We also have:

**Theorem 9.3**

*Assume that $P_j$ are operators such that $0 < \gamma \leq P_j \leq \Gamma$, $j \in \{1,...,n\}$. Then for $p \in (0,1)$*

$$1 \leq \exp \left( \frac{1}{p(1-p)\Gamma^p} \left[ \left( \sum_{j=1}^n \left\langle P_j u_j, u_j \right\rangle \right)^p - \sum_{j=1}^n \left\langle P_j^p u_j, u_j \right\rangle \right] \right) \tag{9.42}$$

$$\leq \frac{\sum_{j=1}^n \left\langle P_j u_j, u_j \right\rangle}{\prod\limits_{j=1}^n \tilde{\Delta}_{u_j} \left( P_j \right)}$$

$$\leq \left( \frac{1}{p(1-p)\gamma^p} \left[ \left( \sum_{j=1}^n \left\langle P_j u_j, u_j \right\rangle \right)^p - \sum_{j=1}^n \left\langle P_j^p u_j, u_j \right\rangle \right] \right)$$

*for each $u_j \in H$, $j \in \{1,...,n\}$ with $\sum_{j=1}^n \|u_j\|^2 = 1$.*

*In particular,*

$$1 \le \exp\left(\frac{4}{\Gamma^{1/2}}\left[\left(\sum_{j=1}^{n}\langle P_j u_j, u_j\rangle\right)^{1/2} - \sum_{j=1}^{n}\left\langle P_j^{1/2} u_j, u_j\right\rangle\right]\right) \tag{9.43}$$

$$\le \frac{\sum_{j=1}^{n}\langle P_j u_j, u_j\rangle}{\prod_{j=1}^{n}\tilde{\Delta}_{u_j}(P_j)}$$

$$\le \exp\left(\frac{4}{\gamma^{1/2}}\left[\left(\sum_{j=1}^{n}\langle P_j u_j, u_j\rangle\right)^{1/2} - \sum_{j=1}^{n}\left\langle P_j^{1/2} u_j, u_j\right\rangle\right]\right)$$

*for each $u_j \in H$, $j \in \{1,...,n\}$ with $\sum_{j=1}^{n}\|u_j\|^2 = 1$.*

**Proof 9.3**   If the following condition is satisfied

$$\delta \le h(t) := \frac{t^{2-p}}{p(1-p)}f''(t) \le \Delta \text{ for any } t \in (\gamma, \Gamma) \tag{9.44}$$

and for some $\delta < \Delta$, where $p \in (0,1)$, then for $p \in (0,1)$, we also have [1]

$$\delta\left[\left(\sum_{j=1}^{n}\langle P_j u_j, u_j\rangle\right)^{p} - \sum_{j=1}^{n}\left\langle P_j^p u_j, u_j\right\rangle\right] \tag{9.45}$$

$$\le \sum_{j=1}^{n}\langle f(P_j)u_j, u_j\rangle - f\left(\sum_{j=1}^{n}\langle P_j u_j, u_j\rangle\right)$$

$$\le \Delta\left[\left(\sum_{j=1}^{n}\langle P_j u_j, u_j\rangle\right)^{p} - \sum_{j=1}^{n}\left\langle P_j^p u_j, u_j\right\rangle\right]$$

for each $u_j \in H$, $j \in \{1,...,n\}$ with $\sum_{j=1}^{n}\|u_j\|^2 = 1$.

If we take $f(t) = -\ln t$, then

$$\begin{aligned}
h(t) &= \frac{t^{2-p}}{p(1-p)}\frac{1}{t^2} \\
&= \frac{1}{p(1-p)t^p} \in \left[\frac{1}{p(1-p)\Gamma^p}, \frac{1}{p(1-p)\gamma^p}\right]
\end{aligned}$$

and by (9.45) we get

$$0 \le \frac{1}{p(1-p)\Gamma^p} \left[ \left( \sum_{j=1}^{n} \langle P_j u_j, u_j \rangle \right)^p - \sum_{j=1}^{n} \left\langle P_j^p u_j, u_j \right\rangle \right]$$

$$\le \ln \left( \sum_{j=1}^{n} \langle P_j u_j, u_j \rangle \right) - \sum_{j=1}^{n} \langle \ln P_j u_j, u_j \rangle$$

$$\le \frac{1}{p(1-p)\gamma^p} \left[ \left( \sum_{j=1}^{n} \langle P_j u_j, u_j \rangle \right)^p - \sum_{j=1}^{n} \left\langle P_j^p u_j, u_j \right\rangle \right]$$

for each $u_j \in H$, $j \in \{1,...,n\}$ with $\sum_{j=1}^{n} \|u_j\|^2 = 1$, which implies, by taking the exponential, the desired result (9.42).

### Corollary 9.4

*Assume that $P_j$ are operators such that $0 < \gamma \le P_j \le \Gamma$, $j \in \{1,...,n\}$ and $p_j \ge 0$, $j \in \{1,...,n\}$ with $\sum_{j=1}^{n} p_j = 1$. Then for $p \in (0,1)$*

$$1 \le \exp\left( \frac{1}{p(1-p)\Gamma^p} \left[ \left( \sum_{j=1}^{n} p_j \langle P_j u, u \rangle \right)^p - \sum_{j=1}^{n} p_j \left\langle P_j^p u, u \right\rangle \right] \right) \qquad (9.46)$$

$$\le \frac{\sum_{j=1}^{n} p_j \langle P_j u, u \rangle}{\prod_{j=1}^{n} [\Delta_u (P_j)]^{p_j}}$$

$$\le \exp\left( \frac{1}{p(1-p)\gamma^p} \left[ \left( \sum_{j=1}^{n} p_j \langle P_j u, u \rangle \right)^p - \sum_{j=1}^{n} p_j \left\langle P_j^p u, u \right\rangle \right] \right)$$

*for each $u \in H$ with $\|u\| = 1$.*

*In particular,*

$$1 \le \exp\left( \frac{4}{\Gamma^{1/2}} \left[ \left( \sum_{j=1}^{n} p_j \langle P_j u, u \rangle \right)^{1/2} - \sum_{j=1}^{n} p_j \left\langle P_j^{1/2} u, u \right\rangle \right] \right) \qquad (9.47)$$

$$\le \frac{\sum_{j=1}^{n} p_j \langle P_j u, u \rangle}{\prod_{j=1}^{n} [\Delta_u (P_j)]^{p_j}}$$

$$\le \exp\left( \frac{4}{\gamma^{1/2}} \left[ \left( \sum_{j=1}^{n} p_j \langle P_j u, u \rangle \right)^{1/2} - \sum_{j=1}^{n} p_j \left\langle P_j^{1/2} u, u \right\rangle \right] \right)$$

*for each $u \in H$ with $\|u\| = 1$.*

**Remark 9.2** If we write the above inequalities for $P_j^{-1}$, then, under the same assumptions,

$$1 \le \exp\left(\frac{\gamma^p}{p(1-p)}\left[\left(\sum_{j=1}^{n}\left\langle P_j^{-1}u_j,u_j\right\rangle\right)^p - \sum_{j=1}^{n}\left\langle P_j^{-p}u_j,u_j\right\rangle\right]\right) \tag{9.48}$$

$$\le \frac{\prod_{j=1}^{n}\tilde{\Delta}_{u_j}(P_j)}{\left(\sum_{j=1}^{n}\left\langle P_j^{-1}u_j,u_j\right\rangle\right)^{-1}}$$

$$\le \exp\left(\frac{\Gamma^p}{p(1-p)}\left[\left(\sum_{j=1}^{n}\left\langle P_j^{-1}u_j,u_j\right\rangle\right)^p - \sum_{j=1}^{n}\left\langle P_j^{-p}u_j,u_j\right\rangle\right]\right)$$

for each $u_j \in H$, $j \in \{1,...,n\}$ with $\sum_{j=1}^{n}\|u_j\|^2 = 1$.

In particular,

$$1 \le \exp\left(4\gamma^{1/2}\left[\left(\sum_{j=1}^{n}\left\langle P_j^{-1}u_j,u_j\right\rangle\right)^{1/2} - \sum_{j=1}^{n}\left\langle P_j^{-1/2}u_j,u_j\right\rangle\right]\right) \tag{9.49}$$

$$\le \frac{\prod_{j=1}^{n}\tilde{\Delta}_{u_j}(P_j)}{\left(\sum_{j=1}^{n}\left\langle P_j^{-1}u_j,u_j\right\rangle\right)^{-1}}$$

$$\le \exp\left(4\Gamma^{1/2}\left[\left(\sum_{j=1}^{n}\left\langle P_ju_j,u_j\right\rangle\right)^{1/2} - \sum_{j=1}^{n}\left\langle P_j^{1/2}u_j,u_j\right\rangle\right]\right)$$

for each $u_j \in H$, $j \in \{1,...,n\}$ with $\sum_{j=1}^{n}\|u_j\|^2 = 1$.

Also, if $p_j \ge 0$, $j \in \{1,...,n\}$ with $\sum_{j=1}^{n}p_j = 1$

$$1 \le \exp\left(\frac{\gamma^p}{p(1-p)}\left[\left(\sum_{j=1}^{n}p_j\left\langle P_j^{-1}u,u\right\rangle\right)^p - \sum_{j=1}^{n}p_j\left\langle P_j^{-p}u,u\right\rangle\right]\right) \tag{9.50}$$

$$\le \frac{\prod_{j=1}^{n}[\Delta_u(P_j)]^{p_j}}{\left(\sum_{j=1}^{n}p_j\left\langle P_j^{-1}u,u\right\rangle\right)^{-1}}$$

$$\le \exp\left(\frac{\Gamma^p}{p(1-p)}\left[\left(\sum_{j=1}^{n}p_j\left\langle P_j^{-1}u,u\right\rangle\right)^p - \sum_{j=1}^{n}p_j\left\langle P_j^{-p}u,u\right\rangle\right]\right)$$

for each $u \in H$ with $\|u\| = 1$.

In particular,

$$1 \leq \exp\left(4\gamma^{1/2}\left[\left(\sum_{j=1}^{n} p_j \left\langle P_j^{-1}u,u\right\rangle\right)^{1/2} - \sum_{j=1}^{n} p_j \left\langle P_j^{-1/2}u,u\right\rangle\right]\right) \tag{9.51}$$

$$\leq \frac{\prod_{j=1}^{n}[\Delta_u(P_j)]^{p_j}}{\left(\sum_{j=1}^{n} p_j \left\langle P_j^{-1}u,u\right\rangle\right)^{-1}}$$

$$\leq \exp\left(4\Gamma^{1/2}\left[\left(\sum_{j=1}^{n} p_j \left\langle P_j^{-1}u,u\right\rangle\right)^{1/2} - \sum_{j=1}^{n} p_j \left\langle P_j^{-1/2}u,u\right\rangle\right]\right)$$

for each $u \in H$ with $\|u\| = 1$.   ∎

Similar particular inequalities may be stated, however we state only the case of one operators, namely, for the operator $P$ satisfying the condition $0 < \gamma \leq P \leq \Gamma$,

$$1 \leq \exp\left(\frac{1}{p(1-p)\Gamma^p}\left[\langle Pu,u\rangle^p - \langle P^p u,u\rangle\right]\right) \tag{9.52}$$

$$\leq \frac{\langle Pu,u\rangle}{\Delta_u(P)} \leq \exp\left(\frac{1}{p(1-p)\gamma^p}\left[\langle Pu,u\rangle^p - \langle P^p u,u\rangle\right]\right)$$

and

$$1 \leq \exp\left(\frac{\gamma^p}{p(1-p)}\left[\langle P^{-1}u,u\rangle^p - \langle P^{-p}u,u\rangle\right]\right) \tag{9.53}$$

$$\leq \frac{\Delta_u(P)}{\langle P^{-1}u,u\rangle^{-1}} \leq \exp\left(\frac{\Gamma^p}{p(1-p)}\left[\langle P^{-1}u,u\rangle^p - \langle P^{-p}u,u\rangle\right]\right)$$

for each $u \in H$ with $\|u\| = 1$, where $p \in (0,1)$.

For $p = 1/2$ we get

$$1 \leq \exp\left(\frac{4}{\Gamma^{1/2}}\left[\langle Pu,u\rangle^{1/2} - \left\langle P^{1/2}u,u\right\rangle\right]\right) \tag{9.54}$$

$$\leq \frac{\langle Pu,u\rangle}{\Delta_u(P)} \leq \exp\left(\frac{4}{\gamma^{1/2}}\left[\langle Pu,u\rangle^{1/2} - \left\langle P^{1/2}u,u\right\rangle\right]\right)$$

and

$$1 \leq \exp\left(4\gamma^{1/2}\left[\left\langle P^{-1}u,u\right\rangle^{1/2} - \left\langle P^{-1/2}u,u\right\rangle\right]\right) \tag{9.55}$$

$$\leq \frac{\Delta_u(P)}{\left\langle P^{-1}u,u\right\rangle^{-1}} \leq \exp\left(4\Gamma^{1/2}\left[\left\langle P^{-1}u,u\right\rangle^{1/2} - \left\langle P^{-1/2}u,u\right\rangle\right]\right)$$

for each $u \in H$ with $\|u\| = 1$.

## Acknowledgement

The author would like the anonymous referee for valuable comments that have been implemented in the final version of the chapter.

# References


[1] Dragomir, S.S. 2009. Some Jensen's type inequalities for twice differentiable functions of selfadjoint operators in Hilbert spaces. Filomat 23(3): 211–222; 2008 Preprint RGMIA Res. Rep. Coll., 11(e): Art. 13.

[2] Furuta, T., Mićić Hot, J., Pečarić J. and Seo, Y. 2005. Mond-Pečarić method in operator inequalities. Inequalities for Bounded Selfadjoint Operators on a Hilbert Space. Element, Zagreb.

[3] Fujii, J.I. and Seo, Y. 1998. Determinant for positive operators. Sci. Math. 1: 153–156.

[4] Fujii, J.I., Izumino, S. and Seo, Y. 1998. Determinant for positive operators and Specht's theorem. Sci. Math. 1: 307–310.

[5] Hiramatsu, S. and Seo, Y. 2021. Determinant for positive operators and Oppenheim's inequality. J. Math. Inequal. 15(4): 1637–1645.

[6] Specht, W. 1960. Zer Theorie der elementaren Mittel. Math. Z. 74: 91–98.

# Chapter 10

# Fuzzy Inference Based Approach of Ant Colony Optimization (ACO) in Fuzzy Transportation Models

*M.K. Sharma, Tarun Kumar*[1],* *Laxmi Rathour*[2],* and *Vishnu Narayan Mishra*[3]

## 10.1 Introduction

The field of transportation is inherently complex and involves an array of variables that are often uncertain or imprecise. Traditional deterministic models, while useful, can fall short when dealing with the nuanced, often imprecise nature of real-world data and circumstances. This limitation has led to the exploration

[1] Department of Mathematics, Ch. Charan Singh University, Meerut, India.

[2] Department of Mathematics, National Institute of Technology, Chaltlang, Aizawl 796 012, Mizoram.

[3] Department of Mathematics, Indira Gandhi National Tribal University, Lalpur, Amarkantak, Anuppur, Madhya Pradesh 484 887, India.

Emails: drmukeshsharma@gmail.com; vishnunarayanmishra@gmail.com

* Corresponding authors: tkvats3@gmail.com; laxmirathour817@gmail.com

of more adaptable, flexible methodologies capable of handling uncertainty and complexity. In this context, fuzzy set theory, pioneered by Zadeh in 1965 [1], and Ant Colony Optimization (ACO), have emerged as influential frameworks that offer new perspectives and innovative solutions in addressing these challenges. This chapter reviews significant research in these domains, highlighting how these methodologies have been successfully utilized and further evolved to tackle transportation problems. Notably, several studies have revealed the synergistic potential when these two methods are combined, which forms the crux of our examination. Zimmermann 2011 [2] expounded on these ideas, illustrating the practical applications of fuzzy set theory. The applicability of fuzzy set theory in transportation problem was first demonstrated by Chanas et al. 1984 [3], who introduced uncertainty into the problem's parameters. The approach was enhanced by subsequent researchers such as Liu and Kao 2004 [4], Gani and Razak 2006 [5], Pandian and Natarajan 2011 [6], Kaur and Kumar 2011 [7], who used extension principles and ranking functions to solve fuzzy transportation problems. More recently, Sujatha et al. 2018 [8] and George et al. 2020 [9] have proposed new methods for solving fuzzy transportation problems, further pushing the boundaries of fuzzy logic in this domain. On another note, Ant Colony Optimization (ACO) has seen a surge in its application in transportation problems [10]. Pioneered by Kazharov and Kureichik 2010 [11], ACO has since been adopted by many, including Kumar et al. 2018 [12] and Ntakolia and Lyridis 2022 [13], to solve complex transport and traffic management problems. The algorithm has also been used in conjunction with fuzzy logic, such as in the work of Di Caprio et al. 2022 [14] and BabaeeTirkolaee et al. 2020 [15], to address issues with uncertainty. Furthermore, advancements in quantum computing inspired Das et al. 2023 [16] to propose a Quantum-inspired ACO for solving a complex traveling salesman problem. Meanwhile, Song et al. 2020 [17] combined fuzzy logic and ACO for dynamic path planning, illustrating the synergistic potential of these methods. ACO has also been applied in optimizing supply chain distribution (Panicker et al. 2013, [18]) and vehicle routing (Li et al. 2019 [19]). These works show the versatility of ACO, which can be enhanced when coupled with fuzzy logic, as shown by Lisangan and Sumarta 2017 [20], Ntakolia [21]. On the topic of traffic control, the integration of fuzzy logic in traffic control systems has gained momentum. Liu et al. 2023 [22], Gamel et al. 2022 [23], Shkera and Patankar 2023 [24], Vechione and Cheu 2023 [25] have shown how fuzzy inference systems can improve the efficiency and robustness of traffic management systems. Additionally, Erdinç et al. 2023 [26] proposed a two-stage fuzzy traffic congestion detector, while Jutury et al. 2023 [27] integrated neuro-fuzzy systems in traffic light control, signifying the continuous evolution and improvement of fuzzy systems in traffic management. As the complexity of contemporary transportation networks escalates, there is a growing demand for innovative, adaptable, and robust methodologies capable of addressing numerous dynamic factors. This research embarked on a mission to integrate a fuzzy inference system and the Ant Colony Optimization (ACO) technique into traditional fuzzy transportation problems—a hybrid approach not previously examined in existing literature, to our knowledge. The methodology we propose offers a renewed outlook on fuzzy transportation problems by encompassing an extensive range of factors. These include distance, fuel price, toll charges, maintenance costs, traffic and weather conditions, and road quality. The holistic integration of these elements presents a more realistic portrayal of the practical challenges inherent in transportation management. Our methodology capitalizes on the power of fuzzy inference systems to effectively manage the intrinsic uncertainties and ambiguities linked with these variables. Concurrently, the ACO technique facilitates the proficient exploration and utilization of potential solutions, thereby boosting the quality of the solution and enhancing the overall performance of the methodology. This research chapter is organized into six distinct sections, including the introduction. The second section provides definitions and a review of existing literature pertinent to our study. The third section delineates the formulation of our techniques. In the fourth section, we apply these techniques to numerical computations and scrutinize the optimality of the problem. The fifth section of the chapter presents and discusses the results obtained from our proposed methodology. These results are derived from numerical computations within our mathematical model. Lastly, the sixth section culminates with the conclusion and a thorough interpretation of our findings.

## 10.2  Preliminaries

### 10.2.1  *Fuzzy Set [2]*

Make $X$ a set that is not empty. A={ (x, $\mu_A$ (x)): x belongs to X } defines a fuzzy set of . Where $\mu_A$ is a function of membership from X to [0, 1].

### 10.2.2  *Fuzzy Number [2]*

Let A be a fuzzy set of X, and let X be a crisp set. If A is normal, fuzzy convex, upper semi-continuous, and support is compact, then A is considered to be a fuzzy number.

### 10.2.3  *Trapezoidal Fuzzy Number (Tr FN) [2]*

If the membership function of a fuzzy number A is provided as follows, then a fuzzy number A is said to be a trapezoidal fuzzy number (TrFN). It is symbolized by $(v_1, v_2, v_3, v_4)$ where $v_1, v_2, v_3$ and $v_4$ are real numbers. The trapezoidal fuzzy

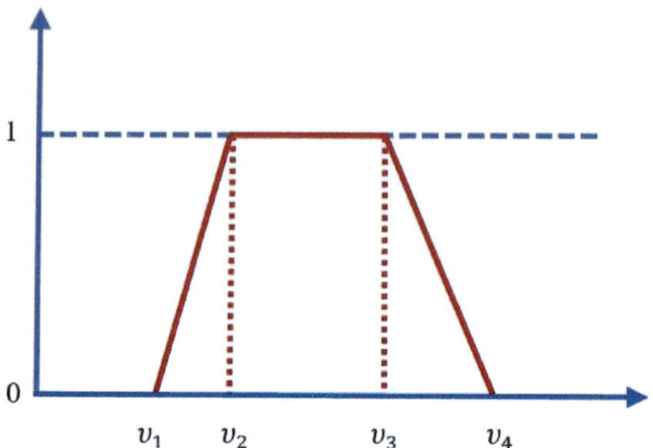


**Figure 10.1:** Trapezoidal Fuzzy number.

number (TrFN) is depicted in Fig. 10.1.

$$\mu_A(x) = \begin{cases} \frac{x-v_1}{v_2-v_1} & v_1 \le v_2 \\ 1 & v_2 \le v_3 \\ \frac{v_4-x}{v_4-v_3} & v_3 \le v_4 \\ 0 & else \end{cases}$$

## 10.2.4   Operations on Trapezoidal Fuzzy Numbers

Let A=$(u_1, u_2, u_3, u_4)$ and B=$(v_1, v_2, v_3, v_4)$ are two trapezoidal fuzzy numbers then

(i) Addition: A$\oplus$B=$(u_1+v_1, u_2+v_2, u_3+v_3, u_4+v_4)$

(ii) Subtraction: A$\ominus$B=$(u_1-v_4, u_2-v_3, u_3-v_2, u_4-v_1)$

(iii) Scalar Multiplication: KA=$(Ku_1, Ku_2, Ku_3, Ku_4)$ if K is positive and KA=$(Ku_4, Ku_3, Ku_2, Ku_1)$ if K is negative.

## 10.2.5   Mathematical Model for Fuzzy Transportation Problem (FTP)

The decision-maker in fuzzy transportation situations is unsure of the precise value of transportation cost, demand, and availability. As a result, the issue might be stated as follows:

Minimize Z=$\sum_{n=1}^{w} \sum_{o=1}^{x} C_{no} \otimes X_{no}$

subject to $\sum_{o=1}^{x} X_{no} \approx a_n$, $\sum_{n=1}^{w} C_{no} \approx b_0$

$X_{no} \ge 0$

Here, Z denotes the whole cost of fuzzy transportation. $a_n$ represents the full amount of fuzzy availability of the commodity at nth source. $b_0$ represents the full amount of fuzzy demand of the commodity at oth source. $C_{no}$ represents the fuzzy transportation cost to each unit from the nth source to the oth destinations. $X_{no}$ represents the quantity of fairly accurate units of the commodity that should be transported from the nth source to oth destinations.

In balanced fuzzy transportation problem, $\sum_{o=1}^{x} a_n = \sum_{n=1}^{w} b_0$ should exists whereas in the case of unbalanced fuzzy transportation problem, $\sum_{o=1}^{x} a_n \neq \sum_{n=1}^{w} b_0$.

Table 10.1  Representation of FTP

|  | 1 | 2 | ... | w | supply |
|---|---|---|---|---|---|
|  | $c_{11}$ | $c_{12}$ | ... | $c_{1w}$ | $a_1$ |
|  | $c_{21}$ | $c_{12}$ | ... | $c_{1w}$ | $a_1$ |
|  | ... | ... | ... | ... | ... |
| x | $c_{x1}$ | $c_{x2}$ | ... | $c_{xw}$ | $a_x$ |
| Demand | $b_1$ | $b_2$ | ... | $b_w$ |  |

### 10.2.6  *Mathematical Model for Ant Colony Optimization (ACO)*

The ant k chases a random step from the vertex i to the next or nearby possible vertices (j,l,q) for set up advantageous solution cumulatively. The next vertex to travel is probabilistically chosen according to the transition probabilities of other vertices such as $P_K$ (i, j), $P_K$ (i, l) and $P_K$ (i, k) with respect to the present vertex i of ant k. This transition probability of ant k to choose the next vertex j from its original vertex i is determined by using the random proportional state transition law, mathematically expressed as:

$$P_k(i,j) = \begin{cases} \dfrac{[\tau(i,j)]^{\alpha}.[\tau(i,j)]^{\alpha}}{\sum_{j\in allowed}[\tau(i,j)]^{\alpha}.[\tau(i,j)]^{\alpha}} & if\, j \in allowed \\ 0 & else \end{cases}$$

Where, $\tau(i, j)$ represents the quantity of pheromone trail on the edge that links the vertex i and i . Further, $\eta(i, j)$ is a heuristic value also called the desirability or visibility of the ant to the building solution on the edge connecting vertices i and j. It is set as the inverse of connection cost or distance between these vertices, i.e., $\eta(i, j) = \frac{1}{d(i,j)}$ It is generally suggested for promoting the cost-effective vertex of the frame which has a large quantity of pheromone concentration. $\alpha \in (0, 1]$ and $\beta \in (0, 1]$ are known as regulating parameters that help to control the relative significance of pheromone versus heuristic values. When all the ants of the ant system have completed their journey, the pheromone concentration is updated on the edges by the pheromone global updating rule, defined as

$$\tau(i, j) \leftarrow (1-\rho).\tau(i, j) + \sum_{k=1}^{n_a} \Delta\tau_k(i, j)$$

Where, $\rho \in (0, 1]$ represents the fractional amount of evaporated pheromone between two steps or iterations of the ACO algorithm. The value of $\rho$ is set to be very small to avoid unlimited pheromone deposition on the path. $\Delta\tau_k(i, j)$

represents the amount of pheromone deposited by ant k on path (i, j), measured in per unit length and expressed as

$$\Delta\tau_k(i,j) = \begin{cases} \frac{Q}{L_k} & if\,the\,ant\,k\,moves\,through\,path(i,j) \\ 0 & else \end{cases}$$

Here, Q is a constant and $L_k$ represents the total travel length of ant k. The value of constant Q is generally set to zero.

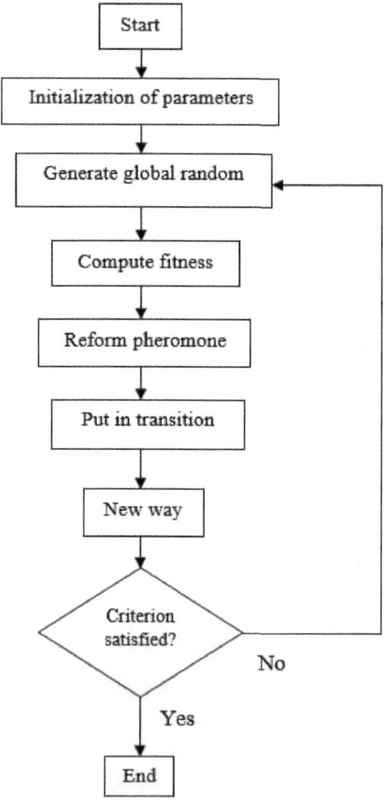


**Figure 10.2:** Flowchart of Ant Colony Optimization.

## 10.3   Our Proposed Plan

The purpose of this study is to introduce fuzzy inference system and ant colony optimization technique in classical fuzzy transportation problems. Now consider the fuzzy transportation problem as defined in Section 10.2. In this problem, according to our knowledge, no such methodology has been found in the literature, in which many factors such as distance, fuel price, toll charges, maintenance cost, traffic conditions, weather conditions and road conditions, etc., coming between the source and the destination are taken into account before giving the allocation, have been considered. Therefore, we have prepared a methodology for classical fuzzy transportation problem with consideration all of these factors with the help of fuzzy inference system and ant colony optimization technique as follows:

Step 1: Consider the fuzzy transportation problem as defined in Section 10.2. Now make a fuzzy inference system (FIS) by taking the factors coming between source and destinations. The FIS constructed with rules that convert these inputs into a fuzzy output representing the cost desirability of a route. The fuzzy output from the FIS (which now represents cost desirability) determine the transition probability levels for the ACO. Paths with lower costs should have higher pheromone concentrations, which will attract more ants. In numerical problem we have taken all these factors as shown in Fig. 10.3.

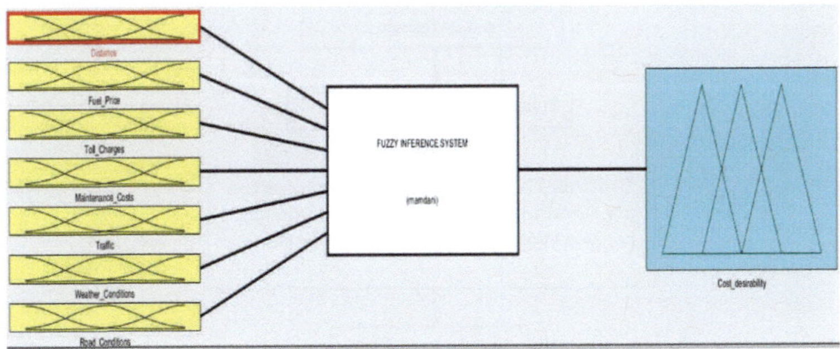


**Figure 10.3:** Fuzzy inference system.

Step 2: From step 1, we will get the transition probability matrix as follows:

Table 10.2 Transition probability values

|         | $D_1$    | $D_2$    | ...  | $D_n$    |
|---------|----------|----------|------|----------|
| $S_1$   | $P_{11}$ | $P_{12}$ | ...  | $P_{1n}$ |
| $S_2$   | $P_{21}$ | $P_{12}$ | ...  | $P_{2n}$ |
| ...     | ...      | ...      | ...  | ...      |
| $S_m$   | $P_{m1}$ | $P_{m2}$ | ...  | $P_{mn}$ |

Now, divide the FTP in pointwise crisp sub-problem and give allocations to the source and destinations which have maximum transition probabilityvalue. Calculate the cost for every crisp sub-problem.

Step 3: The next iterations can be computed in one of two ways: by modifying the rules and updating the transition probabilityor, in the case of equal values in the transition probability table, by adjusting the allocations. Pheromone updating should also reflect the new objective. When ants find a feasible solution (i.e., a path that satisfies transportation demands), they should deposit more pheromone on paths that are less costly. The evaporation rate could also be adjusted based on the cost: paths with higher costs could have a higher evaporation rate, reducing their attractiveness more quickly.

Step 4: This process is repeated until the algorithm converges on an optimal or near-optimal solution.

Flowchart of proposed methodology is depicted in Fig. 10.4.

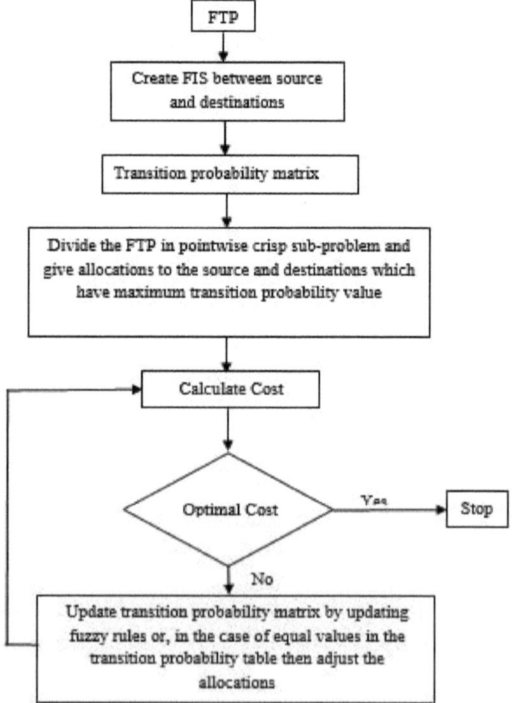


**Figure 10.4:** Flowchart of methodology.

## 10.4   Numerical Computations

Building upon the previous discussions, our next focus will be on the practical application of a fuzzy transportation problem. This problem involves components such as transportation cost, demand, and product availability, all represented using trapezoidal fuzzy numbers. For this purpose, we take into consideration a transportation problem originally presented by Sujatha et al. 2018, as illustrated in Table 10.3. However, an alteration has been introduced to this problem: the value at the (2, 2) position has been changed from (0,0,1,1) to (0,0.5,1,1.5).

Table 10.3  FTP

| | $D_1$ | $D_2$ | $D_3$ | $D_4$ | supply (S) |
|---|---|---|---|---|---|
| $S_1$ | (1, 2,3, 4) | (1,3,4,6) | (9,11,12,14) | (5,7,8,11) | (1,6,7,12) |
| $S_2$ | (0,1, 2, 4) | (0,0.5,1. 1.5) | (5,6,7,8) | (0,1,2,3) | (0,1,2,3) |
| $S_3$ | (3,5,6,8) | (5,8,9,12) | (12,15,16,19) | (7,9,10,12) | (5,10,12,17) |
| Demand (D) | (4,7,8,11) | (0,5,6,11) | (1,3,4,6) | (1,2,3,4) | $\sum S = \sum D = (6,17,21,32)$ |

Table 10.4 Input Factors

| | Linguistic Variables | | |
|---|---|---|---|
| Distance | Short (5-12) | Medium (10-25) | Long (Above 25) |
| Fuel Price | Low (0.20-0.50) | Medium (0.45-0.80) | High (Above 0.80) |
| Toll Charges | Low (0.10-0.30) | Medium (0.25-0.40) | High (Above 0.40) |
| Maintenance Cost | Low (0.40-0.60) | Medium (0.50-0.80) | High (Above 0.80) |
| Traffic Conditions | Low (0-0.4) | Moderate (0.4-0.6) | High (0.6-1) |
| Weather Conditions | Good (0-0.4) | Moderate (0.35-0.65) | Bad (above 0.6) |
| Road Conditions | Good (0-0.4) | Moderate (0.4-0.6) | Bad (0.6-1) |

Now, create fuzzy rules for the FIS. These rules would define how the inputs relate to the output. Fuzzy rules shown in Table 10.5.

### Table 10.5 Fuzzy Rules

| | | Input Variables | | |
|---|---|---|---|---|
| | Distance | Fuel Price | Toll Charges | Maintenance Cost |
| 1 $S_1 \leftrightarrow D_1$ | Short | Low | Medium | Low |
| 2 $S_1 \leftrightarrow D_2$ | Short | Low | Low | Medium |
| 3 $S_1 \leftrightarrow D_3$ | Medium | Low | Medium | Medium |
| 4 $S_1 \leftrightarrow D_4$ | Long | Medium | Medium | Medium |
| 5 $S_2 \leftrightarrow D_1$ | Short | Low | Low | Low |
| 6 $S_2 \leftrightarrow D_2$ | Short | Low | Low | Low |
| 7 $S_2 \leftrightarrow D_3$ | Medi-um | Medium | Low | High |
| 8 $S_2 \leftrightarrow D_4$ | Short | Low | Low | Medium |
| 9 $S_3 \leftrightarrow D_1$ | Medium | Medium | Low | Medium |
| 10 $S_3 \leftrightarrow D_2$ | Long | Medium | Medium | Low |
| 11 $S_3 \leftrightarrow D_3$ | Long | Medium | High | High |
| 12 $S_3 \leftrightarrow D_4$ | Long | High | Medium | Medium |

| Fuzzy Rules | Input Variables | | | Output |
|---|---|---|---|---|
| | Traffic Conditions | Weather Conditions | Road Conditions | Cost Desirability |
| 1 $S_1 \leftrightarrow D_1$ | Moderate | Good | Moderate | Medium |
| 2 $S_1 \leftrightarrow D_2$ | Low | Good | Good | High |
| 3 $S_1 \leftrightarrow D_3$ | Moderate | Good | Good | Medium |
| 4 $S_1 \leftrightarrow D_4$ | Low | Good | Moderate | Medium |
| 5 $S_2 \leftrightarrow D_1$ | Low | Good | Good | High |
| 6 $S_2 \leftrightarrow D_2$ | Moderate | Moderate | Good | High |
| 7 $S_2 \leftrightarrow D_3$ | Low | Good | Bad | Low |
| 8 $S_2 \leftrightarrow D_4$ | Low | Moderate | Good | High |
| 9 $S_3 \leftrightarrow D_1$ | Moderate | Good | Moderate | Medium |
| 10 $S_3 \leftrightarrow D_2$ | Moderate | Moderate | Moderate | Medium |
| 11 $S_3 \leftrightarrow D_3$ | Low | Moderate | Good | Low |
| 12 $S_3 \leftrightarrow D_4$ | Moderate | Good | Good | Low |

By implementing FIS we got the following transition probability value (cost desirability) table:

### Table 10.6 Cost Desirability Values

| | $D_1$ | $D_2$ | $D_3$ | $D_4$ |
|---|---|---|---|---|
| $S_1$ | 0.5 | 0.5 | 0.5 | 0.5 |
| $S_2$ | 0.818 | 0.838 | 0.188 | 0.817 |
| $S_3$ | 0.5 | 0.5 | 0.166 | 0.5 |

Now give the allocations based on above transition probability values in following subproblem-1

### Table 10.7 Allocations for Subproblem-1

| | $D_1$ | $D_2$ | $D_3$ | $D_4$ | Supply |
|---|---|---|---|---|---|
| $S_1$ | 1(1) | 1 | 9 | 5 | 1 |
| $S_2$ | 0 | 0 | 5 | 0 | 0 |
| $S_3$ | 3(3) | 5 | 12(1) | 7(1) | 5 |
| Demand | 4 | 0 | 1 | 1 | |

Cost = 29

Now give the allocations based on above transition probability values in following subproblem-2

Table 10.8 Allocations for Subproblem-2

|  | $D_1$ | $D_2$ | $D_3$ | $D_4$ | Supply |
|---|---|---|---|---|---|
| $S_1$ | 2(6) | 3 | 11 | 7 | 6 |
| $S_2$ | 1 | 0.5 | 6 | 1 | 1 |
| $S_3$ | 5(1) | 8(4) | 15(3) | 9(2) | 10 |
| Demand | 7 | 5 | 3 | 2 | |

Cost = 112.5

Now give the allocations based on above transition probability values in following subproblem-3

Table 10.9 Allocations for Subproblem-3

|  | $D_1$ | $D_2$ | $D_3$ | $D_4$ | Supply |
|---|---|---|---|---|---|
| $S_1$ | 3(7) | 4 | 12 | 8 | 7 |
| $S_2$ | 2 | 1(2) | 7 | 2 | 2 |
| $S_3$ | 6(1) | 9(4) | 16(4) | 10(3) | 12 |
| Demand | 8 | 6 | 4 | 3 | |

Cost = 159

Now give the allocations based on above transition probability values in following subproblem-4

Table 10.10 Allocations for Subproblem-4

|  | $D_1$ | $D_2$ | $D_3$ | $D_4$ | Supply |
|---|---|---|---|---|---|
| $S_1$ | 4(1) | 6(1) | 14 | 11 | 12 |
| $S_2$ | 4 | 1.5(3) | 8 | 3 | 3 |
| $S_3$ | 8 | 12 | 19(6) | 12(4) | 17 |
| Demand | 11 | 11 | 6 | 4 | |

Cost = 300.5

## 10.5   Result and Discussion

The numerical problem under consideration engages with the classical fuzzy transportation problem, wherein factors such as cost, supply, and demand are incorporated as trapezoidal fuzzy numbers. A transition probability matrix was established via a Fuzzy Inference System (FIS) between the source and destination, forming the backbone for transforming the fuzzy transportation problem into a pointwise classical transportation problem. This transformation facilitated the allocation process and subsequent cost calculation. The methodology relies on the creation of a fuzzy rule for each pair of source and destination, resulting in a total of 12 rules. These rules, in turn, generate the preliminary transition probability table. The next iterations can be computed in one of two ways: by modifying the rules and updating the transition probability table or, in the case of equal values in the transition probability, by adjusting the allocations. In the problem addressed here, equal values were present; hence, iterations were performed by altering the allocation of these equal values exclusively.

The continuous reallocation of the equal values in the transition probability matrix led to thegeneration of varied iterations. The computed fuzzy costs and the costs derived from existing methods are tabulated in Table 10.11 and depicted in Fig. 10.5. Figure 10.6 illustrates the process of deriving the cost desirability value using the Fuzzy Inference System (FIS). This procedure requires careful adjustment of the input parameters in accordance with the prescribed rules. By doing so, we can accurately compute the cost desirability value, providing a clear insight into the efficiency of the system. We implimanted this work through the - MATLAB® software, the used fuzzy inference system is applied with the help of MATLAB software.

Table  10.11 Comparison table NWC

|  |  | No (29, 116, 163, 300.5) |
| --- | --- | --- |
| LC | No | (29, 112.5, 159, 300.5) |
| VAM | No | (28, 113.5, 146, 275) |
| Sujatha et al. 2018 | No | (28, 100, 144, 278) |
| Pandianand Natarajan2011 | No | (28, 100, 144, 278) |
| Proposed | FIS with ACO | (29, 112.5, 159, 300.5) (Iteration 1) |
|  |  | (29, 106.5, 154, 298.5) (Iteration 2) |
|  |  | (29, 102, 149, 289) (Iteration 3) |

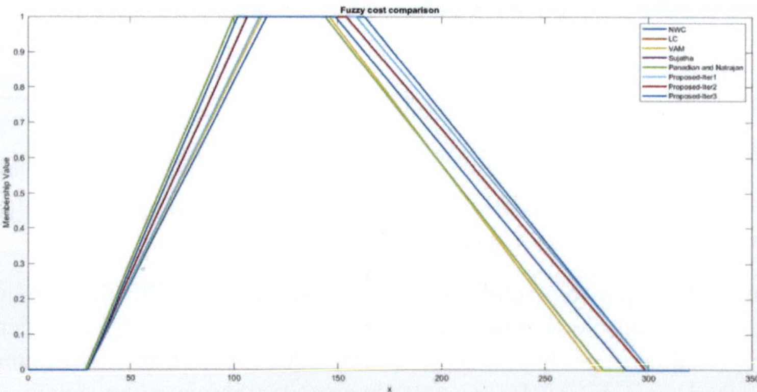


**Figure 10.5:** Fuzzy cost comparison.

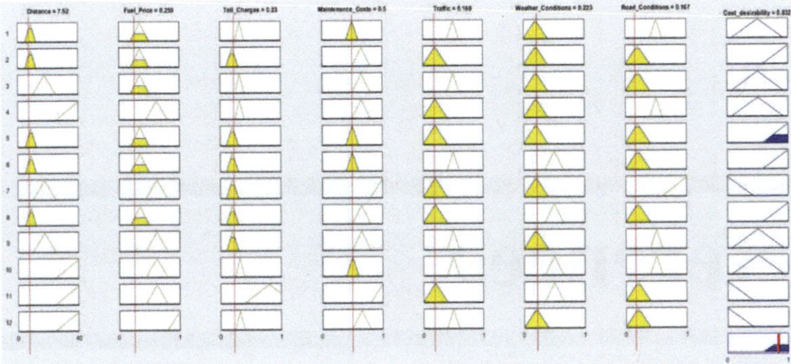

**Figure 10.6:** Deriving the cost desirability value using the Fuzzy Inference System (FIS).

## 10.6 Conclusion

The increasing complexity of today's transportation networks necessitates the development of novel, flexible, and robust methodologies capable of considering multiple dynamic factors. This study set out with the aim of incorporating a fuzzy inference system and the ant colony optimization (ACO) technique into the classical fuzzy transportation problem, a combination that to our knowledge had not been explored in the existing literature. The newly proposed methodology presents a fresh perspective on fuzzy transportation problems by considering a multitude of factors such as distance, fuel price, toll charges, maintenance costs, traffic conditions, weather conditions, and road conditions. The integration of these elements provides a more comprehensive and realistic representation of the real-world challenges faced in transportation management.By leveraging the strength of fuzzy inference systems, our methodology effectively handles the inherent uncertainty and ambiguity associated with these variables. Meanwhile, the ant colony optimization technique allows for the efficient exploration and exploitation of possible solutions, enhancing the solution's quality and the methodology's overall performance. The study, therefore, marks a significant step forward in solving fuzzy transportation problems. The introduced methodology not only broadens the scope of factors considered in the problem but also effectively handles the uncertainty associated with these factors. While the results are promising, it is acknowledged that this is just the beginning, and there is considerable potential for refining and expanding this approach in future research. Ultimately, the study provides a strong foundation for further exploration and development of hybrid models that leverage the combined strengths of fuzzy systems and Ant Colony Optimization, which can contribute to the more robust and efficient management of transportation networks.

# References


[1] Zadeh, L.A. 1965. Fuzzy sets. Inform. and Control. 8: 338–353.

[2] Zimmermann, H.J. 2011. Fuzzy set theory-and its applications. Spr. Sci. & Bus. Med.

[3] Chanas, S., Kołodziejczyk, W. and Machaj, A. 1984. A fuzzy approach to the transportation problem. Fuzzy Sets and Systems 13: 1111–1132.

[4] Liu, S.T. and Kao, C. 2004. Solving fuzzy transportation problems based on extension principle. European J. Oper. Res. 153: 661–674.

[5] Gani, A.N. and Razak, K.A. 2006. Two stage fuzzy transportation problem. J. Phys. Sci. 10: 63–69.

[6] Pandian, P. and Natarajan, G. 2011. An appropriate method for real life fuzzy transportation problems. Int. J. Info. Sci. and Appl. 3: 127–134.

[7] Kaur, A. and Kumar, A. 2011. A new method for solving fuzzy transportation problem using ranking function. Appl. Math. Model. 35: 5652–5661.

[8] Sujatha, L., Vinothini, P. and Jothilakshmi, R. 2018. Solving fuzzy transportation problem using zero point maximum allocation method. Int. J. Curr. Adv. Res. 7: 173–178.

[9] George, G., Maheswari, P.U. and Ganesan, K. 2020. A modified method to solve fuzzy transportation problem involving trapezoidal fuzzy numbers. AIP Conf. Proc. 2277: 090005.

[10] Fidanova, S., Roeva, O. and Ganzha, M. 2020. Ant colony optimization algorithm for fuzzy transport modelling. 2020 15th Conf. on Comp. Sci. and Info. Syst. (FedCSIS), IEEE: 237–240.

[11] Kazharov, A.A. and Kureichik, V.M. 2010. Ant colony optimization algorithms for solving transportation problems. J. Comp. and Syst. Sci. Int. 49: 30–43.

[12] Kumar, P.M., Manogaran, G., Sundarasekar, R., Chilamkurti, N. and Varatharajan, R. 2018. Ant colony optimization algorithm with internet of vehicles for intelligent traffic control system. Comp. Netw. 144: 154–162.

[13] Ntakolia, C. and Lyridis, D.V. 2022. A n-D ant colony optimization with fuzzy logic for air traffic flow management. Oper. Res. 22: 5035–5053.

[14] Di Caprio, D., Ebrahimnejad, A., Alrezaamiri, H. and Santos-Arteaga, F.J. 2022. A novel ant colony algorithm for solving shortest path problems with fuzzy arc weights. Alex. Eng. J. 61: 3403–3415.

[15] BabaeeTirkolaee, E., Mahdavi, I., Seyyed Esfahani, M.M. and Weber, G.W. 2020. A hybrid augmented ant colony optimization for the multi-trip capacitated arc routing problem under fuzzy demands for urban solid waste management. Waste Manag. and Res. 38: 156–172.

[16] Das, M., Roy, A., Maity, S. and Kar, S. 2023. A Quantum-inspired Ant Colony Optimization for solving a sustainable four-dimensional traveling salesman problem under type-2 fuzzy variable. Adv. Eng. Inform. 55: 101816.

[17] Song, Q., Zhao, Q., Wang, S., Liu, Q. and Chen, X. 2020. Dynamic path planning for unmanned vehicles based on fuzzy logic and improved ant colony optimization. IEEE Access 8: 62107–62115.

[18] Panicker, V.V., Vanga, R. and Sridharan, R. 2013. Ant colony optimisation algorithm for distribution-allocation problem in a two-stage supply chain with a fixed transportation charge. Int. J. Prod. Res., 51: 698–717.

[19] Kumar, P.M., Manogaran, G., Sundarasekar, R., Chilamkurti, N. and Varatharajan, R. 2019. Ant colony optimization algorithm with internet of vehicles for intelligent traffic control system. J. Clean. Prod. 227: 1161–1172.

[20] Lisangan, E.A. and Sumarta, S.C. 2017. Route selection based on real time traffic condition using ant colony system and fuzzy inference system. 2017 3rd Int. Conf. on Sci. in Info. Tech. (ICSITech), IEEE: 66–71.

[21] Ntakolia, C. and Lyridis, D.V. 2022. A comparative study on Ant Colony Optimization algorithm approaches for solving multi-objective path planning problems in case of unmanned surface vehicles. Ocean Eng. 255: 111418.

[22] Liu, S., Huang, S., Xu, X., Lloret, J. and Muhammad, K. 2023. Efficient visual tracking based on fuzzy inference for intelligent transportation systems. IEEE Trans. on Intell. Transp. Syst. 1–12.

[23] Gamel, S.A., Saleh, A.I. and Ali, H.A. 2022. A fog-based Traffic Light Management Strategy (TLMS) based on fuzzy inference engine. Neural Comp. and Appl. 1–19.

[24] Shkera, A. and Patankar, V. 2023. Fuzzy inference system for evaluating traffic congestion on urban streets. AIP Conf. Proc., AIP Pub. 2819.

[25] Vechione, M. and Cheu, R.L. 2023. Fault Tolerance analysis of an Adaptive Neuro-Fuzzy Inference System for mandatory lane changing decisions in automated driving. Int. J. Transp. Sci. and Tech. 12: 594–605.

[26] Erdinç, G., Colombaroni, C. and Fusco, G. 2023. Two-stage fuzzy traffic congestion detector. Future Transp.v3: 840–857.

[27] Jutury, D., Kumar, N., Sachan, A., Daultani, Y. and Dhakad, N. 2023. Adaptive neuro-fuzzy enabled multi-mode traffic light control system for urban transport network. Appl. Intell. 53: 7132–7153.

# Index